# Lamps Forever Lit

## A Memorial to
## Kirkland Lake Area Miners

Bernie Jaworsky 2001

# Lamps Forever Lit

## A Memorial to Kirkland Lake Area Miners

by Bernie Jaworsky

Edited by Kerry McArthur
Designed by Sherry Mumford

*"To be remembered
may be the secret to immortality."*
– Anonymous

Cambria Publishing
Calgary, Alberta

CANADIAN CATALOGUING IN PUBLICATION DATA

Jaworsky, Bernie, 1945 –
Lamps forever lit

Includes bibliographical references and index
ISBN 0-9697023-2-9

1. Miners – Ontario – Kirkland Lake Region – Biography. I. Title.
TN27.O5J39 2001 622'.092'2713144 C00-911583-8



Printed and bound in Canada

Published by

CAMBRIA PUBLISHING
Box 61083 Kensington PO
Calgary, Alberta T2N 4S6

E-MAIL: publisher@cambriapublishing.com
Web site: www.cambriapublishing.com

# CONTENTS

# SPONSORS

United Steelworkers – Canadian National Office
United Steelworkers – District 6
Robert J. Kasner
Cliff Lafreniere – Pinewood Park Motors (K.L.) Inc.
Gerry Panagapka CFP, CLU, CH.F.C.
Mervin and John Sullivan – Teck Insurance Brokers
Joe Lalonde – Carpet King
USWA Local 6409 (Kirkland Lake)
Murdoch Group Inc.
Hoggarth Interiors Limited and the O'Connor Family
Kirkland Lake – Economic Development Department
Cheminis Lumber Inc.
Township of Gauthier
Mike McCollough – Northern Pressure Treated Wood
Corporation of the Township of Larder Lake
Township of McGarry (Virginiatown and Kearns)
Tom Drew-Brook
Alex Dumas – Your Independent Grocer
Siemens Westinghouse
Thomas Obradovich Explorations
Michael and Christopher Jaworsky – Waterloo, Ontario
Miners' Memorial Foundation

# FOREWORD

The United Steelworkers is pleased to support this important piece of Northern Ontario – and Canadian – history. The lives and the families honoured by Bernie Jaworsky in this volume are the story of communities and of mining. But more importantly, the story is about individuals, the people who gave up their lives to an industry whose record is filled with obituaries.

Mine safety can never be taken for granted. For the United Steelworkers, it is fundamental to our work as Canada's Mining Union (more unionized miners belong to the Steelworkers than any other union). It is most often the reason miners seek out union representation in the first place. It was why miners at the Westray Coal mine in Nova Scotia were signing union cards in 1992, when the mine exploded and killed 26 men, 11 of whose bodies have never been recovered. Years of public inquiry and court proceedings have only proved what the union and the Westray survivors always knew: that the disaster was the direct result of deliberate management disregard for safety.

Steelworkers have been solidly behind the push for legislation to amend the Criminal Code to hold companies, their directors and senior management accountable for workplace deaths. Otherwise, companies will continue to get away with cutting corners in order to get production out. Deregulation, especially in Ontario, is allowing companies more opportunity to put workers at risk. As the risk increases, the fatalities will surely follow. It makes having a union even more important for the men (and now many women) who work in Canada's mines. With less regulation and enforcement by governments, it is the union that must be more vigilant and forceful in its dealings with mine owners and operators. Indeed, just as this foreword was being prepared, two separate incidents at two Northern Ontario mines killed two men and injured a third – all in one day. There was no union at those mines.

The most important aspect of this book – in which there are no 'happy' endings – is that these men are not merely statistics. Each and every life is a story. Some stories began in other countries. Some included large, extended families. Some were so young, their stories had only just begun when they were killed.

If there is a message beyond this tribute to the dead miners, it is that the killing and the hurt has to stop. As the old saying goes, if we do not learn from history, we are doomed to repeat it. There are lessons within these covers that are worth repeating. We thank Bernie Jaworsky for offering them to us.

Lawrence McBrearty
National Director for Canada
United Steelworkers

Harry Hynd
Director, District 6
United Steelworkers

PRODUCED WITH FINANCIAL ASSISTANCE FROM
THE UNITED STEELWORKERS

# INTRODUCTION

*Kirkland Lake Miners' Memorial. Artists: Sally Lawrence & Rob Moir, Photographer: Lawrence Chrismas*

Located in a park near the Museum of Northern History, the Miners' Sculpture and Memorial stand in proud recognition of the working miner.

The sculpture is a 40-tonne 10-metre-high black granite abstraction of a head frame. On and around the structure are five welded-steel bronze-coated life-size figures of miners at work. One miner is mucking in a drift. Another is scaling loose from the back of a stope. Two miners are machine runners using a jackleg drill to make holes in a rock face, and

*Kirkland Lake Cemetery. Photographer: Lawrence Chrismas*

a stoper drill to drill upwards into a raise. High up, the fifth miner is entering a stope from a manway by means of a ladder.

The Miners' Memorial Foundation also erected four monument additions in close proximity to the sculpture. Three of these, to the left of the sculpture, have inscribed on their black polished surfaces, the names of mine employees who died accidentally while at work. The fourth black monument, located directly in front of the sculpture, is inscribed with an address by Steve Yee, the Chairperson and Founder of the Memorial Foundation and a Macassa miner.

Standing before the monuments with their engraved names, one should be disturbed at the length of the list. How did these workers die? Who were they? Where were they from? How old were they? Did they have families? Where are they buried? It was my wish to answer these questions that provided the motivation for this book.

The book has been compiled as a further memorial to all who accidentally lost their lives in local mines during the past century. Here will be found the biographies of those men who worked in mines at Virginiatown, Larder Lake, Dobie, King Kirkland, Kirkland Lake, Swastika, Boston Creek and Matachewan. During research, other deserving names were discovered and thus 310 biographies in total appear in the book.

The individual biographies were placed in chronological order, usually according to the date of accident. The primary purpose of the book is to create a literary memorial to the deceased, but by placing the accounts in chronological order, a history of the struggle for workplace health and safety evolves. To impart a sense of appreciation for the historical environment surrounding the life and death of the miners, a brief dateline of local events precedes the biographies of each year.

The length of each life story is dependent on the availability of information from research sources including newspapers, accident reports, interviews and inquest reports. Shorter stories are due, of course, to lack of information and no other reason.

Having no first-hand experience working in mines, the author was dependent almost entirely on research for details. Therefore, I wish to acknowledge the following sources: Northern News, Northern Daily News, Kirkland Lake Cemetery monument inscriptions, Ontario Department of Mines Accident Reports, local history texts, Museum of Northern History, Teck Centennial Library, Toronto Star, Toronto Telegram and the Globe and Mail.

Situations could not be explained properly or accurately by excluding mining terms and phrases. Although the stories are written so they can be read easily by the layperson, some technical terms were essential to preserve accuracy. At the back of the book, there is a glossary and a pictorial of the workings of a mine that should be interesting and useful.

In writing, I tried to be sensitive and accurate. Keep the following in mind while reading the accounts: although miners are tough, hard-working men who toil to support themselves and their families, these men also know love, beauty, friendship, kindness, joy, sadness, fear and pain – all the characteristics of being human.

All these stories are sad: in each, a miner dies. An accident in a mine is usually sudden. Hard rock, steel and blasting powder are not considerate of soft flesh and brittle bone. Death in most cases is gruesome, lonely, painful, bloody or suffocating. Death in a mine is not gentle.

Also keep in mind while contemplating a specific accident that underground there is constant darkness – a blackness only a miner knows. What you usually see is what your lamp reveals. Dangers from the side, above or below, are as a consequence unknown to you.

Doing the research has been very enlightening. I formed new opinions about life, death, family, friends, mining, safety and unions, but you will not find my opinions in print. Please read on and form your own.

Bernie Jaworsky

*Three monuments inscribed with names of deceased miners. Photographer: Lawrence Chrismas*

# DEDICATION

This book is dedicated to the memory of the following miners who perished while at work:

Abrams, Max
(a.k.a. Mack Abraham)
Adams, Edward Sidney
Alexander, William
Antuchiewicz, Julian M.
Arthur, Raymond
Baginski, Paul
Banks, James E.
Barabas, Wesel
Barkell, Elmer Paull
Barnett, Turner
Battistone, Antonio
Beaulieu, Romain
Bedforth, John
Beland, Germain Arthur
Bellamy, Arthur Walker
Bellon, Valentino
Bendnsond, Wasil 'William'
Bennett, Leonard Joseph
Bentley, Stoel Smith
Berlingeri, Giovanni
Bizub, John
Booth, T. B.
Bottrill, Jack John Thomas
Boudreau, Gerard
Bourassa, Albert H.
Bowers, Herbert James
Bowman, Alexander Henry
Braaten, George Samuel
Brady, Harry R.
Breakwell, James
Browne-Cadman, George Scott
Brown, John M.
Brown, Johnston
Bulch, Frank
Burton. James
Cameron, Robert, 'Bob' Dickson
Casey, Michael
Castagna, Antonio
Chartrand, Wilfred C.
Chorzepa, Albert
Chouinard, Lucien J. C.
Clark, Tex Malcolm Edward
Clattenburg, Angus
Closs, Alfred Charles
Comptois, Arthur
Conroy, Christopher R.
Cop (Chop), Bozo
Corak, Anton
Coté, Lucien R.
Coulson, John Wesley
Cuthbertson, Thomas James
Daly, Richard 'Scotty'
David, Adolphe
Davis, Roy
de Champlain, Denis Ivan Joseph
Demko, William
Demoline, James Edward
Dologh, Andy
Donaldson, Stephen Leonard
Dool, David Adam
Dragan, Frank
Drul, Michael A.
Dubreuil, James
Duchesne, Edouard
Duda, Karol
(a.k.a. Charles Dooda)
Dufour, Fred
Dunn, Charles David
Eadie, Henry
Eagling, E. H.
Ellis, James Harvey
Elmitt, Garrioch (Garry) George
Englehutt, Kent
Enstiom, Hugo Alfred
Evjent, Johann
Fletcher, Roy
Fortey, William A.
Furber, John Clifford
Furlan, Peter
Gabryelczyk, Joseph
(a.k.a. Joe Gabriel)
Galandzy, M.
Galinoc, Johan
Gallina, Guiseppe
Gibbons, John F.
Gibson, Donald
Gill, Kenneth E.
Glover, Edwin
Golob, Joseph
Gornik, Anton
Gorzalczynski, Stanislaw
(Stanley)
Gosselin, Wilfrid
Greczek, Jan
Greenall, Richard
Gruby, Louis
Guindon, Steven Allan Douglas
Gulbis Janis
Harmala, Arne
Harney, Emmett Joseph
Harrington, Michael
Harris, Tony
Harrop, Frederick
Hemmi, Joseph
Holubec, John
Horbyk, Wasyl
Huard, Gilles
Huhtala, John (Juho) Fredrik
Hulm, Edward
Huncar, Frank
Jendral, Joseph
Jones, Ernest
Jones, Richard 'Radar' T.
Jussila, Reino
Kauppi, Matti Leonard
Kazur, Leonard
Keane, James John
Keller, Alois 'Louis'
Kensy, Leo
Keto, Eric
Klimecki, Jan
Kordic, Eli
Korytko, Stanislaw
Koskela, Gerhard
Kovac, Josef
Kubalak, Jan
Kuros, John
Kurze, Eric William
Kuszneruk, Ignace
Kuzyk, William
Laakkonen, John
Labine, Edward J.
Lafleur, Alfred
Lafond, Fred Joseph
Lalonde, Delphis
Lalonde, William Harry
Lamoureux, Wilfred 'Bert'
Landry, Jean Baptiste
Larmond, James Willard
Latvala, Antti
Lauzon, Fortune
Lenk, Karl
Leverre, Norman E. 'Buzz'
Lewington, Harry
Lindholm, Runor
Linney, Harry
Littlejohns, Ernest
Lloyd, Murdoch C.
Loney, Leslie
Louko, Leo Aatos
Ludwig, Franz
Luoma, William

MacDonald, Alex R.
MacDonald, Archie
MacDonald, Archie A.
MacLeod, Malcolm
Maki, Kalle
Malenoski, Michael
Maloney, Alex
(a.k.a. Alex Malinowsky)
Mantyla, John
Marcouiller, Rolland
McAlinden, Hilary
McCarthy, Patrick Leonard
McDonald, Dean Ellery
McDonald, Duncan H.
McDonald, Joseph B.
McFarlane, Douglas H.
McGillivray, John
McMaster, Hugh
McNaughton, Fred C.
McQuaig, Alexander
McQueen, Robert
Meister, Aksel
Melnyk, Pawlo 'Paul'
Melnyk, Fred (Theodorus)
Melong, Henry
Menard, Henry I.
Michichuik, Joseph
Miciunas, Jonas Vytautus
Micolo, Adriano
Mihelich, Joseph
Miller, Charles M.
Miller, Edward
Mirich, Dane
Moras, Peter (Pietro) Paulo
Morden, Jim 'Irish'
More, Ernest Gordon
Morissette, Jean-Paul
Morrow, Sinklar
Murphy, Leo Wilmot 'Busher'
Murray, Leo Germaine
Mutka, Felix J.
Nadeau, Aurel
Nancarrow, William G.
Nicholl, Alex 'Red'
Nicholson, Robert
Niskanen, David
Nixon, Stephen
Noble, David
Ojala, George
Ollel, Emil (a.k.a. Emil Bartel)
Omelchuk, Harry
O'Reilly, James Sylvester
Orminison, William
Orr, Joseph A.
Ostman, Axel Edwin
Ouellette, Edouard J. B.
Owen, Thomas P.
Owens, Robert 'Robbie' Edward
Paget, Robert
Parker, James Maxwell
Parrott, John Beverly
Patriquin, Weniford V.
Patterson, Howard Robert 'Bob'
Peever, Harvey
Perrier, Donald Joseph
Peterson, Bertel 'Bert'
Piekos, Jan (John)
Pietrzak, Siegfried (Siggy)
Pitcaithley, Robert
Pleban, Joseph
Poloni, Antonio
Pontisso, Albino
(a.k.a. Johnny Albino)
Porre, Nicolai
Porter, Ernest M.
Potkonjak, Sava 'Sam'
Powers, Stephen A.
Purdy, Clifford
Ragus, Stjepan
(a.k.a. Steve Raguk)
Rendes, Niklos
Renick, R.
Rennie, Evan Dougall
Rioux, Fred
Robidas, Armand Gerard
Robinson, James J. 'Red'
Robinson, Percy Michael
Rodgers, Stephen Wayne
Ross, Charles G.
Rossi, Matti
Roy, Leonel
Rubanets, Stive
Rumig, Englebert
Rupert, Archie J.
Ruzick, Branko
Saar, Elmar
St. Gelais, Paul
Samson, Joseph Paul Emile
Sancartier, Joseph
Sandell, Eino
Scott, Robert
Sekulic, Marko
Sheldon, Frederic Arthur Jr.
Sheldon, Robert John Elwood
Sheme, Robbie
Silvola, Henry (a.k.a. Silvala)
Sipprell, Roy Logan
Skinkle, Elwood E.
Slack, Stanley Dorian
Sliva, Albert
Smarz, William
Smith, Royden
Soopalu, Aleksei
Sorle, Peter
Soucie, Anthime Leo
Souriol, Robert Frank
Spiers, Laurence H.
Sragalj, Andrija 'Albert'
Stark, George
Starnio, Carlos
Starr, Thomas Daniel
Stevenson, William
Stinkowji, John
Stinson, Marshall Leslie
Stransky, Rainer
Stupica, Frank Jr.
Sullivan, John Clements
Summers, George 'Scotty'
Susla, William
Szpak, Walter
Szucs, Kalman
Szumelda, Stefan
Taavitsainen, Kalle
Temberg, William
Thibault, Simon M. Joseph
Tomachuk, Alex
Tomc, Frank
Tomljenovich, John
Towtriwsky, Peter John
Trajkowicz, Alexis 'Alex' Joseph
Tremblay, Onide
Trudeau, Fernand
Trudel, Rolland
Turus, Alex
Vachon, Samuel Napoleon
Valitorppa, Jaska
(a.k.a. Jack Kauppa)
Verrier, Léonce
Vincent, Vital Joseph
Vlaisavgevitch, Vujo (Vito)
Walker, Marvin A.
Walton, Joseph S.
Warwick, Roy
Waschuk, Dan
Whalley, Clarence W.
Widdifield, Dale Oakley
Willis, James
Witczak, Richard
Wlasniak, Anthony 'Tony'
Wojcechowski, Leon
Yaworski, Donald
Yergeau, Roger
Young, John
Zalec, Frank
Zarcovitch, Mike

Chapter I

# THE KIRKLAND LAKE AREA

Kirkland Lake is located on the western edge of a 250-kilometer-long gold belt that extends from Northeastern Ontario across the border into Northwestern Quebec. In Ontario, this town became the largest in the area while smaller communities developed around other mines in the belt. These mines began to draw men from all parts of Canada. Many new immigrants from across Europe also came to the area, launching a period of ethnic diversity that exists today.

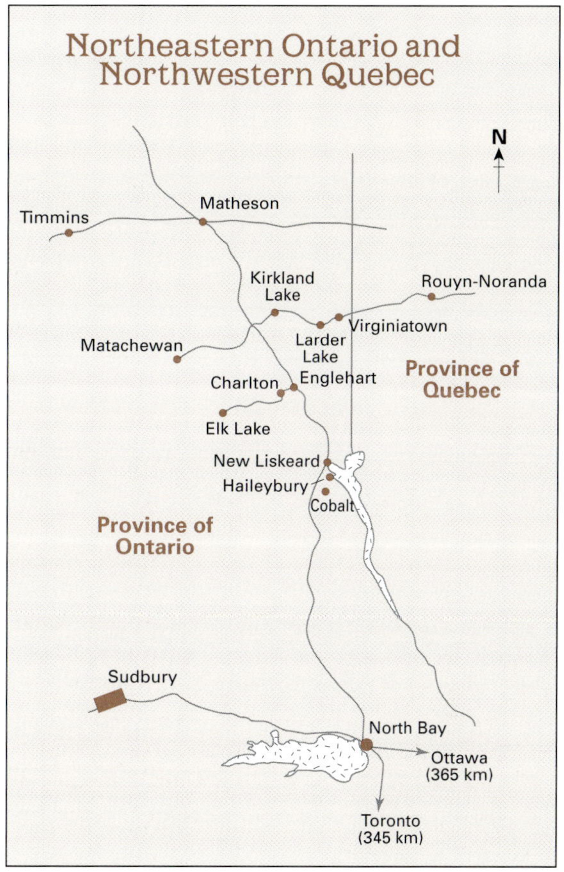


A very short distance from the Quebec border, the communities of Kearns and Virginiatown evolved near their respective mines: Chesterville and Kerr Addison. About 10 kilometers west, on the shoreline of a large lake, developed the town of Larder Lake. Miners from such mines as the Omega, Martin-Bird, Kir-Vit and Fernland lived with their families in homes at Larder Lake. Again progressing a few more kilometers west, one finds the out-of-the-way hamlet of Dobie. This planned townsite, consisting of about seven streets, was the home of miners working at Upper Canada, the Argonaut (Upper Beaver), the smaller Oriole and Brock mines. Along the 'gold break' between Dobie and Kirkland Lake is the village of King Kirkland containing the family dwellings of miners working the nearby Bidgood, Moffat-Hall and Morris Kirkland mines.

A few kilometers to the west of Kirkland Lake is the railroad village of Swastika. Here, the early prospectors disembarked from the train, hoping to discover a fortune in the area. The Kirkland and Larder Lake claims, already staked by prospectors when they arrived by canoe a few years earlier, were now lapsing and could be reclaimed. In Swastika, two small mines developed: the Swastika Mine on the shores of Otto Lake, and the Lucky Cross Mine, along the rail line. The other Swastika area mine, the Baldwin, was located not many kilometers up the rail line. South of Swastika, down the line, the village of Boston Creek developed around the Barry-Hollinger Mine.

About 50 kilometers to the west of Swastika is the community of Matachewan. Here, two gold and base metal mines developed: the Matachewan Consolidated and the Young-Davidson. Matachewan's affinities were at one time to the Elk Lake and the Tri-Town areas further south, but in the '50s, a highway was constructed to Kirkland Lake, allowing the residents and industries of Matachewan to begin their new long-lasting ties with the gold town.

In the gold belt, many other small mines developed and a semblance of community grew about the mines. This consisted of mine buildings, a bunkhouse and a kitchen. The longevity of these mines was often short and the structures were soon abandoned and left to the elements to gradually disappear.

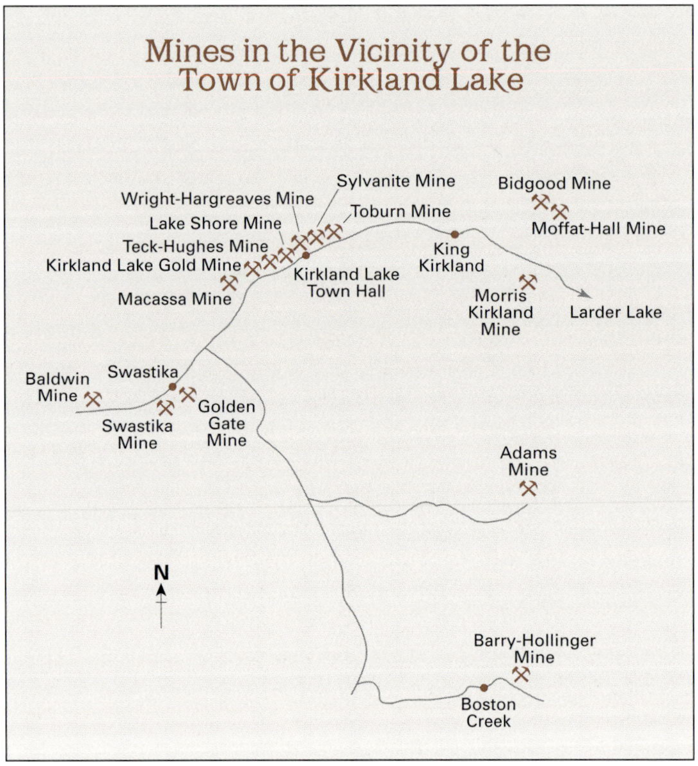


Kirkland Lake developed slowly after 1911 when William Wright and Ed Hargreaves staked their claim on what was to become the Wright-Hargreaves Mine. Soon after, many (including Harry Oakes and the Tough brothers) sank shafts in town in their quest for rich ore. But ultimately as time passed, some could not find the elusive pot of gold and so closed their small operations. Seven mines would remain viable in town; all were located in a crooked row within a short distance of each other. This section of town became known as 'the Mile of Gold'.

A year after Wright and Hargreaves registered their claim, the Tough brothers and Harry Oakes created the Tough-Oakes Mine (later named Toburn) to the east of the Wright-Hargreaves property. Later nearby operations started at the Teck-Hughes Mine in 1917, at Lake Shore Mine in 1918, Kirkland Lake Gold Mine in 1919, Wright-Hargreaves in 1921, Sylvanite in 1927 and finally the Macassa in 1933. Situated in succession from east to west along the Mile of Gold were: the Toburn, Sylvanite, Wright-Hargreaves, Lake Shore, Teck-Hughes, Kirkland Lake Gold and Macassa mines.

Located about 600 kilometers north of Toronto, Kirkland Lake today is a small Northern Ontario town with a population of about 9,000. It has the characteristics of other small towns: a McDonald's, two Tim Hortons, elementary schools, two high schools, a community college, an airport, a downtown commercial area and a small mall. But the elements that give the town its physical character are its geological setting and historical development.

It's important to note that the town was built around the gold mines. In the beginning, there was little planning for the layout of the streets. The main thoroughfare of the downtown area, Government Road, snaked its way left and right, up and down through town. As the mines developed and required a greater workforce, the town added streets to its outskirts to accommodate the building of more homes. Clearly, there was some late planning: some of the streets run parallel to each other and have formed blocks. However, in other areas of town, some odd-angled intersections were inevitable. Building a town around mines situated on Precambrian bedrock caused some unique problems. In some situations, sidewalks were constructed at levels much higher than the streets. Some homes were built on high rock outcrops. In some instances, homes were built without basements; if they had basements, these were blasted out of the bedrock. One can only imagine the problems encountered when infrastructure services for water, sewage and electricity were initiated.

Many buildings along Government Road were constructed quickly for short-term duration. The town's main industry was based on a non-renewable natural resource that could be depleted any day. The mines could go deeper and deeper but one day the gold veins would disappear and so would the town. Wooden frame commercial buildings were hurriedly and inexpensively thrown up to serve the present and not the future. To make the front of the buildings presentable, a flat board façade was constructed on the second storey (to this day, many

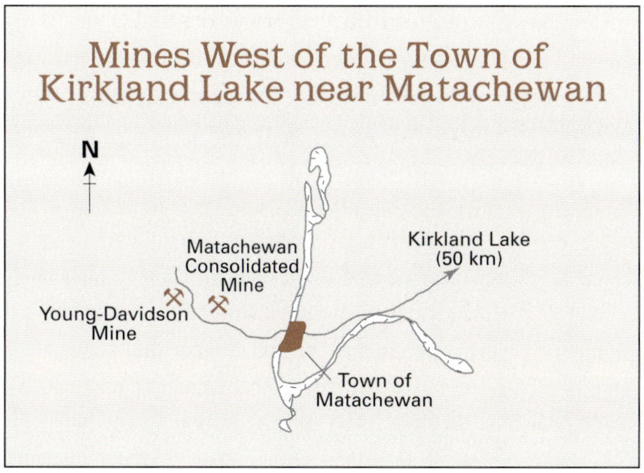


buildings in town retain this affectation). The ravages of downtown fires, prevalent in the '80s and '90s, resulted in some commercial lots on the main drag receiving newly constructed brick buildings. Others, devoid of replacement structures, became horticultural parks financed by concerned local citizens and service clubs.

In the '60s, the Adams Mine was developed south of Kirkland Lake. This was a huge iron-ore-producing mine employing up to 400 men. Due to a good highway and excellent transportation facilities at that time, there was no need for a new community – thus Kirkland Lake derived all the benefits. A similar situation arose in the '90s when two gold mines, Battle Mountain Gold and Holt-McDermott, started up about 60 kilometers northeast of town in the Lightning River area. Employees at these mines now commute each day from communities in Quebec, Matheson, Virginiatown, Larder Lake, Dobie, King Kirkland, Kirkland Lake and Swastika.

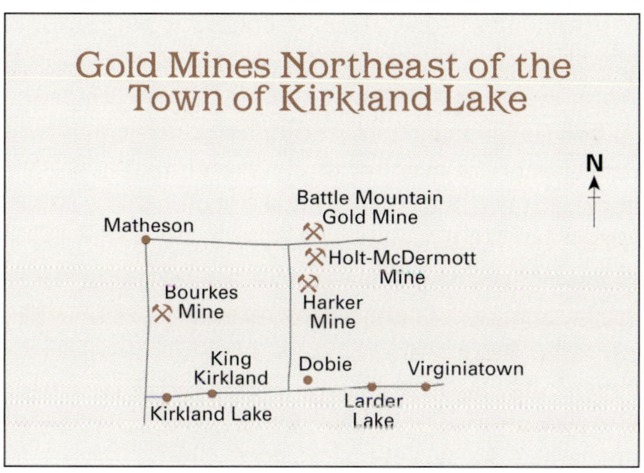


Kirkland Lake has had its good and bad times. On October 24, 1929, the world was plunged into an economic depression and remained there for almost a decade. During these times of uncertainty, gold became the stable currency and the Canadian government sought to bolster its yellow metal treasury. Local gold mines responded with a production boom, causing (ironically) a local economic boom during the early Depression years. Things got even better in 1934 when the U.S. announced it would buy gold at a price near $35 an ounce – within a day, the price of an ounce of gold increased by about $15. This development pushed local production to even higher levels in existing mines and caused investors to open numerous, smaller mines in the area.

The Dirty '30s were the boom years for this area and earned it the title of 'the bright spot of Canada'. Attracting hundreds of job seekers, salesmen, entrepreneurs, gamblers and the like, the population of Kirkland Lake swelled to over 25,000. Over 6,000 men were employed in area mines and were making about five dollars per day. This doesn't seem like much but, compared to the rest of depressed Canada, this was a good wage and the men were thankful they had a job.

The World War II years spelled the end of the gold belt's affluent period and the beginning of the area's decline. Wartime meant base metals were in high demand; by comparison, gold was no longer a precious commodity. The loss of interest in gold, a bitter strike by miners through 1941/42 and a labour shortage caused by the war effort precipitated the closing of many area mines in the '40s. The decline continued in the '50s and '60s: the Toburn, Kirkland Lake Gold, Sylvanite, Wright-Hargreaves, Lake Shore and Teck-Hughes all closed in succession. But there remained some bright spots in the gold belt as the Kerr Addison in Virginiatown had its best production through the '50s and '60s and was one of the biggest gold mines in the world at that time. Also, it was in 1964 that the huge Adams Mine began operations.

All these underground workings are now closed in 2000. The bright refurbished red-sided headframe of the proud Toburn still stands. The Sylvanite site has been completely obliterated and levelled. The Wright-Hargreaves headframe and adjoining buildings have been torn down and replaced with a liquor store and a Royal Canadian Legion building with a bar room appropriately dubbed the 'Hoist Room'. The Lake Shore property is now a mall with a parking lot. A new headframe was built behind the mall in the late '80s – it now stands silent. A few old Teck-Hughes buildings still remain: an assay laboratory of significant industrial architecture and a refurbished Teck-Hughes dry, now a funeral home. Few residents under 50 would be able to give directions to the site of the old Kirkland Lake Gold but if you found the area and looked around, you would still find some obscure remnants and decaying foundations. The Macassa, still employing about 30 men on surface, is doing custom milling for a mine near Matheson.

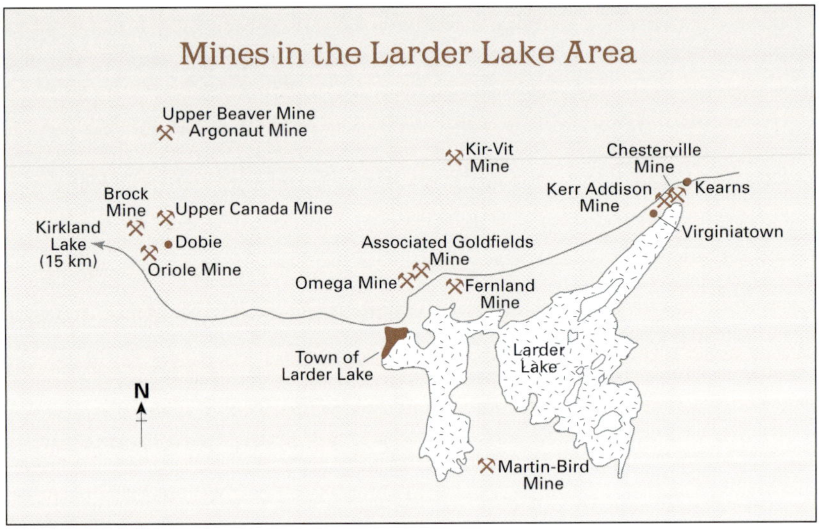


When strangers come into town, one of the first questions they ask is "Where's the lake?" The answer: the beautiful blue Kirkland Lake was filled in with mine tailings during the '20s. Sixty years later, it became profitable for mining companies to dredge and refine these tailings. However, during the '90s, the price of gold dropped and the lake's re-establishment was not completed. The partial lake now has some water in it but no longer has its pristine early-century look.

What is the area like in the new millennium? It would be easy to become negative as the robust mining era is gone and with it went other businesses, leaving our communities with empty buildings and For Sale signs. But Kearns remains a picturesque valley community in the shadow of the grand Mount Cheminis. Virginiatown is a clean, vibrant community involved with its curling rink, an arena, a business section and a large tourist centre. Larder Lake, located on the shoreline of a large deep blue lake of the same name, will always be aesthetically pleasing. Dobie is now smaller, a little more out-of-the-way and much quieter than before. King Kirkland, just east of Kirkland Lake, has held its own or done better. Swastika, near the tumbling rapids of the Blanche River, has a railroad station, a bar-restaurant, a general store, huge community parks and other small businesses. Matachewan, located on the flashing rapids of the Montreal River, now depends on hunting, fishing and tourism. A hotel, school, garage, general store and other small businesses remain viable in this town.

In the largest town of the area – Kirkland Lake – short-term thinking changed to long-term planning in the '80s when politicians and other local futurists decided to construct a large fully serviced industrial park on the south side of town. Kirkland Lake is fast becoming an environmental solutions centre. Presently located in the industrial park is Northland Power, an energy-generating complex utilizing natural gas and wood waste to produce electricity. Westinghouse, a part of the community for many years in the past, has stayed with us and relocated to a new plant at the park to continue rewinding huge industrial motors and generators. An environment-conscious Trans Cycle Industries recycles PCB-contaminated machinery, converting the contents to inert scrap metal and small amounts of liquid PCBs which are then shipped to other areas to be processed and destroyed. New industries, potential residents of the park in the near future, include: Bennett Environmental, a soil de-contaminating industry; Unisphere Waste Conversion, a tire recycling plant; and Alco Power Systems, a battery producing industry. The Kirkland Lake area has a bright future.

Chapter II

# THE EARLY YEARS

## 1900 - 1929

### Dateline 1900 - 1910

- 1902: Construction of a railroad line to Northeastern Ontario begins
- 1903: Silver is discovered in Cobalt; a silver rush ensues
- 1906: A gold rush into the Larder Lake area begins
- 1907: Chesterville Mine on the east shore of Larder Lake is incorporated, Teck Township is surveyed and a nearby lake is named after Winnifred Kirkland of the Surveys Branch, Department of Lands and Forests
- 1909: Gold is discovered in Porcupine; a gold rush into this area ensues
- 1910: In Toronto, Harry Oakes inquires at the Department of Mines about the latest information on geology east of Swastika

### 1911

- Gold is found west of Swastika by a prospector named Baldwin
- Mining claims around Kirkland Lake are staked
- Two new mines have started up in Swastika: Swastika Mine on the north side of Otto Lake and Lucky Cross Mine within view of the railroad
- Harry Oakes, 36, arrives in Swastika by train on June 19

### 1912

- Harry Oakes and the Tough brothers stake claims east of Kirkland Lake at what will become the Tough-Oakes Mine
- Gold ore is bagged and hauled out of the new Tough-Oakes
- Harry Oakes stakes claims on the south shore of Kirkland Lake; later in the year, he purchases adjoining claims and names the property 'Lake Shore'
- A corduroy road is constructed from Swastika to Tough-Oakes

### 1913

- Tough-Oakes produces $66,632 of ore
- Swastika Mine produces $7,172 of ore

### 1914

- Lake Shore Mine is incorporated and shaft sinking begins
- A 26-mile hydro line from Charlton to Kirkland Lake is constructed, providing the new gold camp with 33,000 volts
- Huronia Gold Mine opens south of Beaverhouse Lake, about 15 miles east of Kirkland Lake; much-needed power is generated from a nearby falls
- Tough-Oakes produces $117,644 of gold ore; the mine's master mechanic dies of burns received while on the job and becomes Kirkland Lake's first mining fatality
- World War I begins

### July 12, 1914

## Tough-Oakes Mine

### Murdoch G. Lloyd (1869 - 1914)

Murdoch Lloyd was born in 1869, the year Ulysses S. Grant was inaugurated as the 18th President of the United States and the year the Red River Rebellion began in Canada. Lloyd, 45 and a widower, was employed as a master mechanic at Tough-Oakes.

While repairing a boiler on July 11, he was severely scalded with hot water and steam when a tube suddenly blew out. Suffering extensive burns to his body, he went into shock and died 12 hours later on July 12.

This was the first accidental mining fatality in the Kirkland Lake gold camp.

### 1915

- Kirkland Lake Gold Mine is organized
- Tough-Oakes operates a cyanide mill with a capacity of 125 tons per day
- No fatal mining accidents occur in the area this year

### 1916

- Teck-Hughes builds a mill
- Tough-Oakes employs 300 men and produces $700,000 in gold
- The Ontario Bureau of Mines reports many small gold mines are being established in the local gold camp
- The Roman Catholic Church of St. Jerome is established in town
- No mining accidents in the area gold camps result in deaths this year

### 1917

- Sylvanite Mine is incorporated; an accidental mining fatality occurs here this year, as do mining fatalities at Baldwin and Teck-Hughes
- Teck-Hughes produces $66,722 of ore
- Lake Shore and the newly incorporated Kirkland Lake Gold erect mills
- Northern Ontario Light and Power Company completes an electrical transmission line from Cobalt to Kirkland Lake
- Swastika Mine and Lucky Cross Mine close down
- During the winter, Kirkland Lake's resident doctor Dr. J. P. Wilson is faced with an influenza epidemic

### April 10, 1917

## Sylvanite Mine

## Louis Gruby (1892 - 1917)

Twenty-five-year-old Louis Gruby, an immigrant to Canada from the Austro-Hungarian Empire, was killed at Sylvanite on April 10.

Falling 25 feet from a set of timbers to the bottom of the shaft, he landed on his head and received a severe fracture at the base of his skull. Unconscious, he was taken to the surface bunkhouse where two doctors operated on him. Five hours after the accident, he was pronounced dead by Dr. J. P. Wilson. His family was unknown.

Nine miners would die at Sylvanite during its 48 years of operation; Gruby was the first.

### July 30, 1917

## Baldwin Mine

## Stephen A. Powers (1865 - 1917)

Stephen Powers was born at Perth, Ontario in 1865; his parents were Diven Peters and Catherine Donnelly.

At the age of 52, Powers was married and had seven children. He had been a resident of Cobalt and Haileybury and for three years lived on a farm near Bourkes Station.

Powers was an experienced miner, having worked as a shift boss in many mines in the west as well as the Cobalt camp. He had been in charge of development work at the Baldwin since April 9, 1917. The mine was located along the main line of the Temiskaming and Northern Ontario Railway at mileage 167.

On July 30, he and three assistants were unloading lumber from a rail car at the Baldwin property. He was removing the stakes supporting the load when one broke and flew out with terrific force, striking him in the head. He was knocked backwards and, in falling, struck the back of his head on a boulder. He never regained consciousness.

The railroad locomotive was uncoupled and carried the injured man back to Swastika, two miles away. He died before reaching Swastika where Dr. J. P. Wilson and Dr. Lowery examined the body and found a double fracture of the skull and excessive hemorrhaging.

*Baldwin Mine Property along the Temiskaming and Northern Ontario Railway. Photographer: Unknown*

### August 18, 1917

## Teck-Hughes Mine

## Joseph Hemmi (1881 - 1917)

Born in Finland in 1881, Joseph Hemmi was married and 36 years of age at the time of his accident. While working in a stope at the 200-foot level of Teck-Hughes on August 18, he dislodged a large piece of rock on a muck pile. The 1,000-pound rock rolled on its edge a few feet down the pile and then toppled over on his right leg, causing a severe compound fracture between the knee and the ankle.

As there was no hospital in Kirkland Lake at the time, he was taken to Providence Hospital in Haileybury for treatment and convalescence but he did not recover from his injury. Five weeks after his accident, he died in hospital from shock, following an operation to remove his infected leg.

### 1918

- No mining fatalities occur in the area this year
- Lake Shore mill progresses well and produces $416,414 of gold
- Teck-Hughes continues to produce
- Wright-Hargreaves makes plans to build a new headframe, mill and other mine buildings
- Gold-bearing ore at Tough-Oakes is exhausted and the mine closes
- Patricia Mine in Boston Creek produces $10,114 of ore
- World War I continues for most of the year but ends on November 11; after the war, new countries including Poland, Czechoslovakia, and Yugoslavia are created or re-established

### 1919

- Teck Township (Kirkland Lake) is officially recognized as a townsite, with a population of 675 to 3,000 (due to many transients and backward living accommodations, it was difficult to get an accurate census)

- Kirkland Lake Gold has a new mill and a water system for fire protection; the mine employs 75 men and produces $56,263 of ore
- No accidental deaths are reported this year in area mines
- Burnside Mine constructs a new mill
- Tough-Oakes amalgamates with Burnside Mine
- Work comes to a standstill on June 12 due to a miners' strike that lasts until October 22; miners return to work with no noticeable improvements in working conditions or wages
- Dr. J. P. Wilson sells his practice to Dr. R. H. Armstrong
- The first 'chateau' is built for Harry Oakes. This is a 1.5-storey building constructed on the south shoreline of Kirkland Lake

### 1920

- Three miners die in separate accidents at Kirkland Lake Gold, Teck-Hughes and Goldfields this year
- Huronia Mine becomes the Argonaut
- Kirkland Lake Gold produces $286,901 in ore
- A public school is built in Swastika
- Lyric and Strand theatres are built in town

### January 7, 1920

## Kirkland Lake Gold Mine

## E. H. Eagling (1892 - 1920)

Twenty-eight-year-old E. H. Eagling, an Englishman, was injured on January 7 at Kirkland Lake Gold and died from his injuries in Haileybury Hospital the next day. The unmarried man had been in active service during World War I. He came to the mine on December 1, 1919.

At the time of the accident, he was drilling a hole into the back of a stope on the 400 level. He had nearly completed the hole when a 1,500-pound rock fell from the wall and struck him, fracturing his right arm, right leg and spine, and causing internal injuries.

*Tough-Oakes Property 1920. Photographer: Unknown.*

*Miner drilling hole into back of stope.*
*Artist: Roland Russell*

Mining accidents were to claim the lives of 11 men in Kirkland Lake Gold from 1920 to 1955; Eagling was the first accidental victim.

An inquest was held at Haileybury by Dr. Kane of Cobalt. The ruling was accidental death.

## January 22, 1920
## Associated Goldfields Mine

### Peter Sorle (1881 - 1920)

The summer of 1919 saw a mini-boom in prospecting and development around Larder Lake. The payroll roster at Associated Goldfields now boasted 98 men and the mine had become the largest development proposition in Northern Ontario. While money for development was no problem, obtaining more men to work was. Owner George MacKay and General Manager C. G. Daimpre left Larder Lake to hire additional men; their preference was for Scandinavians. The Temiskaming and Northern Ontario Railroad had some years earlier recruited Scandinavian men to help build Northern Ontario's railway. However, once here, these men were quickly lured away with the promise of larger paycheques from the silver mines in Cobalt. Now Cobalt was being raided for experienced miners for the gold mines. By January 1920, the work force at Associated Goldfields exceeded 200 men.

Thirty-nine-year-old Peter Sorle, single and an employee of Norwegian descent, was injured at about 3:00 p.m. on January 22 and died in hospital on January 24.

He was rolling logs down a skidway near the shore of Larder Lake about four miles from the mine. Struggling to change the course of a log going askew, he caught it from behind with his cant hook. He was thrown over the log and fell with his head resting on one of the skids. While in this position, the log struck his head and fractured the base of his skull.

He was taken to a logging camp about a mile away and later to the mine. A doctor from Kirkland Lake was summoned but it was in the middle of winter and the roads from Kirkland Lake to Larder Lake were not good. The doctor arrived at 4:00 the next morning, 13 hours after the accident.

On January 24, Sorle was taken by stage coach and train to Haileybury Hospital where he died from his injuries.

Coroner Kane held an inquest at Haileybury. The verdict of the jury was as follows: "That Peter Sorle came to his death by accident on Thursday, January 22, 1920 by a log crushing him on the head while (he was) working on a skidway at the Associated Goldfields. We would recommend that companies provide better accommodation for their employees in case of accident."

A few weeks later, a resident physician was hired at the mine.

## April 22, 1920
# Teck-Hughes Mine
### James John Keane (1896 - 1920)

Twenty-four-year-old James Keane was a native of Cork, Ireland; before coming to Kirkland Lake, he was a sailor. He gained employment at Teck-Hughes on October 27, 1919. His duties at the mine were deckman; however, for about three weeks before the accident, he was a cagetender. He was single and a resident of Kirkland Lake for only six months.

At about 4:30 p.m. on April 22, he took a load of sharp drill steel to the 400 level, riding the cage without a lamp. A short time after reaching this level, the hoistman received the 3-1 signal and raised the cage but after it had gone up about 80 feet, he noticed the cable was vibrating in an unusual manner. He stopped the cage for a minute or two, expecting some signal. As he did not receive one, he started the engine slowly but again stopped as the cage was caught in the shaft.

On investigation, it was found that Keane, "in a dying condition and almost stripped of all clothing," was at the 500 level station with his feet in the sump and his head on a rail. The cage, containing 15 or 20 pieces of dull steel, was just below the 300 level. Three or four pieces of steel were caught in the side of the shaft, and eight or ten pieces were scattered on timbers at various positions below the cage. One door of the cage was torn off and the other was still attached but considerably bent.

Keane died before he was brought to surface where Dr. R. H. Armstrong pronounced the miner dead.

On April 23, an inquest ordered by Coroner Jackson was held at the Teck-Hughes offices. Evidence from one of the deckmen showed the deceased was not a careful cagetender and at times rode the cage without giving a signal, or he loaded too much steel on it. Two or three days before the accident, the doors had been torn off the cage because he did not close them properly. The fact that the wall plates in the shaft were scored beneath the cage indicated that the accident was due to one or more pieces of steel projecting beyond the floor of the cage. A verdict of 'accidental death' was rendered.

The undertaker, Thorpe Brothers from Haileybury, handled the funeral arrangements. On April 24, the body was laid to rest in the cemetery at Haileybury.

## 1921

- Wright-Hargreaves mill operates at 100 tons a day and the mine produces $458,780 in gold this year
- Falling rock at Kirkland Lake Gold kills one miner
- The Presbyterian Church is built on Government Road; in 1925, it becomes the United Church

## December 12, 1921
# Kirkland Lake Gold Mine
### Stephen Leonard Donaldson (1888 - 1921)

Stephen Donaldson was born on March 7, 1888 in Glen Almond, Quebec to George Robert Donaldson and Ellen McGuire of Buckingham, Quebec. At the time of his accident, Donaldson, 33, was married with a wife and three children; he also had a brother Gordon living in town.

Donaldson had worked in the Cobalt silver fields from 1908 to 1915, and subsequently lived in Kirkland Lake for six years.

A mine captain at Kirkland Lake Gold on December 12, he and Alfred Johns* were walking the workings. They had completed a stope inspection and were descending a manway to the 700 level when a large mass of rock fell. Donaldson was hit and received a skull fracture and some broken ribs; he died instantly. Johns injured his foot but managed to climb down the manway and get help; he later assisted in the removal of the body from the mine. Donaldson was pronounced dead by Dr. R. H. Armstrong.

An inquest into the accident was held by Dr. Cooper of Englehart. 'Accidental death' was the verdict.

** Alfred Johns would lose a son-in-law, Johnston W. Brown, in an accident at Kirkland Lake Gold in 1948.*

## 1922

- Work begins at Sylvanite but the biggest producer in the Kirkland Lake gold camp is Wright-Hargreaves
- Mines along the shore of Kirkland Lake start to dump their tailings into the lake
- A miner falls to his death in a shaft at Lake Shore
- The Northern News makes its appearance
- Stock market 'boom' begins in America

### October 17, 1922
## Lake Shore Mine
# Kent Englehutt (1883 - 1922)

Kent Englehutt was born in Nova Scotia to Mr. and Mrs. John Englehutt of Moose River on November 14, 1883. At the time of his death, he was 38, married and had two children.

He was an experienced miner, employed at Lake Shore for two years. In 1922, he was a shift boss of an operation to enlarge and retimber a shaft.

On October 17, he was being lowered down the shaft by means of a bucket. At the 100-foot level, he fell from it, plunging 300 feet down and landing in 200 feet of water at the bottom. Men had to use grappling hooks to retrieve his body. His skull was fractured and his body had numerous broken bones. Death would have been almost instantaneous.

Interment occurred in New Liskeard as there was no established cemetery in Kirkland Lake at the time.

Seventy miners were to die accidentally in Lake Shore from 1922 to 1987. Englehutt was the first fatality at this mine.

## 1923

- Biggest producer in the Kirkland Lake gold mining camp is Teck-Hughes, which grosses over $1 million
- Lake Shore treats 100 tons a day
- A miner is killed at Tough-Oakes in January
- Wright-Hargreaves and Sylvanite shaft reach the 1,000-foot level
- The Tough-Oakes and Burnside Mines consolidate to become the Tough-Oakes-Burnside Mine
- Argonaut in Gauthier Township becomes a good producer as a small mine
- The Rouyn rush begins
- Kirkland Lake's population stands at 1,400
- Kirkland Lake Cemetery is established
- Construction of the Nipissing Central Railway east from Swastika begins in April

### January 14, 1923
## Tough-Oakes Mine
# Hilary McAlinden (1902 - 1923)

Two miners narrowly missed death and another died when ore beneath their footing started to move down a chute. Hilary McAlinden, 21, was fatally buried in the avalanche of ore while partners John S. Gillies and Jim Dunagan narrowly escaped a similar fate.

As Gillies and McAlinden were making ready for a blast in a stope, they removed their machine and shut off the water and air lines; then they notified Dunagan, the blaster, that all was ready. They were putting the final work on the fuses while Dunagan was watching them from about 10 feet away. A rumble of rock was heard beneath them – then a second rumble. This time, the rock caved in beneath McAlinden's feet and in an instant he was swallowed in tons of rock. Gillies made a desperate effort to reach him but had to stop when the rocks beneath his feet also started to move. Both Gillies and Dunagan called to their partner and received an answer. Soon muckers gathered at the lower chute to work frantically upwards. Detecting the odor of burning clothing, Gillies called up to entrapped McAlinden. He received this reply: "Pour down some water, I'm burning to death." In his haste to secure some water, Gillies fell down a ladder, breaking two ribs and severely injuring a shoulder. When water was finally secured and poured over the rock above the helpless McAlinden, no answer came from repeated calls. Three hours later, the dead body was retrieved and removed from the mine.

On examination, the doctor found a fracture of the skull above the left eye and burned thighs from the hip to the knee. The fire was attributed to a lamp McAlinden was carrying at the time of the accident.

What had happened to cause the sudden movement of rock downwards? This is a case where little or no settlement of the ore on which the men were standing had occurred in the stope, even though 20 cars had been drawn from below previously. No one had taken notice of this fact until after the ore settled during the accident.

McAlinden had started working in the mine just four months previously, on October 6, 1922. Surviving him were his father and mother, three sisters and six brothers. One of his brothers was a shift boss at Kirkland Lake Gold and two others were employed in mining endeavours north of Kirkland Lake in the Lightning River district. His sisters and parents lived at Mayo, near Buckingham, Quebec. His body was sent there for funeral services and burial.

### 1924

- Tough-Oakes-Burnside catches fire. The staff house, office and one other house are destroyed
- Reverend Cyril Goodier meets Harry Oakes. Oakes gives him permission to build St. Peter's on the Rock Church on one of his lots
- Harry Oakes donates land for a hospital and looks after the debentures
- Kirkland Lake Gold's shaft is 1,600 feet deep but ore values are low
- Five miners lose their lives in four separate accidents at Teck-Hughes
- First steam passenger train comes to Kirkland Lake complete with Pullmans. The Nipissing Central Railroad extends to Larder Lake

### April 29, 1924

## Teck-Hughes Mine

## Carlos Starnio (1890 - 1924)

Thirty-four-year-old Carlos Starnio, Italian and single, met his death at Teck-Hughes on April 29 while working at the 800 level. He was a contract mucker who had been an employee at the mine for some time.

On the morning of the accident, he and partner William Maki were mucking in a narrow drift. Joseph Cuchini handed down a seven-foot steel to Starnio who turned the steel around, end for end, in the confined space. In doing so, he accidentally struck a 10-foot-high muck pile, causing some loose muck to slide. He jumped back behind an ore car to protect himself from the slide when suddenly, two large pieces of rock weighing several tons each came off the side wall and pinned him between the wall and the car, crushing him.

Dr. R. H. Armstrong was called and found Starnio pinned but alive. About two hours later, rescue operations had still not released him and he was pronounced dead by the doctor. The body was finally extricated after another hour.

An inquest into the death was called by Dr. Cooper of Englehart. The verdict of the jury was that the miner met his death by accident, due to falling rock.

### August 24, 1924

## Teck-Hughes Mine

## Two Miners Are Killed In An Accident at Teck-Hughes

Two men working in a stope at the 700 level of Teck-Hughes were killed instantly by falling rock and debris.

Tony Harris and Wesel Barabas were 40 feet below a platform supporting waste rock that would later be used as fill. A slip of a rock from the wall supporting one side of the platform caused rock and other debris to crash down onto them.

It took a day for rescuers to retrieve the body of Barabas; he was found under six feet of rock and debris. Judging by the position of his body, it appeared he had tried to escape the rockfall. He sustained a crushed right side, fractured ribs and a broken collarbone. Death was attributed to pressure on his heart.

There was so much debris and broken rock in the area that it took a blasting operation a week to locate the crushed body of Harris.

An inquest was held by Coroner J. S. McCullough of New Liskeard. The jury brought in a verdict of 'accidental death'.

## Wesel Barabas (1887 - 1924)

Wesel Barabas, 37 and of Polish descent, was an emigrant from the province of Galicia in the Austro-Hungarian Empire.

He worked in the Kirkland Lake gold camp for three or four years, and had been working at Teck-Hughes for only three months. Earlier, he had worked at Beaver Mine and was still residing with his family on the Beaver property at the time of his death.

The funeral arrangements were handled by Allen's Undertaking Parlour; the funeral mass was held at St. Jerome's Catholic Church in town and officiated by Reverend Father Martindale. The new church was filled with friends who later accompanied the body to the cemetery.

Mrs. Wesel Barabas and their six-week-old baby daughter later moved to Timmins to make their new home.

## Tony Harris (1887 - 1924)

Tony Harris, 37 and single, had been an employee at Teck-Hughes for about a year. His funeral service was held at the United Church with Reverend J. F. Anderson officiating. Relatives from Schumacher and local friends attended.

### November 16, 1924

# Teck-Hughes Mine

## Alex Tomachuk (1897 - 1924)

Twenty-seven-year-old Alex Tomachuk, a miner at Teck-Hughes, was brought to Cobalt Mines Hospital after his accident and died there on December 6 from injuries sustained.

On November 16, he descended from a stope and was standing on the eighth level near a chute. It was the practice to throw down dull drill steels from the stope into this chute to the level, a distance of 50 feet. In fact, the chute (measuring 14x16 inches) was designed and constructed for this purpose.

At the time of the accident, P. Conlin, Tomachuk's helper, received the go-ahead from Tomachuk before he threw down two or three dozen steels. Tomachuk was hit in the lower part of the chest by one of the steels that had rebounded when it hit the track or wall or other steels. There were no eyewitnesses to the accident but Tomachuk walked about 300 feet to the station and was taken to surface by the cagetender. He was placed in a hospital room and a doctor was called. As his condition indicated internal injuries of a severe nature, he was sent on the first train to Cobalt. After consulting with the company's surgeon (Dr. R. H. Armstrong), Dr. Mitchell operated. A rupture about four inches long was found in one lobe of the liver. After the operation, the injured miner made good progress towards a complete recovery until December 2, when his condition started to deteriorate. He was operated upon again on December 5 when it was found that he had a malfunctioning bowel. He died the next morning as toxins had entered his system, due to a stoppage of the bowel.

Of Polish descent, Tomachuk had no relatives living in this area; his wife and child lived on a farm at Poplarfield, Manitoba. The train service to their home was infrequent, so it was difficult to notify Mrs. Tomachuk of her husband's death. His body was taken by train to Mileage 104 where it was placed in the vault at the Roman Catholic cemetery.

An inquest was held by Coroner Arnold in Haileybury. According to testimony, the mine captain had issued orders about two weeks before the accident to lower the steel by rope instead of throwing it down the chutes, on account of "the way the steel was being damaged and bent." According to evidence, these instructions were passed on to the men and enforced in a half-hearted manner by Shift Boss Fred Goodman. The drillers' helpers stated at the inquest that they had never heard of these orders.

### November 22, 1924

# Teck-Hughes Mine

## John Bedforth (1881 - 1924)

A prospector in the early days, John Bedforth came to Cobalt where he met and married Amy Louise Deering. Stories have it that he was in Kirkland Lake in 1911 and had staked some claims around Goodfish Lake at the time. He and his wife

*Bedforth was one of eight in the shaft, filling the bucket. Photographer: Unknown*

moved to Kirkland Lake in 1919 where their two children, Francis and George, were born. The family lived on Second Street. Bedforth was 43 when he suffered fatal injuries at the bottom of the Teck-Hughes central shaft while at work on November 22.

He was a shaftman engaged in sinking a shaft, then at a depth of 500 feet. A one-ton bucket filled to within about three inches from the top was being hoisted when the edge of the bucket struck the wall plate of the shaft near the collar. A pail full of small pieces of rock fell to the bottom. Bedforth was one of eight men engaged in filling a second bucket and the only one struck by the falling rock. A piece of rock hit the back of his head and he fell unconscious to the ground.

Both Hoistman Archie Paget and Deckman Roy Robertson stated that hoisting was being done at a regular rate and the bucket was being slowed at the collar. When the bucket struck, the hoistman stopped it. He went to the shaft to see what happened and on the way heard the danger signal. After examining the bucket, he returned to the hoist and raised the bucket for dumping, then lowered the empty bucket to the bottom of the shaft to bring up the injured miner. The doctor examined Bedforth within a few minutes of the accident and found a hole about 1.5-inches in diameter in the top of his head and a possible fracture at the base of his skull. Bedforth died about five hours later as a result of his injuries.

Many floral tributes were delivered to his home where a funeral service was conducted by Reverend

Cyril Goodier. The pall bearers were: J. Hallett, J. Neale, J. Englehart, Boyd Robertson, B. James and J. Gudrie. Interment was conducted in the Kirkland Lake cemetery.

An inquest was held by Dr. Cooper in Englehart. The jury's verdict was 'accidental death'.

The remaining Bedforth family stayed in Kirkland Lake for three more years but on March 14, 1927, Amy Louise died of pneumonia. The two young orphan children moved to North Bay to live with relatives. Francis (Mrs. Marcotte) presently (1999) resides there. George found his way back to Kirkland Lake and was an apprentice machinist at Lake Shore in the late '30s. With the advent of the Second World War, he joined the army. He presently lives in Onaping, Ontario.

Details of the Bedforth family were given by Kirkland Lake's Arthur Turner, nephew of the late John Bedforth.

*St. Peter's On the Rock is built in 1925. Artist: Ann Black*

### 1925

- Miner dies at Teck-Hughes in the 1,000-foot shaft
- Lake Shore Mine produces $1,958,720 of gold and silver but a miner is killed here in May
- Barry Hollinger Mine in Boston Creek produces $65,978 worth of valuable ore but a blasting accident near the property in September claims the life of a miner
- The value of the total production of gold and silver at other local mines this year: Wright-Hargreaves $1,913,401, Teck-Hughes $996,943, Tough-Oakes-Burnside $263,064 and Argonaut $214,143
- Nipissing Central Railroad has its terminus at the Quebec border
- Construction of the new local hospital begins
- One of the new doctors to move into town is Dr. W. C. Rumball

### January 21, 1925

## Teck-Hughes Mine

## Ernest Littlejohns (1901 - 1925)

Twenty-four-year-old Ernest Littlejohns, a single Canadian from Innerkip, Ontario, was an experienced shaftman who had worked in various district mines during the past four years; more than half of this time, he was employed at Teck-Hughes. At the time of the accident, he was working with Edward Gates and Fred Smith, mucking out a round at the bottom of the shaft.

Littlejohns was in the process of sending up a bucket of muck; he kept his left hand on the signal cord and placed his right hand on the bottom of a full bucket in front of him. The end of his right index finger was caught between the full bucket and the top of the empty one. The flesh on the end of his finger – about one inch of it – was ripped off and remained in his glove.

He went to surface and to the office of Dr. R. H. Armstrong for medical attention. The doctor decided to amputate the first joint of the damaged finger. While Littlejohns was administered a chloroform anaesthetic by assistant Dr. Luke, he unexpectedly went into shock and within 10 minutes stopped breathing. Artificial respiration was performed and he was briefly revived, but breathing stopped a second time and a second revival was unsuccessful.

At a post-mortem performed at the coroner's request, Dr. Edis discovered that the spleen and liver of the miner were larger than usual. He also found that the thymus gland, which usually disappears with childhood, was still present. He stated that an adult with a large thymus at maturity is susceptible to shock of any nature and unable to tolerate an anaesthetic.

The jury at the inquest returned a verdict of 'accidental death' while being given an anaesthetic, adding that no censure could be reasonably directed towards the attending physicians.

Littlejohn's father travelled from Southern Ontario to make funeral arrangements. The deceased was a member of the IOOF.

## May 29, 1925
## Lake Shore Mine
### Englebert Rumig (1897 - 1925)

Englebert Rumig, a drilling machine runner employed at Lake Shore, was killed on May 29 when struck by falling rock.

He was working with a partner in a stope and they were raising the arm of a drilling machine when a two-ton mass of rock fell on him, squeezing his head against the machine, breaking his neck and killing him instantly. He was 28 and left a widow but no children.

The funeral service was held at St. Jerome's Roman Catholic Church with Reverend Father Martindale officiating. J. Rumig, brother of the deceased, took charge of the remains after the funeral, and the body was placed on a southbound train. Burial occurred in Mildmay in Bruce County, Ontario.

An inquest was conducted by Coroner Arnold. The jury's verdict was 'accidental death'.

*Drilling machine runner working with partner. Photographer: Unknown*

*McGillivray was at the bottom of the shaft. The other two were at surface. Artist: Unknown*

## September 28, 1925
## Albert Johnson's Prospect
### John McGillivray (1897 - 1925)

John McGillivray, 28 and a native of Nova Scotia, was well known in the mining camps of the district. He died at Cobalt Mines Hospital six hours after being caught in a blast while shaft-sinking at Boston Creek. His skull was fractured, brain damage was physically visible and part of his left arm was torn off. He was brought to Cobalt on the Continental, flagged at Boston Creek station, but died soon after admission to hospital.

An option had been taken at the Boston Creek property by David H. Angus of Haileybury and a contract was let on September 15, 1925 to Albert Johnson, Patrick Finn and McGillivray to sink a 25-foot shaft.

At the time of the accident, the shaft depth had progressed to 14 feet. McGillivray was at the bottom lighting the fuses; the other two men were not in the shaft. He was still in the shaft when the blasts went off.

An inquest into the accident was called by Coroner Jackson of Haileybury. It became complicated and frustrating with unanswered questions about the contract, fuses, matches, dampness, etc. In fact, the jurors adjourned for a period of time because of confused testimony and arguing of witnesses. In time, the jury came to a verdict, finding that McGillivray was killed by a blast; he had used tape fuse that contravened the Mining Act. The jury found negligence on the part of D. H. Angus, Captain Albert Johnson, Patrick Finn and the deceased, McGillivray.

The body was shipped to Nova Scotia for burial. The newspaper named one living relative, Michael McGillivray, the brother of the deceased.

*A 25-bed hospital opened in town in 1926.*
*Photographer: Eddie Duke*

### 1926

- Kirkland Lake Gold locates ore values at the 2,475 level
- The Wright-Hargreaves shaft is at 1,375 feet. An August accident in this shaft claims the life of a miner
- Sylvanite Mine shaft reaches 1,500 feet and a new mill opens
- Total work force in the Kirkland Lake gold camp is 1,460 persons. The mining camp provides the town with a healthy $1,800,000 payroll. Kirkland Lake's population is 3,600 people
- A miner is killed at the Tough-Oakes-Burnside Mill
- Teck-Hughes employs 270 people
- The Lake Shore bunkhouse is built

## August 7, 1926
## Wright-Hargreaves Mine

# Weniford V. Patriquin

## (1891 - 1926)

"In loving memory of Weniford V. Patriquin. He was born on June 13, 1891 and died on August 7, 1926. He was loving and kind in all his ways, upright and just to the end of his days, sincere and true in heart and mind, what a beautiful memory to leave behind." (Epitaph on the stone marking the grave in the Kirkland Lake cemetery.)

Canadian and married with a wife and two children, the 35-year-old Patriquin had been employed at Wright-Hargreaves for over three years. At the time of the accident, he was a cagetender. The hoistman on the shift was his good friend Earl Ripley who initially had asked Patriquin to leave Nova Scotia and come to Kirkland Lake.

Clarence Bonsall and Patriquin were in the process of pushing an ore car onto the cage and were having difficulty doing so. Bonsall pulled the car while Patriquin went around to push with his feet braced on the car and his back to a timber. Unexpectedly, the cage started up with a jerk and Patriquin was taken up about 14 feet. When the cage stopped, he fell to the ground.

According to the 1926 Kirkland District Hospital Records, Patriquin was admitted on August 7 and died the same day.

Dr. Cameron, the mine doctor, testified at the inquest that at 9:20 on the evening of August 7, "I was called to Wright-Hargreaves. When I arrived, Patriquin was in the bunkhouse in a very critical condition and was breathing very heavily. There was a fractured pelvis, bleeding, all sorts of internal injuries and air hunger. He had a very bad injury on his right leg and also on the back of his left leg. He was wrapped in hot water bottles and then transferred to hospital where he was given a stimulant. He was conscious and with his wife until 20 minutes before he died which was at 10:55."

The inquest jury included Foreman J. Ash, Dan Briggs, H. Parker, John Furlong, A. Gosselin, E. E. Doxie and E. Beauschene. Coroner R. H. Armstrong conducted the inquiry and Mining Inspector McMillan examined the various witnesses.*

The jury went out for about 10 minutes and returned with the following verdict: "We the jury find that Weniford Patriquin came to his death by being caught between the cage and the timber, the cage having been raised through signals of some unknown cause."

** One of the witnesses was Deckman Thomas Bentley. His son Stoel would die in a mining accident at Lake Shore in 1959.*

### November 16, 1926

## Tough-Oakes-Burnside Mine

## John Clifford Furber

### (1890 - 1926)

Thirty-six-year-old John Furber came to Kirkland Lake from New Liskeard; he had worked at Tough-Oakes-Burnside about three years. On September 24, he married Violet Ruth Sutton. Two months later he was dead, fatally injured in a mining accident on November 15.

Furber's regular work was to load the buckets of the aerial tram. When there were no buckets to load, it was his duty to assist the crusherman in the crusher house.

On the afternoon of the accident, he had no buckets to load, so he assisted Joseph Potter in cutting steel and nailing a piece of belting on a conveyor. Without saying anything to Potter, he went upstairs to the trommel floor to help Roy Labelle who was clearing away some wet material that had gathered on the casing at the side of the trommel. Labelle went around to the other side and heard something fall. At the feed end of the trommel, he found Furber caught in the gears and being drawn by his clothing between a pinion and a larger unguarded gear wheel. Furber's clothes were being drawn tightly into the gears along with his arms and shoulder. Labelle ran down the stairs and pulled the motor's switch in the crusher room where he gave the alarm. Crusherman Joseph Griffin came at once and threw the belt off the driving pulley; with others from the machine shop, he attempted to release Furber by reversing the movement of the pulley and backing the gears. It took several turns to do this as the clothing was drawn tightly into the gears and wound around the shafting that drives the pinion. By cutting away the clothing and reversing the gears, Furber was released. Both of his arms were broken; his shoulder area was crushed and showed the impression of the cog wheel. His left collar bone was torn away and his chest was crushed, causing internal injuries.

Conscious and in tremendous pain, Furber was asked by Griffin how he got into the gears. He said he'd reached over to lift out a piece of oversize and forgotten about the gear.

The accident happened at about 4:00 p.m. Furber was taken to hospital but died early the next morning. Interment was conducted in the New Liskeard Pioneer Cemetery. The epitaph on the obelisk reads: IN MY FATHER'S HOUSE ARE MANY MANSIONS – "WE WILL MEET AGAIN".

### 1927

- Lake Shore has one fatality and Wright-Hargreaves has three
- The superintendent of Teck-Hughes is killed during a routine underground viewing
- Ferguson Highway reaches Kirkland Lake
- First Baptist Church is built on Tower Street
- Jewish synagogue is built on Station Road
- First Branch 87 Legion Building, the 'Wright House', is constructed on Government Road

### January 17, 1927

## Teck-Hughes Mine

## Joseph A. Orr (1890 - 1927)

Joseph Orr was born to Mr. and Mrs. E. A. Orr in the town of Clarkson, Ontario, located 16 miles west of Toronto.

He had spent time in the Sudbury area, working for Inco for 12 years. He developed many friendships in Copper Cliff, Creighton Mines and Sudbury and was a well-known second baseman on the Sudbury baseball team of 1916/17. He served overseas with the Canadian Engineers Battalion in 1917.

Orr was married to the daughter of Reverend and Mrs. J. Hodgson of Byng Inlet; the couple took up

*Downtown Kirkland Lake 1927. Photographer: Unknown*

residence at Teck-Hughes Property when Orr was hired as superintendent in 1926. The Orrs were members of Trinity United Church in town.

He had a sister, Margaret (a nurse at Toronto General Hospital), and two brothers, F. H. and Howard, both of Clarkson.

Superintendent Orr routinely went underground to inspect operations. On the afternoon of the accident, Orr and Mine Foreman George Gill* entered a cross-cut at the 500 level to make an examination of an area soon to be timbered. To reach the desired area, they had to cross a pile of muck located on a grizzly. Orr, who was in the lead, climbed up the rockpile; it suddenly began to slide, taking him with it. He was caught partway into the grizzly and covered with rock, only his head and one hand left visible. Miners frantically tried to release him but he died within 10 minutes. It took over two hours to extricate his body.

According to the medical evidence of Dr. W. C. Rumball, death was due to shock caused by numerous bruises and by the distortion the body had been subjected to.

The funeral was well attended and there were many floral tributes. The pallbearers were brothers of the Masonic Lodge. Interment took place in Clarkson.

** George Gill's son would die in a mining accident at Teck-Hughes in 1948.*

### February 8, 1927

## Lake Shore Mine

## David Noble (1903 - 1927)

Born in Scotland, David Noble came to Canada in 1919 at 16 years of age. By the age of 20, he was in Kirkland Lake working in the gold mines. In 1925, he married Anne Ryan of Pembroke, Ontario. A joyous result of this marriage was a baby girl born in the Kirkland District Hospital on June 15, 1926.

According to the Northern News, Noble was one of the most popular young men in the gold camp. He had a cheerful disposition which assured him many friends. He was also community-minded: he was a valued member of the local fire department and a dedicated member of the Masonic Lodge.

He had been employed at Lake Shore from June 1923 and was a shaftman, sinking the main shaft from the 1,000-foot level. The shaft had reached a depth of 1,250 feet at the time of the accident.

On this day, he was working with other shaftmen at the bottom. A full bucket of rock was sent up and, while progressing upwards, hit a plank. This caused a piece of rock, about the size of a man's head, to fall from the bucket. The men at the bottom heard the rock striking the timber as it tumbled down and they rushed to the sides of the shaft. Unfortunately, Noble was hit and knocked down. The rock hit his side, causing fractured ribs and a punctured lung. He was rushed to hospital and listed in serious condition. Dr. Mitchell of Cobalt and Dr. G.M. Cameron of Kirkland Lake operated on him and he made favourable progress. But peritonitis set in and his condition deteriorated; he died on February 14.

An inquest was held with Coroner R. H. Armstrong presiding. It was learned that the injuries consisted of a fractured ninth rib: splinters punctured a lung and ruptured the liver and a kidney. The jury brought in the verdict "that David Noble came to his death accidentally by being struck by a piece of rock from a bucket which was overfilled in the shaft of the Lake Shore Mine."

*Inscription on small headstone is simply D. NOBLE. Photographer: Author*

The funeral service was attended by many friends. The funeral procession was headed by members of the Masonic Temple and the local Fire Brigade; the pallbearers were all brother Masons. There were many floral offerings placed as a tribute to the 24-year-old miner. Interment was conducted in the Doric Lodge section of the Kirkland Lake cemetery.

## May 30, 1927
## Wright-Hargreaves Mine

### Branko Ruzick (1900 - 1927)

Branko Ruzick was a Yugoslavian immigrant who came to Canada with the goal to make money and give his family back home a better life. Immigrants like Branko were highly motivated and in large demand: they constantly had their minds on their homes, families and homeland to which they hoped to return one day. The 27-year-old had a wife, parents and two brothers needing support in Yugoslavia.

Ruzick had been working for the mine nearly two years. His supervisors noted that he was a man of more than ordinary ability. First employed as a trammer, then as a driller's helper, he was engaged as a machine driller on May 15, 1927.

At the time of the accident, Ruzick and his helper were preparing to set up a machine to drill a round of holes in the back of a shrinkage stope. As the muck was too close to the roof to allow the bar to be put in place, Ruzick told his helper to go down to level and have the trammers draw off six cars of muck from the chute. Ten days earlier, no rock had been drawn from this chute. There was no reason to believe that the rock had not settled.

When the helper returned, he saw Ruzick standing on the rocks over the chute. As the trammers drew a second car of rock, the rock in the stope caved quickly and without warning. The helper heard Ruzick call and saw him try to grasp an air-line connection while he fell. Unable to rescue Ruzick, the helper called to the trammers for assistance.

When the mine superintendent and shift boss arrived, they found the rock had subsided 25 feet over the chute, leaving almost vertical walls of broken ore on each side. As the conditions left no chance to remove the rock covering Ruzick, it was decided to draw off some rock at another chute to rescue him.

After removing 80 tons of rock, the body was recovered. The attending doctor found that Ruzick's neck was broken and he was dead.

He was given a funeral mass at St. Jerome's Catholic Church where many of his friends gathered. He had two cousins who also worked at Wright-Hargreaves – his only relatives in Canada. Together, they went to the Kirkland Lake Cemetery to view the interment.

To the present day, this grave remains unmarked.

### August 13, 1927

## Wright-Hargreaves Mine

## Nicolai Porre (1894 – 1927)

Nicolai Porre was born in Finland in 1894 and married Lempi there in 1913. Their first child was born in 1915 but died in infancy. Their second child, Irja, was born in Haarma province July 1, 1916; their third, Irma, was born on June 12, 1919.

Porre worked on the railroads in Finland as a section man, cutting and hauling wood, and hauling water. Lempi was an accomplished seamstress.

Porre left his homeland and family in 1925 and came to Northern Ontario where he worked on the construction of the Nipissing Central Railroad between Kirkland Lake and the Quebec border. Soon after, Lempi also left Finland, leaving the children with friends; she and her husband took up residence in Kirkland Lake. Around this time, Porre became employed at Wright-Hargreaves. The farmhouse in Finland was sold and the earnings used to get the children to their parents in Canada.

The two daughters, aged 8 and 11, travelled by themselves by ship from Finland to England and to Halifax. After a seasick journey, they spent a few days waiting alone at the terminal. They spoke no English and, when they got hungry, they would take change from their change purse, dump the coins on the store counter and point at the apples. They left Halifax and travelled by train to Kirkland Lake where they were re-united with their parents in February 1927.

Tragedy struck the family a few months later in May when Mrs. Porre died at home as a result of an abortion performed by a medical layperson. Porre was now left with two very young daughters.

Tragedy would strike again on August 13. Porre, who had almost two full years of experience working as a machine helper at Wright-Hargreaves, was drilling in a stope with machineman John Bohonos. In another area of the stope, Sam Terzich and Clarence Hewson were running another machine and H. G. Woodman was running his machine over the manway. At the beginning of the shift, all the men were informed by Shift Boss Harry Sauvé that No. 25 chute had been hung up and then blasted. Within an hour, two cars of ore were drawn from this chute but it remained hung-up.

Stope Boss George Sygesuk warned the men the chute was still hung up and that soon it would be blasted again. They were told to move to another area of the stope until this was done. Neither Woodman nor Bohonos obeyed these instructions, the former because he was set up over the manway and not over the chute, and the latter because he was not directly over the chute but 15 feet away from it.

The chute was blasted. Bohonos was carried down about 15 feet with the run of muck into the chute. Buried under the rubble, he remained conscious and could see the lanterns of the miners trying frantically to rescue him. His partner Porre was carried further down, about 20 feet. He was also conscious but was buried to a further extent.

After about six hours, Bohonos was freed and, but for a few scratches, bruises and a broken finger, alive. Porre was talking to the rescue crew while being dug out. After 12 hours of being trapped, he was freed and taken to hospital.

There was no evidence of internal injury and he was kept under observation at the hospital. His daughters were brought in through the back door to see him but were not allowed to stay with him for any length of time.

But two days after the accident, he was dead. Dr. C. S. Harris considered death due to shock as a result of exposure.

With their parents gone, Irja and Irma were now orphans. Only memories of a "kind and mellow" father and a "stern and practical" mother would endure.

*Nicolai and Lempi Porre with daughters Irma and Irja (Finland 1921). Photographer: Unknown*

The girls remained temporarily with members of the local Finnish community. A court hearing was held in Swastika where they were asked if they wanted to stay in Canada or go back to Finland. They understood little English but were coached previous to the day of the hearing to say they wanted to stay.

A decision was made and they were allowed to remain. They did get some support from the mine ($15 per month) but saw none of it as it was paid to their guardians.

After the death of their parents, the family's possessions were auctioned off. Mrs. Porre's wonderful linens and Mr. Porre's gold railroad watch were sold. The girls received none of the money. The only things they got to keep were their mother's gold engagement band and wedding ring, and their father's wedding ring.

The girls went to stay at Joki's boarding house. Being older, Irja worked there, sewing, doing the laundry and washing floors – heavy physical work. She was neither paid nor allowed to keep any tips that were given to her. She dropped out of school at grade four to work and support herself and her younger sister. Later, they moved to Hielala's boarding house where the work was easier. However, being an orphan in a strange country was traumatic for Irja and she was forced to overcome many hardships at an early age.

When Irja turned 16, she went to work for Lindfor's Dairy, doing book work as she excelled at arithmetic. Irma attended school for a few more years but eventually quit to work in Butorac's grocery store.

In 1937, Irja married Leo Karen, a Finlander from Larder Lake. She moved there and worked along with her husband as a cook in bush camps, at the Raven Lake Lower House and in a hotel. They adopted a daughter, Milda, who was born in 1951. Leo was a long-time employee at Kerr Addison, first underground, then as a mill electrician. In 1986, Irja and Leo moved to British Columbia to be close to their only daughter and grandson. Leo died there in 1991. Irja, or Irene as she is now more often called, presently (1999) lives in Powell River, B.C. on her own.

Irma married Leo Waldo of Kirkland Lake in 1942 and later moved to Toronto. She now (1999) lives there alone, Leo having died several years ago. She has three children: Bill (now 55), Karen (50) and John.

The personal information for this account was provided by Irene Karen and Milda Karen-Byng. Milda, a lawyer, presently lives in Powell River; she is married to Ted and has one son, Jordan, and two stepsons, John and Dan. Milda remembers well her life in Larder Lake and passes on one anecdote: "I remember one day a rumbling in Larder Lake. My father was at home at the time and told me it was a rockburst from Kerr Addison, seven miles away. He looked at his watch and realized the burst occurred at a time when the miners would be underground at their work places. He said it's a bad one and probably someone has been killed. I remember my feeling of dread of wondering who it was."

While in Larder Lake, Milda became a good friend of Elly Meister. Elly was 12 when she lost her father in a rockburst at Kerr Addison in 1963.

## November 30, 1927
# Wright-Hargreaves Mine
## Fred Dufour (1903 - 1927)

Born in Canada in 1903, Fred Dufour came to Kirkland Lake from Blind River, Ontario and became employed at Wright-Hargreaves in August 1926.

Dufour, single, had a sister, Mrs. J. Campbell, who resided on Tower Street. He boarded a few blocks away at the home of Mr. John Lennon on Third Street. Dufour was a member of the United Church and was highly regarded by his acquaintances. The Lennon children were especially fond of him. The mine captain at Wright-Hargreaves said Dufour was "a first-class workman."

Dufour was killed in a drift on the 800 level on November 30 while assisting the regular scaler, A. Salo. Preparing to scale, the two noticed a piece of loose on the wall and both tried several times without success to bar it out. It was then decided to blast the slab. While waiting for dynamite supplies, they continued to scale the back, thinking there was no danger from the wall. Dufour relieved Salo in scaling and mounted to the second rung of a short ladder to work on the back some nine feet away from the wall. Salo noticed the wall beginning to move and called to Dufour, who jumped from the ladder but was caught by the 20-foot-long slab. He received a fractured skull and a crushed chest, and died instantly.

An inquest was held with Coroner J. F. Edis present. It was evident the slab moved out of position and fell over into the drift. Had Dufour remained on the ladder, he would not have been injured.

Interment took place in the cemetery at Blind River.

## 1928

- Amendment to the Mining Act was made: underground foremen now had to give and receive orders in English. This requirement was soon made mandatory for deckmen, cagetenders and hoistmen
- Tough-Oakes-Burnside Mine goes bankrupt and closes but not before a miner is killed in a shaft accident
- Mining accidents claim lives at Lake Shore, Argonaut, Kirkland Lake Gold, Wright-Hargreaves, Teck-Hughes and Harker Mine
- Number of men employed in the gold mining industry at Kirkland Lake and the adjoining camps is about 2,500
- Ontario Department of Mines reports 43 Ontario miners are killed while working in Ontario mines in 1928. Nine of these are from Kirkland Lake area
- Miners and timbermen earn $4.75 per eight-hour shift. Shaftmen earn $7.00 per eight-hour shift
- New municipal office building is constructed at the corner of Kirkland and Prospect Streets
- Nipissing Central Railroad is finally extended to Rouyn-Noranda after a two-year delay at the border due to interprovincial political haggling

*Miner scaling back of drift. Artist: Roland Russell*

## February 15, 1928
## Lake Shore Mine

### James J. 'Red' Robinson

(1900 - 1928)

James Robinson was a native of Ireland who left his homeland and relatives for Niagara Falls. At 22 years of age, he left Southern Ontario for the gold camp and subsequently became an employee at Lake Shore in 1926.

He was a well-liked single chap, known around the mine as 'Red'. He resided at the Lake Shore Bunkhouse, presently (1999) the Don-Lou Motel. In his capacity as chute blaster, Robinson was in immediate charge of a crew, pulling chutes on the 600 level. The crew, in addition to himself, consisted of two trammers, Emerick Fournier and another miner, and Motorman Jack Kresnik. This crew and others on different levels were under the direction of a mucker boss, Albert Marshall. Before going down on shift, Robinson and Kresnik were instructed by Marshall to pull 50 cars on No. 19 chute. However, on proceeding to work, Robinson ordered five cars loaded at No. 18 chute because it was easier to load there.

When Marshall visited on his rounds, he found the men loading at 18, contrary to his orders. He again ordered Robinson not to pull ore from there. Marshall left to inspect other crews on different levels. Robinson spotted one car at 19 but again loaded a train at No. 18, contrary to orders and to the objections of Kresnik. In all, 15 cars were pulled from 18 and none from 19.

No. 18 had become hung-up a short distance above the brow. Robinson, again acting contrary to mine orders, entered the chute to place a charge of explosive to break it down. The standing order at the mine was that chute blasters must not enter the chutes but must break down any hang-up with a charge attached to a blasting pole. Men who worked with Robinson knew he had been warned many times against the practice of entering the chutes to place a blast.

Robinson took five sticks of powder into the chute with him and, after placing these, asked for and obtained from one of the trammers three sticks more. He then asked for a short piece of wood, presumably to place against the powder to hold it in place. Fournier went to hunt up a suitable piece and Robinson remained in the chute, waiting. Suddenly the muck settled and he was caught in the run of ore. Robinson was dead when he was taken from the chute and, in the opinion of Dr. C. S. Harris who was underground when the body was recovered, had been killed instantly.

An inquest was held before Dr. R. H. Armstrong. The evidence clearly showed that Robinson was deliberately acting contrary to his orders, first in pulling the wrong chute after being told three times not to and again in entering the chute to place the blast. The trammers and motorman testified it was common practice with Robinson to enter a chute and that he rarely used a blasting pole. The workers stated they had warned him of the danger he was incurring in doing so. The jury made a recommendation: "We ask the mine officials to deal severely with any of the mine men who disobey orders and post notices to that effect."

## April 1, 1928
## Kirkland Lake Gold Mine

### Blasting Set Collapses; Five Men Fall to Bottom of Shaft; Two Die

A shift of five men including Joseph Gabryelczyk, Clifford Purdy, J. Cenna, Dave Ackles and Contractor Howard Purdy went on duty at 7:00 p.m., April 1, to near the bottom of an interior shaft at Kirkland Lake Gold. They proceeded to lower the blasting set before putting in a set of timbers. The blasting set was lowered to position by means of a set of chain blocks on each end. The chain blocks were supported by an 8x8 timber that had been 'used' timber.

One side of the set was lowered by the Purdy brothers and the other three men lowered the opposite side. When the first bucket load of planks arrived from above, the men loaded them in position on the set. Suddenly, the 8x8 'used' timber, supporting the one side, broke, allowing the staging to fall about 10 feet to the bottom of the shaft. Cenna was injured by the fall but not seriously.

Ackles and Howard Purdy clung to one side of the blasting set, avoiding a fall. Clifford Purdy and Gabryelczyk fell with the set and were crushed at the bottom of the shaft. Purdy's chest was crushed and Gabryelczyk's neck and skull were fractured. Both probably died instantly.

The practice of using timber previously subjected to blasting for supporting a weight is not a safe one, given the difficulty of detecting cracks in the timber. Normally, such pieces of 'good' timber would support many times the weight carried in this instance.

## Joseph Gabryelczyk (a.k.a. Joe Gabriel) (1889 - 1928)

Joseph Gabryelczyk, a recent Polish immigrant, lived with his wife and five children in Kirkland Lake. The family had moved into town in 1928. One sister, Agnes* (Mrs. Stanley Kensy) was already here. Her husband operated the local blacksmith shop on the corner of Kirkpatrick Street and Teck Avenue.

Left surviving were Gabryelczyk's wife Rose and their children, Bruno, Ted, Chester, Steve and Emily. Interment was conducted in the local cemetery, presently (1999) an unmarked grave.

Some of the information for this account was provided by the grandson of Joseph Gabryelczyk who presently (1999) lives in the Kirkland Lake area.

** Agnes Kensy would lose a son in an accident at Lake Shore in 1952.*

## Clifford Purdy (1890 - 1928)

Cliff Purdy came to town in 1924 and first worked at Sylvanite and later at Kirkland Lake Gold. He, his wife and two sons, Edmund and Arthur, lived on Duncan Avenue.

Purdy was originally from Spry Harbour, Nova Scotia and his aging mother still lived there. He had five brothers – Walter of Toronto, Thomas of Thedford Mines, William of Cobalt, George of Spry Harbour and Howard of Kirkland Lake.

The body was transported by rail to Spry Harbour for burial.

### April 18, 1928

# Argonaut Mine

## Joseph Pleban (1900 - 1928)

Joseph Pleban, of Polish descent, was born in Canada and raised in Barry's Bay, Ontario. At the time of the accident he was 28 years old and single.

Employed as a mucker at Argonaut since August 1927, he previously had worked at the Hollinger in Timmins as a machine helper.

During the night shift of April 17, he and partner Frank Virtan were tramming ore from chutes beneath stope No. 1105 at midnight, then quit for lunch. After lunch, they entered the stope from the 10th level to muck ore into two empty chutes. They worked until 3:30 when Virtan left the stope after telling Pleban to meet him at the level station in about 10 minutes.

Virtan was waiting at the station when the cage stopped at the level but Pleban did not arrive. Virtan went to the ladderway that led to the stope and shouted to his partner. Receiving no answer, he went down the ladders and found Pleban lying dead on top of the muck. A mass of rock had fallen from the hanging wall and broken in two on the footwall. The lower piece evidently passed over Pleban when it slid down the pile of broken muck. His skull and left leg were fractured and his chest was crushed.

The body was sent to Barry's Bay by train. Funeral services and burial occurred there.

### May 25, 1928

# Wright-Hargreaves Mine

## Frank Bulch (1899 - 1928)

Frank Bulch, 29, had a wife and three children living in Yugoslavia. As was common among Eastern European immigrants in those days, the husband left for the new world to earn money to bring the rest of the family to Canada at a later date. It was Bulch's intention to build a small home in town in the summer of 1928.

Bulch was employed at the mine for 14 months as a trammer on the grizzly, and later as a chute blaster. At Christmas time, he served as a substitute

*Motorman in charge of small storage battery locomotive and two cars. Photographer: Unknown*

motorman for the odd shift. On the morning of May 25, he was placed in charge of a small storage-battery locomotive on the 1,250 level, a position that he desired and deserved as he had quickly become a very capable worker. On this day, he was instructed on his new duties by Eugene Jodouin, the mucker boss on that level. Bulch made a trip alone about 1,600 feet on the level and brought out two ore cars. Jodouin assisted him in dumping the contents of the first car into the ore pass and stood beside the locomotive while Bulch drove to move the second car into position for dumping. At the time, Jodouin was turned to allow the locomotive to pass by when he heard a sound and noticed Bulch was not in the locomotive but lying on top of the ore car behind it. Jodouin immediately reached for the controller of the locomotive and stopped the motor. He lifted Bulch to a position between the locomotive and the car and went to the station to get assistance from three men, John Lowes, Dawson Hoadley* and John A. McDonald. Together, they pushed the locomotive to the station and there lifted out the injured miner. He was taken by cage to surface where he was examined by Dr. Harris and sent to hospital. It was found he had sustained severe abdominal injuries including a rupture of the bladder and a fractured pelvis. He died at the hospital about one-and-a-half hours after the accident. Death was due to shock from injuries and internal hemorrhage.

An inquest was held by Coroner J. F. Edis of Swastika. The placing of the locomotive and the two cars under the chute gave a minimum of 15 inches where Jodouin stood opposite the chute. It seemed Bulch stood up when he started the motor and was caught between the empty car and the chute. The jury brought in the verdict of accidental death.

Bulch's brother, who worked at Coniston, Ontario, came to town to make the funeral arrangements. Interment was conducted in the Kirkland Lake cemetery. The site is presently (1999) an unmarked grave.

** Dawson Hoadley's brother-in-law, Robert Nicholson, would die in a mining accident at Lake Shore in 1930.*

## June 5, 1928
## Tough-Oakes-Burnside Mine

### M. Galandzy (1900 - 1928)

Galandzy, 28, a recent immigrant to Canada, was born in the Austro-Hungarian Empire in 1900. He came to Kirkland Lake from Winnipeg where his wife lived at the time of his death. He had worked at the Tough-Oakes-Burnside for about a year employed as a timberman's helper.

On the morning of June 5, he and T. Piipo were told by the shift boss to place a timber and snatch block to serve as a sheave in the manway at the top of the station on the 1,600 level. The two went down in the bucket to the level and proceeded to rig up the block and cable. While installing the apparatus, they decided to go up about 16 feet. Piipo wound 40 or 50 feet of cable around an empty wooden box and then handed the snatch block, with an equal amount of slack cable, to Galandzy. They entered the bucket and rang one bell, thinking they were going to be hoisted slowly and they could stop the bucket at the desired height. Hoistman Harry Barkell*, not having received a three-bell signal, thought the bucket was released and began hoisting it to the 1,100 level. When Piipo noticed they were being hoisted past the top of the station, he endeavoured twice to ring one bell to stop, then he shouted to Galandzy to immediately drop the block and cable down the shaft. In all probability, the block caught on the edge of the still-rising bucket and, when all the slack had been used up, the bucket tipped at about 1,350 feet, causing Galandzy to fall out and down the shaft. Piipo was thrown behind the timbers and caught himself on a piece of blocking. Men at the bottom of the shaft heard the noise of the falling cable that eventually got hung up on a hanging rod; within seconds, Galandzy's body fell to the bottom, followed by his boots and lamp.

When the bucket reached the 1,100, it was right-side-up but damaged. The contractor, Donald McLellan and the deckman removed the damaged bucket and replaced it with another, and McLellan went down the shaft to investigate. There was some damage and obstruction in the shaft, almost causing another spill. McLellan found Piipo standing in a manway at 1,350 feet. By means of ladders, and again the bucket, they got to the bottom of the shaft and found Galandzy.

The dead man had fallen a distance of 320 feet. His body was placed in the bucket and sent to the 1,100, with the men riding on the edge. The body was examined by a doctor who found instant death was due to a badly fractured skull.

** Harry Barkell's father would die in 1935 due to an injury received while working at Teck-Hughes.*

### June 27, 1928

## Harker Mine

## Matti Rossi (1891 - 1928)

Matti Rossi, 37, Finnish, with a wife and one child living in Kirkland Lake, was fatally injured by scalding. He had fallen into the sump of the shaft at the 500 level of the Harker Mine, Harker Twp., at 8:30 a.m. on June 27. The injuries resulted in death on July 5 at the Presbyterian Hospital in Matheson where Rossi had been moved after a severe trip over miles of rough, wet trails and a further journey of 25 miles by car.

A miner with 18 years experience, Rossi had been employed at the mine since May 15, 1928 following its re-opening. The mine had been closed down since March 1925. Work resumed on the 500 level four days before the accident when Rossi had contracted to drive a crosscut from the shaft. On the morning of the accident, he and partner Saraho were drilling but experienced trouble with the machine drill. The defective part was removed and Rossi left the workplace intending to have the necessary repairs made to the part at the surface.

Shortly after leaving (it was 150 feet to the shaft), Saraho heard cries for help. Rushing to the shaft, he found Rossi had fallen into the sump and was unable to get himself back onto the level. The water in the sump was hot due to an exhaust line from a steam pump working at the 250 level. Saraho pulled Rossi out of the hot water and immediately ripped off his rubber clothing.

When brought to surface, Rossi was found to be scalded on his hands and from his waist to his feet. He was quickly prepared to make a most trying trip, carried by stretcher through the bush to a point 14 miles away where he was met by a doctor who later testified the first aid rendered was excellent and every comfort that could have been provided under such circumstances had been given. The injured miner reached the hospital on June 17, about 15 hours after the accident.

From the evidence gained at the inquiry held at Matheson by Coroner W. J. Grummet of Ansonville, it was learned the steam pump was working at the 250 and exhausting into the sump at the 500. Normally, the exhaust would not be in the water but above it; this time, however, there was a great rise in the sump water owing to recent heavy rains.

When Rossi was brought to surface, he made this statement: "I signalled for the cage." It was the practice at this newly reopened mine to rap on the pipeline when signalling for the movement of the cage. At the time, the cagetender had not responded. "I saw a bar that I thought was the cage bar," he continued. "I made a mistake." Examined after the accident, the shaft proved to be barred on the south side but the bar on the north side from which Rossi approached was standing vertical.

Apparently, conditions at the shaft station when Rossi arrived there were cloudy and steamy, owing to the exhaust of the steam pump. He mistook the bar at the south side as being the bar on the north side and stepped to get onto a cage but fell into the sump filled with heated water. That the bar was not guarding the shaft on the north side seems to have been Rossi's own fault as he had unloaded steels from the cage for his shift and had not lowered the bar when finished. At the time of the accident, he and his partner were the only persons working on the 500.

### July 28, 1928

## Wright-Hargreaves Mine

### John Bizub (1901 - 1928)

Originally of Czechoslovakia, John Bizub came to Canada to find employment in the local gold camp. With mining experience from his native land, he had been employed at the Wright-Hargreaves since April 10, 1928. Aged 27 and single, he was killed in this mine on July 28.

On the night of the accident, he was working as a helper with Arno Ranta, a driller – a strange partnership this was as Bizub spoke only Czechoslovakian and Ranta spoke only Finnish. They were drilling a breast in No. 1751 stope and their drill was set up on a pile of ore about 14 feet deep. Located below the ore was No. 2 chute. The trammers on the day shift had drawn seven cars from the No. 1 chute, fifteen from No. 2 and ten from No. 3, under the direction of Shift Boss Harry Sauvé. The ore drawn from No. 2 only caused a settlement in the stope near the first chute. The odd juxtaposition of the settlement to the chutes was noted neither by the shift boss nor by the two miners who were now located above the No. 2 chute.

After scaling their work place, the two drilled a hole and were half done when they were suddenly carried down by a run of ore into No. 2 chute. Ranta was buried to his middle but remained in a standing position, with the muck to one side of him rising about five feet higher than his head. Bizub was completely buried under about four feet of muck.

John Myelda and his helper heard Ranta's calls for help and came to his assistance. Word was sent out to Lyman McKenzie and his shaft crew at the station, and work of digging out the men began. Ranta was released alive in an hour but it was about two hours before Bizub could be uncovered, as considerable timbering was necessary to hold up the broken ore. He showed no signs of life when uncovered, though Ranta claimed he was alive for at least the first half hour after being buried.

An inquest was held by Coroner R. H. Armstrong, and the jury brought in the verdict of accidental death with a recommendation that the mine should in future provide a list of cars drawn from each chute.

### August 27, 1928

## Teck-Hughes Mine

### Max Abrams (a.k.a. Mack Abraham) (1898 - 1928)

Thirty-year-old Max Abrams, single and a native of Egypt, was injured in a sand blast on the fifth level of Teck-Hughes on August 27 and died 10.5 hours later. He was an employee at the mine for only a month and had four years of previous underground experience in other Kirkland Lake mines.

On the night of the accident, he was mucking off the floor of a drift. About half an hour before the accident, Shift Boss Arthur Parker* and Mucker Boss Frank Butler visited the place and blasted down some rock from a stope. About five minutes before the accident, John Vidas, the deckman at the winze, saw Abrams take two sticks of Forcite and a capped fuse from a cupboard in the station. A few minutes later, a blast occurred without warning. Vidas found Abrams lying between the rails a few feet from a rock shattered by a sand blast. Parker and Butler were called and took the injured miner to the first aid room where Dr. W. C. Rumball arrived a few minutes later. The injuries included a frontal fracture of the skull, a broken arm and shoulder, and terrible injuries to the face. No explanation could be given as to how the accident occurred. The deceased miner's lamp and an open knife were found undamaged on the floor of the drift.

An inquest was held at the mine conducted by Dr. J. F. Edis; the coroner's jury brought in a verdict of death by accident.

** Arthur Parker's brother would die as a result of a mining accident at Teck-Hughes in 1957.*

### 1929

- At Lake Shore, three employees are killed
- Ontario Department of Mines reports 48 miners in Ontario lose their lives while working in Ontario mines in 1929. The Kirkland Lake area claims nine
- There are 2,140 miners in Kirkland Lake; the total population of the town is 7,200
- 'Black Friday' occurs in New York as the U.S. stock exchange collapses on October 28. The world economic crisis begins

### February 23, 1929

## Lake Shore Mine

## Duncan H. McDonald

### (1896 - 1929)

Duncan McDonald, 33 and single, came to town from Alexandria, Ontario. He was employed as a pipefitter at Lake Shore.

On the day of the accident, he and M. Hanlin were installing an airline in a shaft at the 1,400 level. McDonald was delivering lag screws to the workers when he slipped and fell through an opening between the manway and the cageway. He landed 24 feet below on a three-inch-diameter pipe that prevented him from falling another 106 feet to the bottom of the shaft. He was rescued from this precarious position by fellow workers, taken to surface and then admitted to hospital.

Medical examination showed he suffered a crushed chest, severe abdominal injuries and a broken right leg. Three days later, on February 26, he died in hospital. His body was sent by train to Alexandria for funeral services and burial.

An inquest into the death was held. The jury returned a statement of accidental death and made a recommendation: "That guard rails be put in all compartments open to a manway."

### May 31, 1929

## Teck-Hughes Mine

## R. Renick (1891 - 1929)

Renick, 38 and a Yugoslavian immigrant, lived in town with his wife and five children. On the day of his accident, he was working with a crew of men to remove old timbers from the Teck-Hughes south shaft. This work was preparatory to the placing of permanent guides into the shaft.

The men were about 24 feet from the bottom of the shaft. Renick came off the skip and got onto the divider; he had a wrench in his hand and was reaching over to loosen a nut when the crew boss told him to get off the divider so he could raise the east skip. Just as Renick was straightening himself, the west skip came down. There were only a few inches between the divider and the skip, and he was caught between.

Dr. R. H. Armstrong was called and found Renick dying. He had a fractured arm, broken ribs, a crushed pelvis, internal and external hemorrhages and a severe crushing over the heart. He died before reaching hospital.

At the inquest conducted by Dr. J. F. Edis, fellow workmen stated Renick was always a careful worker. "The accident was no doubt caused by the lack of co-operation between the shaft leader, deckman and hoistman," said Edis.

### August 9, 1929

## Bidgood Mine

## Blasting Accident Claims Two Lives at Bidgood

Ignace Kuszneruk and Dan Waschuk were good friends because they had a lot in common: both were recent immigrants, both had a wife and family back in Eastern Europe and they worked as partners in the same mine.

Unfortunately, they both made the same fatal mistake and died on the same day, August 9. A double funeral for them was held at the Kirkland Lake cemetery.

On this fateful day, the two were working together in Bidgood at routine activities involving blasting. After a blast, they miscalculated the time and returned to the blasting area too soon. Waschuk had a wrench in his hand as he was in the process of turning on the air, a routine procedure to clear the area of smoke and dust. Kuszneruk was following close behind when the unexpected blast occurred. Waschuk was killed instantly and Kuszneruk died the same day in the Kirkland Lake hospital.

The two were buried side by side. Their caskets were adorned with wreaths placed by fellow miners from Bidgood. The pall bearers were fellow immigrants from Eastern Europe.

## Ignace Kuszneruk (1904 - 1929)

Ignace Kuszneruk was a Ukrainian immigrant with a wife and child in Ukraine. He had been an employee at Bidgood for only three months. His gravesite at the cemetery is unmarked.

## Dan Waschuk (1892 - 1929)

Dan Waschuk was a recent immigrant to Canada from Poland, where he had a wife and child. He had been an employee at Bidgood for only five months. His grave in the cemetery is also unmarked.

### August 28, 1929

# Oriole Mine

## Laurence H. Spiers (1907 - 1929)

Laurence Spiers, 22 and single, came to the local gold camp from Burk's Falls, Ontario. At the time of his death, he worked at the Oriole Mine in Gauthier Twp., about 17 kilometers east of Kirkland Lake. (The Oriole Mine later became the Anoki Mine that later became part of the Queenston properties.)

It may be that his death was due to lack of experience. After setting an incomplete or malfunctioning blast, he went back to the area. Some witnesses said that he was having difficulty relighting the fuse, he dropped a lit match or assembled a fuse improperly – whatever. The result was fatal but not instantaneous.

The blast struck him in the head. Workers coming to his aid found him bleeding from a macerated face. He was taken from the mine to the hospital in Kirkland Lake. He was alive but doctors held no hope of recovery. He died five hours later.

At the inquest, a peculiar type of evidence was offered. Evidently, before Spiers died, he was conscious and gave a report to one of the doctors in the hospital. According to the evidence, he stated he was having difficulty lighting the fuse when all of a sudden – he did not know what happened. The jury ruled that the fatal injury was accidental.

### September 21, 1929

# Wright-Hargreaves Mine

## Henry Melong (1884 - 1929)

The Melong name is well known in town today. The family members are descendants of Henry Melong who came here from Antigonish, Nova Scotia.

Melong and his wife, the former Mary McGinnis (1884 – 1958), lived on Comfort Street with their nine children: Donald, Katherine, Mary, John, Sadie, Martha, Henry Jr., Theresa and Frank. Henry Sr. had been employed at Wright-Hargreaves since 1922. Before the accident, he had been working in the crusher house for eight months.

*Henry Melong. Photographer: Unknown*

He died under mysterious circumstances on September 21. His partner, Iver Oberson, left the crusher house on a routine break, leaving Melong alone in charge of operations. Oberson came back after his break and found all systems running but Melong was lying on the rolls with his head bumping on the hub of the wheel. Oberson immediately shut down the power and rescued the still-conscious Melong. He was bleeding from the head due to cuts on the scalp and a fractured skull. He was taken to Kirkland and District Hospital where he died the next day. Death was due to head injuries sustained in the accident.

Although there were no witnesses, the inquiry jury concluded Melong had left everything running while he went to inspect the feed chute. He was wearing a new pair of boots at the time and the jury suspected he slipped while doing the inspection and fell backwards, striking his head on the hub of the revolving wheel.

He was buried in the Kirkland Lake cemetery.

The personal details for this memorial account were given by Henry Melong's daughter-in-law, Mrs. Henry Jr. Melong. Grandsons David and Henry found a tiny photograph of their grandfather on a cuff link and made a concerted effort to have it reproduced for this book.

### December 3, 1929

## Wright-Hargreaves Mine

### Kalle Taavitsainen (1892 - 1929)

Kalle Taavitsainen, a native of Finland, was married and lived with his wife and two children in Kirkland Lake.

He was injured at Wright-Hargreaves on September 8, 1928 when a rockburst blew out on the 2,600 level and struck him. He was quickly removed from the mine and placed in Kirkland District Hospital where he lay totally paralyzed from the hips down due to a broken back and a ruptured spinal cord.

He remained in the local hospital 279 days before being transferred to the Toronto General Hospital in June 1929. He died there December 3.

### December 12, 1929

## Lake Shore Mine

### James Willis (1889 - 1929)

James Willis came to the local gold mining camp early. History has it he was an early prospector in these parts along with Harry Oakes. Oakes struck it rich but it would seem Willis was not so lucky.

He came to Kirkland Lake from Depot Harbour, a town near Parry Sound, to do some prospecting. He remained single and stayed to work in the mines after his prospecting days were over.

On December 12, he was sent out to the Diamond Lake gravel pit about 15 miles east of town. The pit was owned by Lake Shore as the mine required large amounts of stone backfill for its underground workings.

On a cold December day, Willis, being responsible for loosening up the frozen fill, decided to thaw out five sticks of dynamite by a warm stove. This was a fatal mistake. The dynamite blew up during the thawing process and he was killed instantly, his body badly mutilated by the explosion.

The remains were transported back to town where his two brothers later arrived from Montreal to make burial arrangements. A funeral was held and the body was placed in the storage vault until spring interment in the Kirkland Lake cemetery. Presently (1999), it is an unmarked grave.

### December 29, 1929

## Lake Shore Mine

### James Burton (1901 - 1929)

Jim Burton, 28, single, was hit on the head by a bucket while underground in Lake Shore at the 1,200 level on December 29; his helper, Hilyard Liscumb, was unhurt. A cable attached to a bucket used for conveying wrenches and steel up and down an inclined raise had failed, allowing the free bucket to tumble down the raise and strike Burton's head.

Dr. G. M. Cameron went down to the level to aid the injured miner and found him alive but comatose and sustaining severe head injuries. He was given first aid, taken to surface and rushed to hospital.

An operation was performed to remove ten small pieces of fractured bone from Burton's meninges, the linings of the brain. Eight hours after the accident, Burton succumbed to his injuries and died.

An inquest was held where jurors C. J. Sweet, H. J. Gibson, H. McDonald, H. C. Waldron, C. Lintlop, F. W. Haynes and Albert Robertson were selected. Foreman Haynes returned the jurors' verdict of 'accidental death'.

*Kirkland Lake Gold Miners (wall mural in town). Artist: Eric Meyer*

Chapter III

# THE THIRTIES

## 1930

- $17 million worth of gold is extracted from Lake Shore this year
- Tough-Oakes-Burnside Mine is re-opened as the Toburn
- Nine fatalities occur in local mines: one is killed in Kirkland Lake Gold; five die in Lake Shore; Teck-Hughes loses a miner and Wright-Hargreaves has two accidents that claim one life each
- The lake called Kirkland Lake is now completely filled with mine tailings
- West End Public School opens for students living in the area of Macassa and Kirkland Lake Gold

## February 25, 1930
## Wright-Hargreaves Mine

### Gerhard Koskela (1895 - 1930)

Twenty-eight-year-old unmarried Gerhard Koskela, a native of Finland, of 88 Fifth Street, was fatally injured at Wright-Hargreaves on February 25.

*Drilling machine runner and helper. Photographer: Peak Studios Kirkland Lake*

He was running a drilling machine at the time of the accident. Helper Frank Sterbenach went a short distance for more drilling steels when Koskela drilled into a missed hole containing about three sticks of powder. The resulting explosion caused a fractured skull, a broken left arm, and numerous cuts and bruises to his face and body. Removed from the mine and taken to Kirkland District Hospital, he received medical attention that did not save his life.

At the inquest conducted by Coroner J. F. Edis, Shift Boss H. McDonald said he visited the face when the men were starting up but did not notice the missed hole. Sterbenach stated that the total face was not washed down before the drilling. The jury returned a verdict of accidental death.

The body was buried in the local cemetery. The inscription on the headstone over the gravesite reads: GERHARD KOSKELA – BORN NOVEMBER 29, 1895 – DIED MARCH 1, 1930.

## March 19, 1930
## Lake Shore Mine

### William Orminison (1899 - 1930)

William Orminison, 31, unmarried and of Romanian nationality, was fatally injured at Lake Shore while drilling into a missed hole on March 19. Helper John Jamieson was badly shocked and lost the sight of his right eye.

Orminison received severe injuries to his left side; his left leg and wrist were broken, and he received undetermined internal injuries. He was rushed to hospital but died there. According to hospital records, his address was 26 Goodfish Road.

At the inquest held by Coroner J. F. Edis, it was learned that at the time of the accident, the two were starting their first hole on a breast they'd failed to wash down. They drilled into a missed hole containing three sticks of powder. The jury returned a verdict that Orminison came to his death through his own carelessness – that is, by not washing down the face before drilling.

*Author's note: In some instances, there is no record of where some of the men were buried or they were buried in unmarked graves. Tough financial times existed then. Families couldn't afford headstones and in some cases could not afford the price of a plot. Many were buried in the cemetery by friends and fellow workers during the darkness of night and, the next day, an observing cemetery caretaker would note in his records that an unscheduled burial had taken place. Perhaps the remains of William Orminison were handled this way. Also, many of the graves are presently unmarked. It may be they were never marked or they were marked with a simple wooden cross that over time rotted away.*

*It is ironic that these men broke rock for a living but a rock was not placed over their grave. Fortunately, a monument, the Miners' Memorial, has now been erected in their memory.*

### April 9, 1930

## Kirkland Lake Gold Mine

### Wilfrid Gosselin (1906 - 1930)

Wilfrid Gosselin, 24, French Canadian, residing in town with his wife and two children, died April 25 from injuries received April 9 while working in a winze below the 4,000 level at Kirkland Lake Gold.

At the time of the accident, he and three others were mucking a cut in the winze. The shaft leader had gone to the 4,000 level to make up the primers to blast the remainder of the round when he heard an explosion coming from the winze. About three tons of rock came off the wall, part of it falling on Gosselin while he was in a stooping position. The other men apparently escaped serious injury.

At the inquest held by Coroner J. F. Edis, all witnesses stated that the walls of the winze were perfectly sound before the accident. One theory was that water and air had penetrated along a fault in the rock forming a gas which, combined with pressure, caused the rock to break away and fall. A verdict of accidental death was returned by the jury.

### June 7, 1930

## Teck-Hughes Mine

### John Mantyla (1896 - 1930)

John Mantyla, 34, was injured in the winze at the 25th level of Teck-Hughes June 7 and died in hospital June 13. Mantyla, of Finnish descent, lived in town with his wife and three children.

The doors of the deck, where the bucket is dumped, when closed, completely cover the winze. Ray Davis, deckman at the time of the accident, closed the doors but did not notice that one door wasn't closed completely. The bucket was dumped and rock fell through the opening into the winze where Mantyla was working.

At the inquest held by Coroner J. F. Edis, Shaft Leader Maki said the rock that struck Mantyla on the shoulder was about six inches square. The jury returned a verdict of accidental death.

### August 2, 1930

## Lake Shore Mine

### Harry Lewington (1870 - 1930)

There is a headstone in the Kirkland Lake cemetery engraved with the following inscription: FATHER AT REST – HARRY LEWINGTON – DIED AUGUST 2, 1930 – AGED 60 YEARS – 'ROCK OF AGES CLEFT FOR ME'.

Harry Lewington was instantly killed on the 600 level of Lake Shore when he walked into an open sand pass raise and fell 75 feet to the muck below.

The raise had been broken through with the firing of the final round at 3:30 a.m. The morning shift, which included Lewington, was apparently told of this breakthrough. Fifteen minutes into the shift, he was walking through the crosscut on the level with an unlit carbide lamp in his hand when he stumbled into the open raise. Normally, this area is well lit by a string of electric lights; however, the blasting had broken two lights near the raise with the result that visibility was not good. An inquest held before Coroner J. F. Edis returned a verdict of accidental death.

Born in England, Lewington came to Canada in 1875 when he was five years old. He had been an employee at Lake Shore for six years. He was survived by his wife, two sons and a daughter.

The funeral took place at the family residence at 68 Third Street. The pastor of the Baptist Church, Pastor Manktelow, conducted a service that was attended by a large number of employees from Lake Shore. There were many floral tributes presented. The pallbearers were: George Craig, Dave Ramsay, H. McPhail, Thomas Pinder, A. Whittle and J. Meath.

### September 18, 1930

## Wright-Hargreaves Mine

### Alexander McQuaig (1899 - 1930)

Alexander McQuaig was killed at the transformer house of Wright-Hargreaves on September 18. He was Canadian, 31 years of age and unmarried.

At the time of the accident, workmen were removing a 4x4 cross-arm from the transformer house without having disconnected the source of power. They were connecting one end of a steel cable to the insulator and intended to attach the other end to the building. The old cable would then be cut off and the crossarm removed.

Mr. Hunter, who was in charge of the work, left the men for a short time. He had asked McQuaig and Apprentice Murphy to take measurements, get the material for the change but not to do any work on the line.

The two men prepared the cable for one line and then proceeded to complete the work. Wearing gloves, Murphy ascended a ladder about three feet above the floor. McQuaig stood on the concrete floor holding one end of the cable in his bare hands. Murphy, attempting to pass the other end through the loop of the insulator, accidentally touched the cable to the bare line circuit that was conducting 2,200 volts. McQuaig fell to the floor.

Artificial respiration began at once and a doctor was called. Efforts to revive McQuaig continued for about three hours but failed.

### October 9, 1930

## Lake Shore Mine

### Robert Nicholson (1902 - 1930)

On October 9, 28-year-old Robert Nicholson, a sampler at Lake Shore, apparently stepped upon and disturbed the hung-up ore in a chute. The 100 tons of ore moved down the chute along with Nicholson and he was later found in the rubble at the 1,200 level. Apparently, he was killed almost instantly.

There were no witnesses to what actually happened, for he was working alone. He was not missed until the end of the shift; when checking the tag board in the dry room on surface, the foreman realized that Nicholson was still underground. A search was started and, five hours later, the body was found.

Born at Gateshead on Tyne in England, Nicholson came to Canada in 1924 and was employed at Lake Shore from 1925. Miss Evelyn Brooks, formerly of Lancashire, England, became his wife in 1927 when they were married in Toronto. They had three children: George (2), 14-month-old Robert and nine-week-old baby Margaret.

Surviving Nicholson besides his immediate family were his parents, Mr. and Mrs. G. Nicholson, brother George and a sister, Mrs. D. Hoadley, all of Kirkland Lake.

According to the Northern News, many floral tributes were received before the funeral service. The funeral at Trinity United Church was officiated by Reverend A. Gillies. A male choir, composed of fellow workers of the deceased, sang hymns at the church that was filled to capacity. Interment was conducted in the local cemetery.

An inquest was held before Coroner R. H. Armstrong and the jury returned the following verdict: "We the jury find that Robert Nicholson came to his death at the Lake Shore Mine on October 9, 1930 from injuries received, namely, a fractured skull caused by accidentally falling down No. 2 chute. We consider negligence has been shown on part of the men."

A flat stone marks the grave. The inscription simply reads: ROBERT NICHOLSON – 1902 - 1930.

## November 7, 1930
## Lake Shore Mine

### Fred Rioux (1890 - 1930)

Fred Rioux, 40, single and Canadian, was instantly killed at the 2,200 shaft station of Lake Shore by a one-ton slab of rock. The accident occurred at 12:30 p.m. on November 7.

Rioux was already dead when Dr. C. S. Harris arrived at the mine and, on examination, found the deceased had a broken neck, crushed ribs and severe internal injuries.

That day, a crew was at work in the shaft bottom where muck was loaded into cars on the level and transferred to No. 3 shaft for hoisting. Rioux and another mucker, Olsen, and two trammers were handling the muck on the level.

Just before the accident, the trammers were taking a full car to the shaft, so Rioux and Olsen sat down on a plank under an untimbered portion of the station and waited for the trammers to return with the empty car. At this instant, a slab of rock fell off the wall, toppled over and crushed Rioux beneath it.

At the inquest held by Coroner R. H. Armstrong, the jury returned a verdict of accidental death.

Interment was in the Kirkland Lake cemetery. The inscription on the headstone reads: FRED RIOUX – 1890 - 1930 – 'AT REST'.

## November 12, 1930
## Lake Shore Mine

### Raymond Arthur (1898 - 1930)

Raymond Arthur was an experienced 32-year-old Canadian miner who, before his employment at Lake Shore, had worked at Tough-Oakes and the Dome and McIntyre Mines in the Porcupine District.

At Lake Shore, he was a drill helper on a machine operated by Pat Bertram. On the day of the accident, the two were working in a five-foot-high stope located between the 1,400 and 1,600 levels. William Allen, their shift boss, instructed them to scale as he had found some loose debris. The two worked with their scaling bars for about 15 minutes when Bertram went around to the other side of the loose to pick up a machine bar. He suddenly heard a rock fall and a yell. He looked around and saw a three-foot-square rock on his partner. He immediately got help to remove the rock.

Arthur was alive but in severe shock. He was taken to surface and to hospital where he was observed by Dr. G. M. Cameron. X-rays revealed multiple fractures of the pelvis and a swelling in his back. At 8:00 a.m. the next day, free gas accumulated in his abdomen. Surgery revealed a tear in the intestine and early stages of gangrene. His condition deteriorated throughout the day and he passed away at 8:00 p.m.

An inquest was held before Coroner R. H. Armstrong; the jury returned a verdict of 'accidental death'.

*Men crowd around Wright-Hargreaves gates looking for work. Photographer: Eddie Duke*

## 1931

- Many transient workers come to the area looking for employment during this Depression year. Men live in 'hobo jungles' around the community
- Toburn mills 125 tons of ore per day
- Macassa is successfully developed
- Lake Shore produces 533,757 ounces of gold
- Eleven miners are killed in the Kirkland Lake gold mining camp

- The 2,475 level of Kirkland Lake Gold is extended 1,000 feet into Macassa ground before good ore is found
- Kirkland Lake Mine Rescue Station building is constructed on Government Road
- Two hundred and fifty new homes are built in town this year
- Golf enthusiasts purchase a farm south of town for their future golf course
- An eight-room high school is built for local students
- A 'red' gathering of a crowd of 2,000 people downtown is dispersed with difficulty by police. Ten men are arrested and six receive two-month jail terms
- Dominion of Canada census states that 17 nationalities are represented in Kirkland Lake

*Silence and inactivity surround the gates now located in a quiet park. Photographer: Author*

### January 7, 1931
## Lake Shore Mine

## Marvin A. Walker (1909 - 1931)

Marvin Walker, born in Renfrew County in 1909, was the eldest son of Mr. and Mrs. Anson Walker of Lombardy, Ontario. While in his hometown, he was a well known, popular young man.

He was single and 22 when he came to Kirkland Lake. His cousin, Mrs. Walter Sylvester, was a resident of the town at that time. She would later accompany the body to its final resting place in Smiths Falls cemetery.

On January 7, Walker was riveting on surface in the construction of the Lake Shore mill. He was on a scaffold along with Sam Verday, when the ram plank, which supported one end, snapped and caused the structure to collapse. Verday's arm caught on a steel bar which saved his life, as he was rescued in this position. Walker fell a distance of 27 feet and landed on a classifier. His injuries included a fractured skull and serious internal trauma. He lived for only a few minutes after the fall.

Many friends and relatives attended the funeral service held at the United Church in Lombardy. Reverend R. Wylie, pastor, officiated the ceremony.

Walker was survived by his parents; three sisters, Mrs. Jno Rankin of Ottawa and Hazel and Julia at home; and four brothers, Jack, Howard, Cecil and Gerald, all at home.

### January 8, 1931
## Lake Shore Mine

## Adriano Micolo (1890 - 1931)

Forty-one-year-old Adriano Micolo was Italian and married, his wife and child living in Italy. He was an experienced miner, having worked at Wright-Hargreaves for four years before being hired at Lake Shore in July 1930.

On January 8, one day after the accidental death of Marvin Walker on surface, Micolo was scaling underground in a Lake Shore stope with partner Billok when a 10-ton rock fell on him. He was killed instantly, while Billok escaped unhurt.

### March 9, 1931
## Wright-Hargreaves Mine

## Michael Casey (1897 - 1931)

A coroner's jury absolved all concerned from any blame in connection with the death of Michael Casey in the Wright-Hargreaves mill on March 9. The verdict stated that he came to his death from a fractured skull while he was oiling a bearing from the wrong side of a bearing hanger. It was thought that Casey put an arm through the hanger to reach a grease cup when his sleeve caught in a gear close by.

Mill Superintendent Malcolm Black said that no one had ever tried to grease this bearing from the wrong side before. When F. J. Baillie and Wilfred Halley found the body on the revolving shaft, they turned off the switch and tried to get the body down. Dr. C. S. Harris was called to the mine where he found Casey with an arm through two gears and the side of his face beyond recognition. Death was attributed to a fractured skull.

Casey was born in Larkhall, Lanarkshire, Scotland in 1897. Interment was conducted in the Kirkland Lake cemetery; the grave is unmarked.

After the fatal accident, a preventive wall was built to prevent workers from making the same mistake.

### May 10, 1931

## Sylvanite Mine

### Hugo Alfred Enstiom (1908 - 1931)

Hugo Enstiom, 23, was born on June 14, 1908; he was of Finnish and Swedish descent.

He had been working at Sylvanite for two years. Single, he boarded at the home of his sister, Mrs. H. Ganson at 121 Pollock Avenue. He was a timberman working in the main shaft about 100 feet from surface when the accident occurred.

A crew was in the process of installing new guide brackets for the cages. Enstiom was working in the lower regions of the shaft when a plank from the upper region fell. A later inquest could not determine the origin of this plank. It struck Enstiom on the head, fracturing his skull although he was wearing a black safety hat at the time. Examination of the hat afterwards showed it had lost its original rigidity and was quite soft, affording little protection. Dr. Harris arrived at the mine dry to attend to the injured miner but Enstiom lived for only five minutes.

Interment of the body took place in the local cemetery.

### May 22, 1931

## Wright-Hargreaves Mine

### Stjepan Ragus (a.k.a. Steve Raguk) (1905 - 1931)

Drilling into a missed hole in a crosscut on the 1200 level at Wright-Hargreaves on May 22 was the cause of an accident in which Stjepan Ragus, 26, machine helper, was killed and S. Backland, the operator of the machine, lost the sight of one eye and received severe face and chest injuries. It was believed that only a small amount of powder remained in the hole as the explosion only blew off a small amount of rock. The drilling steel was slightly bent and the machine was left standing after the blast.

In 1926, Ragus came from Yugoslavia, leaving his parents. He was a member of the local branch of the Croation Fraternal Society which did the arrangements for his funeral.

The service for the single man was held at St. Jerome's Roman Catholic Church where Father Jodoin conducted the mass. At the graveside, S. Kalember, John Velacich and Ivan Marohnich, three officials of the Society, conducted the last rights of the Croation Fraternity. The funeral was a large one, as close to 100 cars were in the funeral cortege.

Present at the inquest were Coroner J. F. Edis, Crown Attorney F. L. Smiley, Mine Inspector E. C. Keeley and jurymen J. Edwards, A. J. Elliott, H. Maxwell, D. Briggs, A. Jack, W. Louden and F. Ferguson. Several witnesses, including Mine Superintendent L. B. Smith, were called.

At the inquest, Dr. G. M. Cameron testified that he was called to the mine on May 22 where he proceeded at once to the shaft house. Half a minute later, the injured miner was brought to surface and lived but a few moments. Death was attributed to an internal hemorrhage caused by a small stone severing a blood vessel just above the heart. Two ribs and the breast bone were fractured as the stone entered the body.

The gravestone at the local cemetery states: STJEPAN RAGUS – OVDJE POCIVA STJEPAN RAGUS – ROD 1 FEBRUARY 1905 – UMRO 22 MAY 1931 – OVAJ SPOMEN PODIZUMI ATAC DURO I MATIANDA – POCIVAS U MIRU

Translation: Stjepan Ragus – Here rests Stjepan Ragus – Born 1 February 1905 – Died 22 May 1931 – This gravestone erected by his father and mother – Rest in peace.

### June 19, 1931

## Kirkland Lake Gold Mine

### William G. Nancarrow (1905 - 1931)

The Northern News described this fatal accident as "the most distressing and gruesome fatality in the history of mining in Kirkland Lake."

Twenty-six-year-old William Nancarrow was a skiptender at Kirkland Lake Gold on June 19. He was in a winze, riding a skip loaded with 80 loose drilling steels. It was proper procedure to operate a skip with the door closed but this time the door was left open. While ascending rapidly in the shaft, the skip became lopsided and the drill steels slid, striking the timbers of the shaft and causing the steels to fly about in all directions. Nancarrow must have been killed instantly in this frenzy of flying steel. Also by some manner, his body was ejected out of the skip and fell several hundred feet to the bottom of the shaft where it was found in the sump, mangled and mutilated beyond recognition.

Nancarrow was married; at the time of his death, his wife and baby were visiting Cornwall, England. He had an aunt and uncle in town, Mr. and Mrs. G. Cousins.

He was survived by his immediate family and a mother and sister in England. The funeral service was held at Trinity United Church with Reverend A. Gillies officiating. Interment took place in the local cemetery; the grave is unmarked.

### July 7, 1931

## Wright-Hargreaves Mine

### Royden Smith (1911 - 1931)

Roy Smith was born on June 6, 1911 to Joseph Smith (1811 – 1931) and Maria Smith (1885 – 1948). He was a youthful employee at Wright-Hargreaves; on July 7, he was working in the powerhouse as an electrician with three years experience. The beam on which he was working gave way, causing him to fall a distance of 10 feet. A witness to the accident, Stanley Duggan, said he did not think Smith touched any live wires before his fall. Remarkably, Smith picked himself off the ground in a dazed condition and then collapsed, unconscious. Artificial respiration was administered by the employees present but resuscitation failed. A doctor on the scene stated, "discolouration of the face may have indicated electrocution but it was impossible to say if this was a factor in the death."

At home at 96 First Street, Mr. and Mrs. Smith suffered the anguish of losing their 20-year-old son. Interment took place in the local cemetery. Roy Smith is buried with his parents.

### July 28, 1931

## Teck-Hughes Mine

### Sinklar Morrow (1905 - 1931)

Twenty-six-year-old Sinklar Morrow, born in Norfolk County, Ontario in 1905, was the son of Mr. and Mrs. James Morrow, who in 1931, lived in New Liskeard.

Tragedy came to the Morrows twice that year. Two months before Sinklar's death, their other son Eldred died in the Cobalt Municipal Hospital as a result of electric shock he received at nearby Hound Chutes while in the employ of the Northern Ontario Power Company.

Sinklar Morrow was injured at Teck-Hughes and died about 15 minutes following the accident. On July 28, he and P. Bertrand were drawing muck from a chute on the 16th level, when the chute became hung up. After working at the plugged area for some time, they decided to blast it. Morrow placed the powder between two large pieces of rock and was about to light the fuse when Bertrand told him to wait until he turned off the air. Bertrand descended the ladder and had just taken two steps away from it when he heard Morrow call. On going back, he found his partner pinned between a large rock and a post. His chest was badly crushed and he died before he could be released. Dr. W. C. Rumball pronounced Morrow dead at the site.

The funeral service was held in New Liskeard and interment took place in the New Liskeard cemetery.

### August 8, 1931
## Lake Shore Mine

# Antonio Battistone (1900 - 1931)

Antonio Battistone, a 31-year-old Italian immigrant whose wife and two children were still living in Italy, lived at 41 Government Road East. His only relative in Canada was a brother-in-law, Leonard Cicci, who lived in Schumacher.

On August 8, Battistone and partner Frank Perusek were pushing an ore car of muck when a rockburst suddenly blew 25 tons from the side of the drift. This fall of rock at the 1,800 level partially buried the two men. Perusek, who was at the far side of the car, received three fractures in his right arm and several minor cuts; he was hospitalized. Battistone, at the near side, received the full brunt of the burst and was killed instantly.

The funeral, handled by Wray and Simm's Undertaking Parlour, was held at St. Jerome's Roman Catholic Church. Interment was in the local cemetery. The grave is unmarked.

### August 29, 1931
## Lake Shore Mine

# Edward Sidney Adams
### (1907 - 1931)

Twenty-four-year-old Ted Adams, born in Ludlow, England in January 1907, came to Kirkland Lake and was employed at Lake Shore for two years. He and his newly arrived girlfriend from England, Violet Mann, were married in town on July 28, 1931. A month later, Adams was killed in a mining accident.

Working in the headframe as a top deckman, he was disconnecting the hoisting apparatus from the bucket of ore when suddenly he slipped but

*Ted Adams (on left), and his good friend Jack Bottrill. Both killed in separate accidents at Lake Shore. Photographer: Esme Hurd*

managed to prevent a long fall down the shaft by catching onto the timbers. Because Adams was alone, the hoistman did not realize that he was clinging to the side of the shaft when the bucket was sent down for more ore. The descending bucket crushed his right shoulder, fractured ribs and crushed his pelvis. After workmen heard his cries for help, he was rescued and brought to surface where he died.

Surviving Adams were his wife, his parents and eight brothers and sisters in Ludlow, and a brother William living in Kirkland Lake. Also surviving was his good friend, Jack Bottrill, who would be killed five years later in an accident at the same mine.

The funeral service was held at St. Peter's Anglican Church. Adams was buried in the local cemetery. The inscription on the headstone reads: IN LOVING MEMORY OF EDWARD SIDNEY ADAMS – BORN JANUARY 1, 1907 – DIED AUGUST 29, 1931 – 'NOT DEAD, ONLY GONE BEFORE'.

## November 13, 1931
## Sylvanite Mine

### Robert McQueen (1896 - 1931)

Robert McQueen was born in 1896 in Dalmeany, near Kirkcaldy in Scotland. In 1928, he married in Kirkland Lake; his bride was from Dalmeany. They later left to live in their hometown in Scotland for a year and returned in March 1931, to reside at 110 Duncan Avenue. One of McQueen's sisters, Mrs. W. Easdale, lived in town.

He was initially employed at Kirkland Lake Gold but took new employment at Sylvanite on October 1, 1931, becoming a mucker boss.

On the date of his death, he was near a 100-foot-deep vertical stope that was being prepared for backfilling. Its entrance was barricaded. He had not been seen for two hours until found at the bottom of this stope, dead, with a crushed skull and a broken leg. He had entered the forbidden passageway and had fallen a hundred feet to his death.

McQueen was survived by his mother, a sister and two brothers, all in Scotland, his sister in Kirkland Lake, and a brother in India. The funeral mass, attended by a large number of friends, was held at St. Jerome's Roman Catholic Church and conducted by Reverend Father D. Jodoin. Interment was in the local cemetery; the grave is unmarked.

### 1932

- Eleven miners die, tragically and accidentally, in area mines this year
- Unemployed men line up at the mine gates, hoping to be hired
- Annual payroll in the mines of Kirkland Lake is $4 million
- John Martin and sons start up the Martin-Bird Mine, south of Larder Lake; their financier is S. J. Bird of Toronto
- Church of the Holy Name and the Assumption Church are built
- Kirkland Lake Curling Club opens
- Polish Hall is constructed in town
- Seven people from Lake Shore drown when a sudden storm tips their boat on Larder Lake on Thanksgiving week-end
- Kirkland Lake waterworks project begins

## January 9, 1932
## Lake Shore Mine

### Robert Scott (1893 - 1932)

Robert Scott was born in 1893 at Montrose, Scotland. In his youth, he took an active part in athletics and was a keen student of mining. He first worked in a Scottish coal mine and later in a shale mine. On one occasion, a Danish company chose him to go on a mining expedition to Iceland. While in Scotland, he was the assistant manager of a plant owned by the Scottish Shale Oil Company. He left his native Scotland in 1929 and came to Kirkland Lake where he began work at Lake Shore.

"He had a quiet and unassuming nature and his geniality made him many friends in town," the Northern News stated. He was a loyal member of the Rechabites, a Scottish temperance society.

On January 9, on the 1,800 level, three helpers, J. Hogg, Jock Burns and W. Carswell, were coming out of a drift with a car load of sand near the charging station. The train backed up to let them on the siding and then moved on. The men and their car came out onto the main line but they forgot to close the switch. A second train was loading sand in the crosscut at that time.

Scott, the shift boss, spoke to the three as they passed by but did not notice the open switch. He decided to eat his lunch at the charging station and sat down on a powder box located on the track between a five-ton locomotive and the wall. It was against the rules to eat lunches in such places. The second train, loaded with sand, was soon to pass this charging station. Its locomotive was behind the train of cars moving towards the station with O. Oddenholm walking ahead. Oddenholm did not notice the open switch until he saw the cars taking the wrong track. He blew his whistle but not in time to permit the motorman to stop his locomotive. The train hit the parked locomotive in the station and drove it forward, crushing Scott against the wall. He received a crushed chest, broken collar bone and severe lacerations to the head; he was killed instantly.

Surviving were his wife and mother, both of Luthermuir, West Lothian, near Edinburgh, Scotland, and two sisters, one in Scotland and one in France.

The funeral services, held at Wray's Undertaking Parlors, were conducted by Pastor Manktelow of the Baptist Church. The body was shipped to Scotland for burial.

## March 22, 1932
## Wright-Hargreaves Mine

### Alex R. MacDonald (1897 - 1932)

One man was instantly killed, four were seriously injured and four more suffered minor injuries in a mine accident at Wright-Hargreaves at 7:00 p.m. on March 22.

The nine men were lowered on a steel blasting set near the bottom of No. 4 shaft when the lowering chains broke, causing them to fall a distance of 28 feet.

Killed in the accident was Alex MacDonald, 35, who received a fractured skull, a broken leg and severe facial injuries. Another miner, Howard Purdy, was in a similar accident at Kirkland Lake Gold in 1928, during which he lost his brother; Purdy survived this fall but received a fractured right femur and a broken collarbone. J. Vouri received a fractured leg and cuts. Thomas Bullock broke two ribs and had bruises to his body. Lauri Laakso fractured his shoulder and received lacerations on his cheek. The other injured men were: Sydney Laird with a cut lip, cut nose and leg abrasions; Livingstone Brydges with a fractured wrist; H. Wennerholm with a bruised forehead; and E. Stewart with bruises on his hip and shoulder.

Alex MacDonald, born in Judique, Nova Scotia, had a brother living in Sudbury. He was single and lived at the home of Mr. and Mrs. E. R. Ostrom, at 62 Second Street.

At the inquest held by Coroner R. H. Armstrong, Master Mechanic William Smith stated in his evidence that when he built the set and specified chain blocks, he had no idea that men would ride on the set while it was lowered. Shaft Leader Bullock and Shaft Captain J. McGuire said they thought the chains were safe enough to handle the set with men on it.

The chains were sent to Toronto for testing. One chain broke at 12,500 pounds, another broke at 15,000; these two were classified as 'flawed'. Normal similar chains without flaws broke at 20,150 and 17,475 pounds.

As a result of the tests, the following changes occurred at the mine in April 1932: (1) the new chain blocks on blasting sets were now of greater capacity, (2) to prevent strain, the chains were taken off the set when timbering was finished, and (3) men were not allowed to ride the set while it was lowered.

## April 5, 1932
## Lake Shore Mine

### Two Miners Die When a Missed Hole is Drilled

A 12-hole slash had been drilled on the 2,200 at a Lake Shore drift face on a previous shift and then blasted. The round did not break well.

Joseph Michichuik was ordered by Shift Boss Mr. Postle to drill more holes in the face and reblast it. He drilled four new holes down the centre but instead of drilling a new lifter hole, he put a seven-foot steel into an old hole five-and-a-half feet deep. The powder in the hole exploded, blowing the collar off the hole. The fractured rock from the collar hit Michichuik in the neck and head, killing him instantly.

Marko Sekulic, his helper, was cranking the drilling machine at the time of the blast and received the charge in the face and chest. Both his eyes were blown out and he died in hospital about 10 hours after the accident. Death was attributed to secondary shock.

### Joseph Michichuik (1896 - 1932)

Thirty-six-year-old Michichuik, single and a native of Romania, immigrated to Canada in 1910. He came to Kirkland Lake in 1930, gained employment at Lake Shore and made his home with Mr. and Mrs. Joe Laka at 18 McKelvie Avenue. He had an earlier accident at Lake Shore and lost a couple of fingers in that incident. In 1931, he purchased a farm near Hamilton, Ontario. It was his intention to give up mining and soon become a farmer. He was a member of the local Moose Lodge.

The funeral was held at St. Peter's Anglican Church and services were conducted by Reverend Paul Markovich, a Serbian Orthodox priest from Hamilton.

## Marko Sekulic (1902 - 1932)

Born in Ribeniku Lika, Croatia, on October 22, 1902, Marko Sekulic was a married man who came to Canada alone in 1926. While in Kirkland Lake, he lived at 135 Pollock Avenue. At the time of his death, his wife, son and parents were living in Croatia.

The inscription on the headstone over the grave in the Kirkland Lake cemetery reads: MARKO SEKULIC – ROTEN U RIBNIKU LIKA BOATIAN – GOD 1902 OKT 22 – UMRO 1932 APRIL 5 – AT KIRKLAND LAKE, ONT. – UPOIEVA O BIOZEIME.

### April 15, 1932

# Wright-Hargreaves Mine

## Thomas P. Owen (1895 - 1932)

Thomas Owen, born in 1895 at Penrhydendrath, North Wales, left his native country in 1924 and came to Canada, settling in Kirkland Lake where he made many friends. He was a member of the local Doric Lodge A.F. and A.M. and a member of the Kirkland Moose Lodge. He was regarded as a good worker at the mine.

Owen died on April 15 when he, Steve Bougie, J. Hallett and three others were engaged in shaft sinking operations, timbering an underground ore pocket at Wright-Hargreaves.

To lower the long timbers into the pocket, the trap door and part of the covering were removed; this left an exposed hole into the shaft. The size of the hole was five feet square but the 34-inch-diameter bucket hung in the hole, flush with the top of the level. A 22-foot length of 12x12 timber was moved into the drift with an end lying about seven feet from the edge of the hole. Using a cant hook, Owen tried to move the end closer to the hole by giving it a sudden jerk. The hook slipped, he went backwards into the space between the bucket and the hole, and fell 60 feet to the bottom of the shaft where his head hit solid rock. He was killed instantly.

He was survived by his wife, Olwen, a five-year-old son, and his parents who were living in Wales.

The funeral service was held at the Masonic Hall where a large number of friends and about 100 lodge members gathered to pay their last respects. Reverend M. C. Tait of Trinity United Church conducted the service. Six members of the Masonic Order acted as pallbearers. At the graveside, several Welsh comrades sang an emotional Welsh hymn as part of the ceremony. The grave is located in the Doric Lodge section of the local cemetery. The inscription on the stone is: T. OWEN – OLWEN OWEN – 1895 - 1990.

### May 5, 1932

# Lake Shore Mine

## Paul Baginski (1892 - 1932)

Paul Baginski, born in Poland in 1892, served in the Polish Army during World War I. He was married and had three children.

He left his family in Europe and came to Canada in 1929 during the world-wide Depression to get work. His first job was at the Abana Mine; later, he came to Kirkland Lake to work at Lake Shore where he quickly gained a reputation of being a careful and faithful employee.

He was an active member in the Polish Catholic Society and is remembered as playing the part of Santa Claus at the Society's Christmas Party.

On May 5, Baginski and his partner, Stanley Pogoda, were underground at Lake Shore engaged in loading holes when suddenly an explosion occurred. Baginski, standing directly in front of the exploding hole, received the full impact in his face and left arm. Pogoda, nearby, received a broken right arm and was suffering from shock when fellow miners came to his aid. Both were taken immediately to hospital where Baginski received treatment for his badly mutilated face. He was conscious but succumbed to his injuries and died

*Paul Baginski was Santa Claus at local Polish Club (Christmas 1931). Photographer: Unknown*

the next day. Pogoda survived the incident and was released from the hospital after 15 days.

Funeral arrangements were handled by Wray's Funeral Parlour. The funeral mass was held at St. Jerome's Roman Catholic Church with Reverend Father Jodoin officiating. Six fellow members of the Polish Society acted as pallbearers. Other members and friends accompanied the body to the church and then to the local cemetery for burial. The grave is unmarked.

### May 25, 1932
## Barry-Hollinger Mine
## Andrija 'Albert' Sragalj
### (1897 - 1932)

Many Croatian immigrants came to Canada to work the mines of Val d'Or, Noranda, Porcupine, Sudbury and Kirkland Lake. One Croatian immigrant, S. Mahovlich, presented working conditions in the McIntyre Mine at Schumacher in his poem:

### The Peasant and the Worker

*And now let your thoughts go,*
*and let them see the sights*
*Of the underground, where in holes stir.*
*Like worms through tunnels they crawl*
*And heavy rocks pull to the daily sun.*
*During their life they dig their grave*
*Passing their entire life in darkness.*
*No one cares for the poor people,*
*Especially when they are miners.*
*Only when someone loses his life,*
*The newspapers make a little noise.*
*But everything subsides, and so does this lament,*
*As does anything human pertaining to the poor.*
*And when crying widows*
*Count the bodies lifted from the dark caverns.*
*For this no one's head aches*
*And the world sleeps quietly as before.*

Albert Sragalj was an honest hard-working Croatian immigrant seeking a new life in Canada. Employed at the Barry-Hollinger, located about a mile from the mining community of Boston Creek, he lived alone but had a good friend and neighbour, John Mavrinac.

In Christmas of 1931, Albert was reunited with his wife and two children, Navy and Andy, when they left Yugoslavia and came to Canada. A few months later, he died while working at the mine.

Thirty-five-year-old Sragalj was on work duty at the Barry-Hollinger when suddenly he became ill and died a few minutes after reaching the surface. It was thought that death was likely due to heart failure.

The funeral mass took place in the Roman Catholic Church at Kirkland Lake and interment was in the Kirkland Lake cemetery. A large number of neighbours and fellow workers came to pay their last respects to their friend. At the cemetery the grave is now marked by a recently placed headstone (2000).

His children presently (1999) live in London, Ontario.

*Andrija Sragalj with hand-held carbide lamp.*
*Photographer: Unknown*

## June 25, 1932
## Teck-Hughes Mine

### Garrioch (Garry) George Elmitt (1908 – 1932)

At the coroner's inquest there was testimony that Garry Elmitt, 24 and a cagetender at Teck-Hughes, had sometimes jumped across an open shaft in the face of repeated warnings. This evidence was given by several employees at the mine who were questioned about the tragedy. The jury's verdict was: "We the jury, find that Garrioch George Elmitt came to his death by a fracture of a skull caused by the deceased falling down the shaft on June 25th, 1932."

Dr. R. H. Armstrong reported that Elmitt was alive when brought to surface after falling 26 feet into the sump of a shaft from the 3,600 level but that he ceased breathing about 20 minutes later. Death, he told the jury, was caused by multiple fractures of the skull.

Mine Superintendent George Gill testified that Elmitt had plunged into the sump after striking some planking during the fall, and the fall ended on a pile of rocky spillage. "No one saw the accident," the mine superintendent said, "and the only possible way he could get hurt was by crossing the shaft. He had no business crossing whatsoever, had no occasion to jump and I understand he had been warned against this practice before his death."

Frank Dafoe, a witness, was asked by Mine Inspector Keeley if he had ever seen Elmitt jump across the shaft. In reply, Dafoe said, "Elmitt told me that he sometimes leaps the hole."

Robert Parker, a shift boss, said there was no reason for anyone to jump across from one side to another unless it was to avoid walking about 100 feet.

Elmitt had been very popular in the community where he had lived for some years. He had been employed at the mine since 1929. The 24-year-old gained a wide circle of friends and his death was particularly sad as he was married only a little more than two weeks to Marjorie Rachel Lee of North Bay. (Marriage date was June 8, 1932.)

A jury statement was released at the end of the inquest: "There were no eyewitnesses, hence the cause of the accident is not definitely known. In travelling from side to side of the shaft as a cagetender's duties require, two safe means are possible: either to cross through the cage while it is spotted at the level or to travel around the shaftpillar, a distance of about 80 feet. There is no doubt that the accident occurred through Elmitt's falling into the shaft from the 3,600 level but it cannot be determined whether his fall was a result of a silly practice or some other unknown cause."

## July 8, 1932
## Lake Shore Mine

### John F. Gibbons (1902 - 1932)

Born in England in 1902, John Gibbons came to Canada in 1928. He lived in Kirkland Lake for a few years and roomed at the home of Mr. and Mrs. William Moran at 54 Tower Street.

He was active in soccer and, at the time of his death, was a member of the local football club executive. In England, he was a rugby player of note.

Gibbons was killed at Lake Shore in the early morning of July 8 while working in a stope on the 27th level. He was engaged in boarding up the stope wall when the mishap occurred.

Until this time, the stope was worked as a shrinkage stope but was prepared for cut-and-fill. In preparation for the change, the stope was pulled of muck. On the night shift of July 7, two muckers were sent to the stope to shovel the remaining muck into the chutes. Shift Boss C. L. Jerome then instructed Gibbons, a timberman, to scale the walls and to place sprags where necessary to make the place secure. Jerome returned later and was talking to Gibbons through the chute when suddenly a large slab weighing well over 20 tons slid off the footwall. Gibbons was struck down and completely buried by smaller pieces that had come off the wall. By the time fellow workers removed the rocks from his body, he was dead.

His injuries consisted of a fractured leg, a crushed chest and injuries to the skull. Mine Physician Dr. G. M. Cameron and Dr. Ernest Ratcliffe responded to the call for medical aid; however, nothing could be done.

Thirty-year-old Gibbons was single and a member of a family residing at Cumberland, England. His family included his parents and two sisters.

Funeral services held at Trinity United Church were conducted by Reverend M. C. Tait. Many friends attended; among the floral tributes was a wreath from fellow employees and a wreath from Lake Shore Mines Limited.

The interment occurred in the local cemetery; it's possible that the body was later exhumed and shipped back to England.

October 20, 1932

## Lake Shore Mine

### Charles G. Ross (1904 - 1932)

A sister's joy over a long-planned visit with her brother changed to grief on October 21 when she arrived in Kirkland Lake to find him dead.

Charles Ross was killed in a rockfall at Lake Shore a few hours before his sister, Miss Dora Ross of Toronto, stepped off the train at Swastika to spend several days with him. They had made plans of the visit some weeks previously and had spoken of moving their widowed mother in Toronto to this community to live with her son.

Ross, 28 and unmarried, was a stope boss at the mine. He died shortly after 11:00 p.m., a few minutes after coming on shift with his men. He was in charge of slashing, taking down backs and timbering. Earlier, the afternoon shift had blasted a nine-hole slash just before coming off shift at 11:00. The blasting had knocked out many props and posts. Ross and his crew of four men started their shift by replacing these when suddenly 50 tons of loose fell. Ross was killed and the other four were injured.

Gordon Howes (21) and Nick Pochopsky (41) were taken to Kirkland District Hospital where Howes was treated for a crushed leg and Pochopsky for bruises and lacerations.

Witnesses said that Pochopsky escaped death by a miracle. When the falling loose fell over him, it formed a protective arch that saved him from serious injury. The other two men were given first-aid treatment. Dr. G. M. Cameron and Dr. C. S. Harris were at the scene, attending to the injured. It took hours before the rescue crew found the body of Ross beneath the fallen material.

Ross was born in Rosshire, Scotland and lived with his family in Wales for some time before coming to Canada. He had been employed at Lake Shore for almost 18 months and was highly popular with his fellow employees.

His sister stayed in town with her friend, Miss Anna W. Johnson, while funeral arrangements were made. The service was held at Trinity United Church with Reverend A. Gillies officiating. The pallbearers were Murdoch McLeod, Clarence Whalley*, James McDonald, William Allen, Frank Wysocki and Ralph Hurd, all employees at Lake Shore. Many floral tributes covered the casket; sprays and wreaths came from the Lake Shore Mine, shift bosses at the mine, F. Grierson, friends from Lake Shore, the Johnson Family and Mr. and Mrs. Hector McPhail. Miss Ross accompanied the body as it was shipped to Toronto by train. Another funeral service and interment occurred there.

His mother, three sisters, Eva, Dora and Sybil, and three brothers, Alton, Colin and Paul all survived him in Toronto.

** Clarence Whalley would die in a 1954 accident at Lake Shore.*

December 2, 1932

## Lake Shore Mine

### Howard Robert 'Bob' Patterson (1904 - 1932)

When Scotty O'Donnell, the undertaker's assistant at Wray and Thorpe's Undertaking Parlor, was called in from the bush to report for duty, he found a sad task awaiting him. It was his responsibility to help prepare for burial the body of his closest pal, 28-year-old Bob Patterson, who was fatally injured in an accident at Lake Shore on December 2.

Patterson was a former United States Marine who had spent 1931 and 1932 in and around Kirkland Lake. At one time, he was living in a bush camp near the King-Kirkland Mine with his pal O'Donnell and several others. He had been employed at Lake Shore for almost two months and was living at Allan Jack's boarding house at 32 Prospect Avenue.

Patterson was married and had two children; the family lived in Sault Ste. Marie, Ontario. His father was a well-known mining man in Northern Ontario and for years had interests in the Michipicoten and Goudreau areas. Patterson was a member of a shaft crew working the No. 3 shaft at the 3,325 level on December 2. A two-ton tugger was installed at this level to lower timber to the shaftmen working 80 feet below. A small bucket was attached to the tugger cable.

At about 9:10 p.m., he told his fellow shaftmen that he was not feeling well and that he was going up to the level for a drink. He got on the bucket and rang one bell. Previous instructions were given to not ride the bucket under any circumstances. As the bucket was hoisted, it began to spin very fast. The hoistman saw Patterson's head and shoulders come through the deck; he then saw him fall through the hole at one side of the bucket. He plummeted 80 feet down the shaft. He was rescued, taken to surface and then to hospital where he died at 11:40 p.m. His death was attributed to a fractured skull.

He was survived by his immediate family, parents Mr. and Mrs. R. H. Patterson, a brother Wilfred, and a married sister, all living in Sault Ste. Marie.

### 1933

- Three miners are killed in accidents at Lake Shore; two are killed at Teck-Hughes and a miner dies at the Bourkes Mine
- Macassa constructs a 200-ton mill while Wright-Hargreaves is milling 1,000 tons per day
- Kirkland Lake Gold pays shareholders their first dividend in 19 years
- Toburn struggles to stay open
- A passable road is constructed from Kirkland Lake to Larder Lake and on to Rouyn
- Mine managers are well known to their employees. The following include the managers at some local mines: Colin A. Campbell (Barry-Hollinger), Oscar Knutson (Bidgood), William M. Sixt (Kirkland Lake Gold), E. W. Todd (Lake Shore), G. A. Howes (Macassa), Thomas L. Wells (Matachewan Consolidated), C. E. Rodgers (Sylvanite), R. J. Henry (Teck-Hughes), M. W. Summerhayes (Wright-Hargreaves), M. W. Hotchkin (Toburn)

### January 9, 1933
## Lake Shore Mine

## Jan Klimecki (1898 - 1933)

Members of the local Polish Society expressed their sorrow at the funeral of Jan Klimecki, 35, who died of injuries after falling down a raise into a stope at Lake Shore.

At the time of the accident, John Stefanich was lowering planks by means of a rope to Klimecki in the raise. When this job was completed, Stefanich came down to help and at that time realized that his partner had fallen into the stope.

Klimecki came to Kirkland Lake in 1926, leaving his wife and three children in Poland. He was a very active member in the local Polish Mutual Aid Society and was one who lent his aid in the construction in 1932 of the new Polish Hall* on Woods Street. At the time of the construction, he was a director and shareholder of the Kirkland Lake Construction Company, the firm that built the hall. In amateur theatricals sponsored by the Wanda Society, he took an active part and was treasurer of that organization.

Funeral services for Klimecki were well attended. The body was conveyed from the O. J. Thorpe's undertaking rooms on Duncan Avenue to the Polish Hall where it lay in state while many friends paid their respects to his memory. The funeral was held at Assumption Church where a newly arrived Polish priest, Reverend Father Callistus Winiarz, conducted the mass. The funeral cortege then proceeded to the cemetery where interment in the winter vault took place. Floral offerings accompanied the body to its resting place.

** As of 1999, the Polish Hall is the last active original ethnic hall in Kirkland Lake.*

### January 12, 1933
## Bourkes Mine

## Runor Lindholm (1909 - 1933)

Twenty-four-year-old Runor Lindholm, single and of Finnish ancestry, was killed instantly when he fell down a shaft from a point near the surface into 50 feet of water below the 250-foot level. The accident occurred at the Bourkes Mine on January 12.

Lindholm and fellow worker Fred Burke were climbing from the bottom of the shaft to the surface when the accident happened. About two or three ladder lengths from surface, Burke, who was leading, told Lindholm to wait on a landing until he got to the landing above, as he considered the ladder there too weak to carry both of them. Lindholm did this and, as Burke reached the surface, he heard his partner fall.

It was determined later that Lindholm fell off the ladder into the cage compartment. The partition between this compartment and the manway had been removed earlier to gain access to the pumps carried on the cage during a dewatering activity in the shaft. The partitions were not yet replaced on the top four sets.

An inquest was held at Swastika on January 18 before Coroner J. F. Edis. The jury returned a verdict of accidental death.

Interment occurred in the Kirkland Lake cemetery.

### February 25, 1933

## Teck-Hughes Mine

### Joseph S. Walton (1893 - 1933)

Joseph Walton, 40, an ore car sampler at Teck-Hughes, died from injuries he received in a fall down a ladder. His family – wife Mary and three children – lived on the Teck-Hughes property.

Walton was a native of Newsham, Northumberland, England. He was an experienced underground miner who'd worked at Teck-Hughes for six years and previously for 25 years in other mines in Canada and England. At one time, he was a five-year resident of Nanaimo, B. C. He was exceedingly popular among his employees at the mine and was held in high esteem by members of the Loyal Order of the Moose, having been a valued member for three years. He was an ex-serviceman who served in the British Army during World War I after enlisting in England.

He was alone at the time of the accident and was proceeding from the 30th to the 31st level when he apparently missed his step on a ladder and tumbled 50 feet down an inclined winze. He was rescued and taken to hospital where he died of a concussion, a week later, on March 4.

Surviving him were his wife, son Stanley (6), and two daughters, Mamie (19), and Edna (14). His mother survived him in England; he had four brothers, three living in Africa and another in England.

The funeral service was preceded by rites under the auspices of the Moose Order at the residence of the deceased. Reverend A. Gillies officiated the service at Trinity United Church. The bearers were R. West, Alex Scott, A. Skne, G. Connell, S. Gamble and W. Padget. Many people attended the services. Interment was in the local cemetery.

It is believed that Mrs. Walton and her three children returned to England to live.

### April 15, 1933

## Lake Shore Mine

### Henry I. Menard (1898 - 1933)

Henry Menard, a native of Gracefield, Quebec, was born to Mr. and Mrs. Oliver Menard in 1898 and spent his boyhood and early youth there. During the World War, he enlisted for overseas service; subsequently, he went to France with the 24th Infantry Battalion and was wounded at the Battle of Vimy Ridge. Having the rank of Sergeant, he was honourably discharged at the conclusion of the war. Returning to Canada in 1918, he lived for a while in Hull, Quebec before coming to Kirkland Lake in October of 1920. It was also this year that he married his wife, Laura. He was first employed at Wright-Hargreaves, then Sylvanite and later at Lake Shore where he was regarded as a highly capable employee. He kept alive his World War memories and friendships by keeping active membership in the local post of the Canadian Legion.

Thirty-five-year-old Menard of 40 Duncan Avenue lost his life when he and his crew of shaftmen were scaling loose rock just above the 3,950 station in No. 3 shaft at Lake Shore. A broken rock brow carried with it three sets of timbers and Menard was instantly killed when one of the heavy timbers crushed his chest. His companions, L. Miller and G. Lamoreaux, were slightly injured in the mishap.

The funeral rites were very impressive as full military honours were accorded to the deceased. Highland pipers played and a guard of honour from the local post of the Canadian Legion was present. From his late residence, the funeral cortege with its flag-covered casket proceeded to Assumption Church; there, a Requiem High Mass was sung by

Reverend Father D. Jodoin assisted by Reverend Father H. Leduc of Ramore and Reverend Father H. Quenneville. The church was filled with many friends and the casket was adorned with many floral tributes to this popular man. Burial rites at the local cemetery were completed with strains of the 'Last Post' and 'Reveille' sounded by Jerry Smith. The casket was carried by Jean Danis, Jack Neale, George Cope, Tom Murphy, William Johnston and Joe Meath. The pipers at the funeral were Donald Graham and Robert Hannah. Menard was survived by his wife, his parents, and a sister, Mrs. Joseph Tessier of 13 Brookbank Avenue.

### November 23, 1933

## Lake Shore Mine

### Frank Dragan (1907 - 1933)

Frank Dragan, a native of Yugoslavia, was an employee at Lake Shore who met instant death on November 23. His family – his wife, the former Pearl Burling and their two children – lived at 49 Hudson Bay Avenue. Before coming to Kirkland Lake in 1930, he lived for a time in Toronto and Sudbury. He was a personable 26-year-old who was very popular in town. At the time of the accident, he and Frank Mihelic, both machine runners, and their helper, Mike Vrbos, were driving a raise. One hole in the centre of the raise had already kicked through to the 400 level and Mine Captain William Allen spoke to them through this hole. At about 2:00 a.m. Mihelic told Dragan to go to the top of the raise at the 400 to guard and to see if it was safe to blast; they would communicate to each other through the hole. At about 3:00, Mihelic completed the loading of the holes and he called to Dragan up the raise, telling him that he was going to light the fuses if all was clear on the 400. Dragan called down through the hole, giving the all-clear. The fuses were lit and the shots went off at 3:20.

Dragan's broken body was found at about 3:45 at the top of the raise. Was there some miscommunication between him and Mihelic? Apparently, Dragan was communicating down the hole when the shots went off. He received practically the full force of the charge and was partially buried under a resulting fall of rock.

The funeral rites were held at O. J. Thorpe's undertaking rooms with Reverend E. G. Dymond of St. Peter's Church officiating. Many fellow countrymen were present at the funeral and accompanied the body to the cemetery for burial. The pallbearers were John Belej, Martin Clobucar, Andrej Kovac, John Stefanich, Nick Dejanovic and Frank Tomc*. Surviving were his immediate family in Kirkland Lake and relatives in the United States and Europe.

** Frank Tomc would be killed a month later in a mining accident at Teck-Hughes.*

*Headstone photograph of Frank Tomc. Photographer: Author*

### December 24, 1933

## Teck-Hughes Mine

### Frank Tomc (1896 - 1933)

Born at Metlili, Yugoslavia in 1896, 37-year-old Frank Tomc left his wife and family in 1926 to work in Canada's gold mines. He returned to his native country for a visit with his family in 1932 but again returned in March 1933 to Kirkland Lake where he took up residence at 48 Main Street.

On the day before Christmas, he was handling an empty ore car across an ore pass on the 23rd level at Teck-Hughes when he lost his footing and fell down a pass, a distance of 70 feet. Fellow workers immediately lowered ladders in a rescue attempt but found a lifeless body. He probably died instantly.

Many friends and co-workers attended the funeral service at Holy Name Church. The body was placed in the local cemetery vault awaiting spring interment.

A lengthy Yugoslavian inscription is found on the stone marking the grave: TUJA POCIVA FRANK TOMC – ROJEN LETA 1896 – SUHOR U PRI – METLILI YUGOSLAVIA – PONESRECIL SE JE 24 DECEMBRA 1933 TECK-HUGHES – RUDWIKUIN JE ISTEGA DIIE TUDI UMRL BODI MU LAHKA – KANASKA ZEML JICA.

## 1934

- Joseph Golob, Delphis Lalonde, Archie MacDonald, Emil Ollel, Eric Keto, Walter Szpak, Henry Eadie, Joe Mihelich and Thomas Cuthbertson are all fatal victims of mining accidents in the Kirkland Lake district in 1934
- U.S. sets the gold price at $35.00 an ounce; this is good news for the gold mining camps
- Bidgood Mine is incorporated
- Golden Gate Mine is incorporated
- Moffat-Hall Mine, which closed in 1932, reopens this year in February
- CJKL begins broadcasting
- Dionne quintuplets are born near Callander, Ontario

## January 26, 1934
## Teck-Hughes Mine

### Joseph Golob (1900 - 1934)

Joseph Golob, 34, a machine helper at Teck-Hughes, met death while at work on January 26. He was an employee with more than six years of underground mining experience.

Born in Yugoslavia on January 1, 1900, he was a single man who in 1927 came to Kirkland Lake where he lived with his brother Tony at 62 Woods Street. Other members of his family including his parents remained in Yugoslavia.

On the day of the accident, he and partner Javo Popovich were working in a stope 23 feet above the 3,200 level. The previous shift had blasted and left one missed hole and two cut-off holes. It was now the duty of these two men to reblast the three holes. Golob prepared the holes for blasting and sent Popovich down to the level to turn off the air. He did so and began to walk back when the first shot went off. Coming into the stope, he found Golob lying directly under the holes. He was dead; killed by the blast.

An inquest was held before Coroner J. F. Edis. The jury concluded this was an unexplained blasting accident evidently caused by a premature blast.

Many fellow countrymen attended the funeral mass conducted by Reverend Father Callistus Winiarz at Holy Name Church. Burial took place in the local cemetery.

## February 10, 1934
## Lake Shore Mine

### Delphis Lalonde (1905 - 1934)

Delphis Lalonde, a native of Lemieux, Ontario, was born on April 24, 1905. In 1915, the family moved to the lumbering town of Charlton, Ontario. Lalonde left the town in 1929 and came to Kirkland Lake where he got a job at Lake Shore.

On April 29, 1929, he and Valerie Caza (1905 – 1988) of St. Joachim, Ontario, were married in Englehart. The marriage produced four children: Leo (1930 – 1946), Thérèse (1931 – 1931), Albert (b. 1933) and Roger (1934 – 1953). The family lived at 130 Carter Avenue.

On February 10, he and Mike Tonuik were engaged in scaling operations in a stope. The two were standing on a muck pile when suddenly a large slab of rock broke off the wall and precipitated a slide of muck down towards the millhole. Tonuik jumped to safety but Lalonde was carried towards the hole. A grizzly covering the millhole had a bar missing leaving a space large enough for Lalonde to pass through. He was drawn into the hole to a depth of 20 feet and covered with finely broken ore; his body was not recovered until six hours later. He had died of suffocation.

*Marriage of Delphis Lalonde and Valerie Caza (1929). Photographer: Unknown.*

He was predeceased by a daughter in 1931. Surviving were his wife and three young sons, parents Joseph Lalonde and Azelie Lauzone of Charlton, four brothers: Fred, John and Napoleon of Charlton and Julien of Kirkland Lake, and three sisters: Dilema (Mrs. Adelard Lauzon), Lucia (Mrs. Dennis Desjardins) and Louise (Mrs. R. Barryall).

The funeral mass officiated by Reverend Father D. Jodoin was held at Assumption Church. A large number of friends and relatives attended the service for the 28-year-old miner. Interment took place in the local cemetery.

At the inquest held by Coroner J. F. Edis, the jury returned a verdict of accidental death.

Alice Laronde, niece of Delphis Lalonde, researched some details for this account. The photograph and other details were supplied by Albert Lalonde who was only a baby when his father, Delphis, was killed. Albert, now 67, presently (1999) lives with his family in Val d'Or, Quebec. Joseph Lalonde, nephew of Delphis, lives in the Kirkland Lake area. Many other relatives remain in this area and the grandchildren reside in Quebec.

### May 9, 1934
## Teck-Hughes Mine
## Archie A. MacDonald
### (1900 - 1934)

Archie MacDonald, 34, lived on Federal Street with his wife and their five children. He was injured at Teck-Hughes at 8:30 p.m. on May 9 and died in the hospital the same night.

MacDonald, a machine runner, left a stope to examine an air line in a manway. He was in a leaning position, doing the examination, when suddenly a 150-pound piece of rock fell and struck him in the middle of the back. Two other machinemen in the area came to his aid. He was taken to surface and sent to hospital where it was found that he had sustained a broken leg, broken arm and severe chest injuries. He died three hours after the accident.

He was interred in the Kirkland Lake cemetery. The inscription on the stone over his grave reads: ARCHIE MACDONALD – 1934.

### May 28, 1934
## Martin-Bird Mine
## Emil Ollel (a.k.a. Emil Bartel)
### (1899 - 1934)

Thirty-five-year-old Emil Ollel, single and born in Norway in 1899, worked as a blacksmith and mechanic at Martin-Bird near Larder Lake. On May 28, while working underground, he was overcome by carbon monoxide and subsequently fell down the shaft.

Mine Manager John Campbell had placed a three-horsepower gasoline pump on a floating raft, 35 feet down from the collar of the shaft, to wash down an area on surface for geological inspection. He started the pump and intended to let it run until the gasoline was used up. The pump ran for only 10 minutes and stopped before he had completed the wash down. He and Ollel decided to go down the shaft to see what had happened to the pump. Suspecting a blockage, they removed the suction pipe. Ollel ascended to surface to get a wrench and returned. The pipe was taken off, examined and replaced. They couldn't find the problem so Campbell decided to leave it until the next day. He climbed the ladders to the surface and Ollel followed. Campbell arrived at the surface and turned around just as Ollel let go of the ladder and fell backwards down the shaft. Oscar Otterson, who was on surface at the scene, saw Ollel fall and immediately ran down the ladders to rescue him. He tied a rope around and under Ollel's arms and Campbell and two others hoisted Ollel up to surface. He lived for about five minutes and then expired.

A post mortem examination was made by Dr. Shipley, who stated that Ollel died of carbon monoxide poisoning.

Interment occurred in the Kirkland cemetery.

An inquest was held before Coroner J. F. Edis at Kirkland Lake. At the inquest, Campbell stated he felt no ill effects from the exhaust gas.

According to a Department of Mines Report, a prosecution resulted after the accident: "A charge was laid against John Campbell, Manager of Martin-Bird Syndicate, for operating contrary to Subsection 27, Section 143, of the Mining Act, in that a gasoline pump was operated in a shaft at that property."

A plea of 'guilty' was entered before Magistrate Atkinson at Kirkland Lake on June 14, 1934 and a fine of $100 plus costs was imposed.

### June 24, 1934

## Kirkland Lake Gold Mine

### Eric Keto (1905 - 1934)

Eric Keto was born in Finland in 1905. At the age of 21, he left his native land and came to Ontario where he worked in the Frood Mine at Sudbury before coming to Kirkland Lake in 1933. He was employed at Kirkland Lake Gold at the time of the accident. Keto, 29, was married. He and his wife, Hilda, lived at 54 Lebel Avenue.

On June 24, Machine Runner Sam Vachon* and helper Keto had been on shift for only an hour, scaling in a stope. Vachon stopped scaling and handed the scaling bar to Keto who then proceeded to take down more loose rock. A minute or two later, the hanging wall suddenly collapsed, crushing Keto. He was killed instantly.

The funeral service was held at O. J. Thorpe's undertaking parlours. Interment occurred in the local cemetery.

** Sam Vachon would die accidentally in the same mine, seven years later.*

*Headstone photograph of Walter Szpak.*
*Photographer: Unknown*

### June 27, 1934

## Wright-Hargreaves Mine

### Walter Szpak (1903 - 1934)

In Poland on December 28, 1903, Walter Szpak was born to Joseph and Mary Szpak. As a young adult, Walter left his native country and came to Canada for a new start in life. After a number of years in Kirkland Lake, he was an experienced miner at Wright-Hargreaves. In 1932, he married the former Anna Zawojska and the couple took up residence on Woods Street.

On June 27, 1934, Szpak was working underground at Wright-Hargreaves with his partner, Jackymek. They had drilled off the first round of 16 holes in a raise and were in the process of lighting it. They lit ten holes but had difficulty lighting the 11th as the fuse would not light. The fuses to the other holes were lit and burning. Szpak cut the fuse three times before he finally succeeded in lighting it. Jackymek repeatedly asked Szpak to leave but to no avail.

Jackymek jumped off the staging and swung under a brow of rock. Then Szpak jumped but it was too late, as the first hole exploded at the same time. He received the full force of the explosion and was killed instantly.

An inquest was held and the jury returned with the following verdict: "We, the undersigned jurymen, find that Walter Szpak came to his death by an accident attributed to misadventure on the part of the deceased."

The interment of 30-year-old Walter Szpak took place in the Kirkland Lake cemetery.

## November 8, 1934
## Kirkland Lake Gold Mine

### Henry Eadie (1909 - 1934)

Henry Eadie , a native of Glasgow, Scotland, had been in Canada for nine years before coming to Kirkland Lake from Windsor in January 1932.

In a June wedding in 1934, he married the former Bertha Watts of Dane. The couple took up residence in Chaput Hughes. Five months later, Eadie was dead, killed in an underground accident at Kirkland Lake Gold.

He had been an employee at the mine initially on surface and later as a cagetender. There were no witnesses to the accident occurring on November 8 as Eadie was working alone, operating a cage in a winze. He was found crushed beneath the cage at the 5,850 level.

He was predeceased by a brother, Jim, who drowned in Winnipeg in the summer of 1934. Surviving Eadie were his wife, a sister (Mrs. E. Horne) of 93 Second Street, and a brother and sister living in Scotland.

The funeral service for the popular 25 year-old was held at the Baptist Church on Tower Street with Reverend M. S. Barr officiating. Interment was conducted in the local cemetery.

The inquest jury's verdict was 'accidental death.'

## December 5, 1934
## Sylvanite Mine

### Joseph Mihelich (1905 - 1934)

Born in Yugoslavia in 1905, Joseph Mihelich came to Canada without his wife and son during the Depression, seeking employment at the local gold mines.

While here, he professed to being an atheist and became involved with the Communist Party of Canada.

On December 5, Mihelich, a three-year employee at Sylvanite, was working in a stope with Machineman Jack Hardy. A fall of about two or three tons of loose suddenly occurred, pinning Mihelich but not injuring Hardy. When Mihelich was released from the pile of rubble, he was dead due to a fractured skull.

A hasty inquest into the death was conducted by Coroner R. H. Armstrong. The jury returned a verdict of accidental death.

Present at the funeral service held at the Ukrainian Hall was Mihelich's closest relative in Canada, a brother-in-law from Sudbury. Mihelich was given a Communistic funeral; no clergy were present since the service was not religious but very political. Representative politicians of the local Communist Party made speeches. Later, the casket containing the body was paraded through the various streets in town. Interment occurred in the local cemetery. The grave is presently unmarked.

The photograph was donated by Anthony Paver, a staunch retired union man from Hamilton, a distant relative of Joseph Mihelich. Paver was born in Kirkland Lake but he and his family left the area during the 1941/42 miners' strike. His father, Anton Paver, was a miner at Sylvanite in the '30s.

Anecdotal unverified stories are passed on from generation to generation; the following one was given to Anthony by his father: "As the mine rescue crew were trying to free Mihelich, he was still alive but pinned and in terrific pain. To help him ease the pain, they sent him a bottle of whiskey on the end of a rope."

At first glance, the funeral photo may give a morbid impression. However, it's important to remember that Joseph Mihelich's wife and son lived

*Funeral of Joseph Mihelich (Ukrainian Hall – Kirkland Lake). Photographer: Unknown*

in Yugoslavia at the time of his death. To demonstrate to his family that he died surrounded by friends, this photograph was taken and sent to Europe. To show his everlasting love and devotion to his wife, their wedding photograph was placed in front of him, in the casket.

### December 6, 1934
## Macassa Mine
# Thomas James Cuthbertson
### (1902 - 1934)

Thomas James Cuthbertson was born in 1902 to Mr. and Mrs. Thomas Cuthbertson of the farming area in the Township of Kerns, a few miles northwest of New Liskeard.

He had worked in the Beattie Gold Mine in Quebec and in the Kirkland Lake gold camp. His fraternal twin brother William was also an employee at Macassa.

On December 6, a day after an accident at Sylvanite that killed Joseph Mihelich, Thomas Cuthbertson was injured in an accident at Macassa and was taken to Kirkland Lake Hospital where he put up a gallant battle for life. His friends were hopeful, but pneumonia that so often accompanies a serious injury ended his life on December 9.

At the time of the accident, Machineman R. Hudson and helper Ely Hill were working in a stope, 95 feet above the 2,475 level. Hudson had finished drilling and told Hill that he was going down the manway to the level to shut off the air and water. He also told him that when he was ready, he would rap on a pipe as a signal for Hill to lower the steels to him.

*Thomas Cuthbertson. Photographer: Unknown*

At level, Hudson had difficulty in closing the air line. While he was working at it, Shift Boss George Smith and Cuthbertson arrived. Cuthbertson, a pipefitter, told Hudson he would close the valve of the air line for him. He loosened the nut on the plug by tapping it. Hill, 95 feet above, hearing this tapping, mistook it as a signal to lower the steel. He tied three pieces of six-foot-long steel to a new hemp rope and began to lower them. About halfway down, the knot became loose and a steel dropped, followed by the other two. Cuthbertson, below, was pierced through the left side of his back by the first steel; the other two landed near Hudson.

Cuthbertson was immediately taken to hospital for surgery. According to medical reports, he was progressing favourably until he developed pneumonia and died. At the time of his death, he was 31 and single.

The body was taken to Thorpe's Funeral Chapel for services. Many wreaths adorned the casket, tributes from relatives, friends and associates. The pallbearers were John Woodruff, A. C. Perry, Cameron Fraser, Dan Morrison, Dan Abrams and Patrick Purcell, all fellow workers from Macassa. The body was taken by train to Uno Park and interment occurred in the Milberta, Ontario cemetery.

Thomas Cuthbertson was survived by his father of Milberta, his brother, three sisters, Viola (Mrs. Gorden Larett) of Falconbridge, Emma (Mrs. Leo Simpson) and Ethel (Mrs. A. Larouche), both of Kirkland Lake.

An inquest was held by Coroner R. H. Armstrong. The jury returned a verdict of accidental death and recommended that a safer method of lowering steel be used.

Kay Cuthbertson, wife of the late William Cuthbertson, presently (1999) resides in Kirkland Lake. She provided the photograph and some personal information for this memorial to her brother-in-law. Kay, by the way, is the sister of the late Robert Carlin, an historical union organizer at Kirkland Lake and Sudbury.

## 1935

- Considering all the mining activity in the area, the incidence of accidental deaths seems to decrease. Even so, seven miners die on the job
- Kirkland Lake gold mining camp of 3,490 working men is bustling with activity due to the $35-an-ounce price of gold
- Seven mills in town operate 24 hours a day
- Lake Shore is now one of the largest gold mines in the world and mills 2,000 tons of ore a day
- Teck-Hughes mills 1,000 tons a day
- Sylvanite mills 432 tons a day
- Wright-Hargreaves operates underground on 28 levels
- Omega Mine employs 175 men
- Morris Kirkland Mine is incorporated
- Beaverhouse Lake Gold Mine (formerly the Argonaut Mine) is incorporated
- Harry Oakes resigns as President of Lake Shore and plans to move to the Bahamas because the government taxes him too heavily
- The new Trinity United Church is built in town

## January 15, 1935
## Toburn Mine
### William Kuzyk (1904 - 1935)

Thirty-year-old William Kuzyk, Polish and single, was killed in No. 549 stope at Toburn on January 15. He and partner E. Lundquist were mucking when Shift Boss V. Risberg came into the stope for inspection and noticed no loose ground. During the shift, the two miners continued to muck and uncovered a 2.5-ton piece of loose on the hanging wall. This piece broke away suddenly and struck Kuzyk, pinning him against the footwall. He died standing up with only his head visible. It took a crew an hour and a half to extricate the body. Instant death was attributed to severe crushing injuries to the chest and abdomen.

Interment occurred in the Kirkland Lake cemetery; the burial site is unmarked.

## January 16, 1935
## Lake Shore Mine
### Anthony 'Tony' Wlasniak (1902 - 1935)

Polish immigrant Tony Wlasniak, 33 and single, was killed January 16 at Lake Shore. The accident occurred a day after fellow countryman William Kuzyk was killed at Toburn on the other side of town.

Wlasniak was working on the 2,400 level, barring out a large rock from a 35-degree sloped muck pile. For about half an hour, he strenuously tried to dislodge this rock. Finally, he was successful and, when the muck started to move, he either collapsed or lost his balance and fell into the run of muck towards the millhole. He was drawn into the millhole and the large rock then crashed through the split lagging lining the millhole. It momentarily held this position and then fell into the millhole, crushing Wlasniak about the head and chest, causing instant death.

Interment occurred in the Kirkland lake cemetery. The gravesite is unmarked.

An inquest was held before Coroner J. F. Edis. The jury returned a verdict of accidental death attributed to internal hemorrhage and multiple chest injuries.

*Elmer Paull Barkell. Photographer: Unknown*

## April 12, 1935
## Teck-Hughes Mine
### Elmer Paull Barkell (1878 - 1935)

Elmer Barkell was born at Lake Linden, Michigan on September 20, 1878. His parents, Bartholomew Barkell (1854 – 1922) and the former Elizabeth Paull (1854 – 1934), at one time lived in Wisconsin but moved to Lake Linden before Elmer was born.

At Lake Linden, Elmer married the former Mary Jane Nicholls (1881 – 1938). The marriage produced three children: Elmer Lloyd, Alferetta Ruth and Alfred Harry, all born at Lake Linden.

At 35 years of age, Elmer came with his family to Canada where they settled in the silver boom town of Cobalt. This town had its beginnings in 1904 and by 1913, when the family arrived, it was a going concern, a town famous around the world.

Cobalt hit its peak in 1919 and after that began to slow down. Two of the children, Alferetta and Harry, now young adults, saw a promising future a little further north and moved to the gold boom town of Kirkland Lake. Lloyd left but went south to Toronto.

Elmer and Mary Jane remained in Cobalt where they owned and operated an automobile dealership and garage. By 1934, the business was not doing well so they sold out and moved to Kirkland Lake to be closer to their two children.

At 56 years of age, Elmer Barkell had a high degree of mechanical experience; he was hired on at Teck-Hughes. His mechanical experience made him a valuable employee as an underground pumpman.

On April 3, while tightening a gland-nut on a pump on the 3000 level, the wrench slipped, causing an abrasion on the third finger of his right hand. Thinking that the wound was slight, he did not take the time for first-aid treatment and continued to work regularly for three more days. However, on April 7 he found that his right arm was considerably swollen and painful. He was treated by a doctor but efforts to check the infection were unsuccessful and on April 12, he died. A post-mortem examination revealed the cause of death as septic poisoning.

Interment occurred in the Kirkland Lake cemetery where he is now with his parents, his wife and his grandson who drowned at the age of 13.

The personal history and the photograph of his grandfather is credited to Donald Barkell who lives with his family in Kirkland Lake. He is the son of Alfred Harry Barkell.

### June 11, 1935
## Teck-Hughes Mine
### Sava 'Sam' Potkonjak
### (1901 - 1935)

Thirty-four-year-old Sam Potkonjak, a Yugoslavian immigrant, married for only a year and a half, lived with his wife Milena at 31 Hudson Bay Avenue. He was killed in a Teck-Hughes stope June 11.

*Headstone photograph of Sava Potkonjak.*
*Photographer: Author*

He and his partner, A. Kosicki, were scaling and had spent considerable time trying to bring down one particular stubborn piece of loose. When lunchtime arrived, they decided to leave the piece and go to the level to eat. Potkonjak had to pass under the piece to reach the manway and when he did so, the eight-ton rock fell and crushed him, causing instant death.

Interment was in the Kirkland Lake cemetery. The inscription on the headstone over his grave reads: OVDJE POCIVA – SAVA POTKONJAK – RODGEN 9 JAN GOD 1901 – POGINUO 11 JUNA GOD 1935 – USPOMEN PODIZE – SUPRUGA MILENA. Translation: Here lies Sava Potkonjak – Born 9 Jan year 1901 – Killed 11 June year 1935 – Remembering the resurrection – wife Milena.

### July 26, 1935
## Lake Shore Mine
### Albino Pontisso (a.k.a. Johnny Albino)
### (1890 - 1935)

Forty-five-year-old Albino Pontisso, Italian and with a wife and two children living in Italy, boarded at 54 McKelvie Avenue. His occupation at Lake Shore was listed as mucker.

On July 26, Pontisso and J. Reitveld were working in a stope at the 2,700 level. They were tramming sand, used for back-fill, from a raise to the end of the stope. Before the accident, there was some difficulty in keeping the sand moving in the 40-degree raise. In fact, during the July 25th shift, the sand was not moving so water was added at the top of the raise and allowed to run for two hours.

Since the sand was still not moving well on their shift, Rietveld set up a blast in the raise to get it moving. The blast caused about seven tons of wet sand to come down. Pontisso went to the bottom of the raise and started shovelling sand into the tram. Unexpectedly, 10 more tons of sand came loose, crashed down through the raise, carried out a bulkhead, knocked Pontisso off his feet and covered him slightly. Bert Malloy and Oli Langerid, who were working about 40 feet away, heard Pontisso's cries for help. Malloy rushed to his aid and uncovered sand from his face but another rush of sand came down the raise causing the rescuers to retreat to a safe distance. They called for more help and tried to block the opening where the bulkhead had been carried away. They went back and uncovered Pontisso the second time but again another rush of sand let loose and buried his head to a depth of three feet. He was dug out the third time but by this time he had died of suffocation.

The funeral mass was held at Holy Name Church and interment took place in the local cemetery. The inscription on the headstone reads: ALBINO PONTISSO – BORN JULY 1, 1890 – DIED JULY 26, 1935 – 'FRIENDS REMEMBER'.

### August 2, 1935
## Bidgood Mine

### Andy Dologh (1904 - 1935)

Thirty-one-year-old Andy Dologh, married and Czechoslovakian, was killed August 2 at Bidgood Mine, about 10 miles east of Kirkland Lake.

Dologh, a machine runner, was working with his partner, Ed Nelson, in a drift on the 500 level. They drilled a round of holes and loaded each hole with about six sticks of powder. Due to the dampness of the rock face, they were slightly delayed in lighting the round but left quickly. They were about 20 feet from the face when the first shot went off. Both were knocked down by the blast but Nelson regained consciousness and walked to the shaft where he met Nils Saari and P. Tukanen. They immediately took Nelson to surface and returned to the drift with Mine Manager Oscar Knutson. The three men were unable to reach Dologh because of the smoke and gases from the blasts. Saari, using a wrench, broke the pipe line and turned on the air while Knutson went back to surface to call the mine rescue station in Kirkland Lake.

Mine Rescue Superintendent P. C. Smith arrived and, along with two mine employees, went underground wearing masks. In the drift, they found Dologh bleeding from many small wounds in his back and from a compound fracture of his left arm. He was taken to surface but he was dead. Dr. Shipley later stated that death was attributed to shock and hemorrhage.

Interment was in the Kirkland Lake cemetery. The epitaph on the headstone states: 'DEATH IS CERTAIN, THE HOUR UNSEEN'.

### September 11, 1935
## Lake Shore Mine

### John Clements Sullivan
(1912 - 1935)

John Sullivan and his twin brother Daniel were born in Argentite, Ontario in 1912 to a northern pioneer family, Mr. and Mrs. Daniel J. Sullivan, who came to Cobalt in 1905.

In the spring of 1934, John left his hometown and came to Kirkland Lake where he got a job at Lake Shore as a driller. On May 5, 1935 he married the former Mary Brezenski of Cobalt and the couple took up residence in town. Four months later on September 11, at the age of 23, he was fatally injured in a stope on the 2,000 level at Lake Shore.

Sullivan and his helper, John Burnett, received instructions from Shift Boss M. McLeod to take down an overhang. Sullivan was drilling the fourth hole into the overhang when suddenly about a ton of rock fell on him. He sustained a badly crushed chest and abdomen and died within five minutes.

Sullivan was survived by his wife, three brothers, Daniel, Clarence and Wilbert, five sisters, Mrs. Fred Davidson, Mrs. Hazel Peters, Mrs. William Cochrane of Duparquet P. Q., and Geraldine and Marguerite, both at home.

He was buried at the Roman Catholic Cemetery at North Cobalt.

## 1936

- Twenty-one miners in the Kirkland Lake area are killed in mining accidents this year
- There are 4,640 mine employees in Kirkland Lake
- This year would become the highest production year in the Kirkland Lake gold camp; declines would occur from now on
- Kirkland Lake Mine, Mill and Smelters Union, Local 240, is granted a charter on May 1
- Wright-Hargreaves sinks another new shaft
- Lake Shore is the best producer in Kirkland Lake
- Bidgood Mine in Lebel township operates successfully
- The Moffat-Hall Mine in Lebel closes; one person is killed there during salvage operations
- Fernland Mine, located between Larder Lake and Virginiatown, is incorporated
- Omega Mine near Larder Lake mills 300 tons of ore daily
- Kerr Addison Gold Mines Limited is organized
- Harry Oakes now lives in the Bahamas
- Wright-Hargreaves Ball Park opens (currently the Civic Service Stadium)
- Kirkland Lake Police Department declares war on gambling by conducting a Saturday night raid on two downtown clubs. Fifty-two men are charged and are taken to the police station in a bus specially rented for this action
- Symington Funeral Parlour opens
- S. A. Pain is the manager of Golden Gate Mine. In the '60s, he will write two well-known local books: *Three Miles of Gold* and *The Way North.*

## February 1, 1936
## Barry-Hollinger Mine

### Angus Clattenburg (1898 - 1936)

Angus Clattenburg was born at Wolfville, Nova Scotia. His family, including his brothers and sisters, left the Maritimes and came to Cobalt where his father worked in the silver mines.

The 38-year-old Clattenburg was married to the former Louella Cheaney. The family of two children lived in Boston Creek near the Barry Hollinger Mine where their father was employed and killed on February 1.

Mining Inspector Ed Keely of Swastika conducted an investigation into the fatal accident that occurred during the closing-down operations at the mine. He learned that Clattenburg and two others, Eric Ritchie and John Mulvihill, were working in the main shaft, 85 feet below the 200 level. They were disconnecting pipes and sending them to surface for salvage. A pipe broke loose and struck Clattenburg on the head and arm, killing him instantly. Ritchie, who was working above him and Mulvihill, working below, were not injured.

*Miner cleaning out hole with scraper.*
*Photographer: Unknown*

February 20, 1936

## Teck-Hughes Mine

### Two Drillers are Killed in a Blasting Accident

Two drillers, Matti Kauppi and Axel Ostman, were fatally injured on the 43rd level of Teck-Hughes on February 20.

The accident occurred as they were drilling the 13th hole in the face of a crosscut. This hole was drilled to a depth of about three feet when the drill intersected the remnant of another hole from the previous round and an explosion resulted.

Shift Boss Frank Dafoe had visited the men four hours earlier and watched them wash down the face and clean out the old holes using a scraper.

Due to the explosion, both received fractured skulls and multiple body wounds from flying pieces of rock. Kauppi was treated at the first-aid station and then rushed to hospital but died about an hour after the accident. Ostman lost both eyes and was taken to hospital where he died 20 hours later.

"A possible explanation for the failure to discover the missed hole may have been attributed to the sheared nature of the rock in this face. This occasionally results in the remains of old holes squeezed to such an extent that their detection and location is difficult. Unless extreme care is exercised, such a hole and its contents may not be discovered in a new face." – a statement made at the inquest.

*Two victims of drilling into missed hole are buried side by side. Photographer: Author*

### Matti Leonard Kauppi

(1903 - 1936)

Matti Kauppi, 33, born in Finland in 1903, was married and living at 49 McKelvie Avenue.

### Axel Edwin Ostman

(1903 - 1936)

Axel Ostman, 33, born in Finland on November 18, 1903, was married and living at 15 McKelvie Avenue.

The funerals were held simultaneously at the Finnish Evangelical Lutheran Church on Sunday, February 23, with Reverend Martin Halinen officiating.

March 1936

## Barry-Hollinger Mine

### Robert 'Bob' Dickson Cameron (1884 - 1936)

Bob Cameron, born at Montague Gold Mines, Nova Scotia in 1884, grew up with mining, working in the northwest fields at Cobalt, Silver Centre and in other mining districts. He was a mine captain at the Keeley in the Cobalt camp and became captain at Barry Hollinger in Boston Creek where he worked for six years after making a trip west.

In March of 1932, he fell 160 feet, 80 of them vertically, in Barry Hollinger. To the amazement of friends, 48-year-old Bob was sitting up in his bed the next morning, complaining only of a sore back. After medical treatment, he was able to work again for a short period of time but his injuries were serious and they gradually caused him to become an invalid.

Death came to him in 1936 due to his injuries. He died in Cairn's Hospital located at 22 Woods Street in Kirkland Lake. The Northern News at that time stated, "He was a beloved character of the mining world, who, for four-and-a-half years, put up a gallant fight against injuries that would have killed a man of lesser physique."

*Bob Cameron.*
*Photographer: Unknown*

Many 'old-timers' in the gold camp made their last tribute to Cameron at the Symington's Chapel where he rested. His funeral service was in Trinity United Church and interment took place in the Kirkland Lake cemetery.

He was survived by his wife, two daughters, Olive (Mrs. Tom Webster) and Hattie (Mrs. Calverly), and three sons: Jim, Colin and Bill.

Presently (1999), Olive lives in Kirkland Lake. She verified this account of her father. Three of her children, Leonard Webster, Chuck Webster and Freda Ross, reside in Kirkland Lake.

### April 1, 1936

## Lake Shore Mine

### John Laakkonen (1897 - 1936)

John Laakkonen, a married man, was living at 108A Woods Street in Kirkland Lake.

On April 1, he was working in a stope at the 2,900 level of Lake Shore with his partner, J. Mickelson. Laakkonen was stooped over, pushing pieces of rock through the grizzly by hand, when a rock fragment weighing about 50 pounds suddenly bounded down the muck pile and hit him, breaking his neck. Mickelson lifted him into the manway, considering him only stunned. He obtained assistance and Laakkonen was taken to the shaft and then to surface but he was already dead.

The funeral service was held at the Finnish Evangelical Lutheran Church with Reverend Martin Halinen officiating. Interment occurred in the local cemetery. The inscription on the headstone over his grave reads: IN MEMORY OF – JOHN LAAKKONEN – DIED APRIL 1, 1936 – NUUKKUAS RAUHASSA – RAKKAHIN – LAAKKONEN.

### June 11, 1936

## Matachewan Consolidated Mine

### William Stevenson (1877 - 1936)

Fifty-nine-year-old William Stevenson, British and married, was killed in a stope at the 160-foot level of Matachewan Consolidated on June 11.

No. 1 E stope was a shrinkage stope with five chutes. On the morning of June 11, B. Sutherland and D. Ainsley were instructed to pull the No. 5 chute. Ernest Bodin and Angus McDonell were drilling the bench in the stope near this chute. Stevenson, the mine captain, met the chute-pullers tramming their cars to the station; he then entered the stope. He climbed up the muck pile to where Bodin was drilling and began to examine some ore on the wall.

Suddenly, the muck pile dropped about eight feet, causing Stevenson to fall backwards and plunge head first into the running funnel of muck. When the movement of muck finally ceased, he was almost completely buried with only one of his feet left visible. An attempt was made to extricate him but the exercise proved futile as the funnel-shaped muck pile around him could not be held back. After two-and-a-half hours of rescue work, it was decided to pull the muck from the chute and extricate the body that way. The body was released after pulling six cars of ore. Stevenson had died of suffocation.

### July 21, 1936

## Wright-Hargreaves Mine

### William Temberg (1904 - 1936)

William Temberg, 32, single and Russian, was employed as a drill helper at Wright-Hargreaves. He was fatally injured on the 550 level July 21 and died in the Kirkland Lake District Hospital August 2.

On the day of the accident, Temberg and his partner N. Jackowic had finished drilling in a stope and were bringing their dull steels out to the top of a steel-slide, 90 feet above the level. They stacked them before lowering them to the level.

The regular method of lowering steels in a raise is as follows: One of the men goes to the level and, when clear of the steel-slide chute, shouts to his partner at the top of the raise; the partner places the steel into an apparatus called the 'carrying finger'. The steels are then lowered in the raise using this finger.

Jackowic instructed Temberg to go down to the level to unload the steel from the carrying finger when it was lowered. As the finger was at the bottom of the steel-slide, Temberg was instructed to pull on one end of the rope to signal to Jackowic to pull the finger up.

Jackowic received the pull on the rope shortly after Temberg shouted 'okay', signalling he was in the clear. Jackowic then pulled the carrying finger up but just as the finger got to the top, he accidentally dislodged some steels, causing them to fall down the raise. Hearing a shout from Temberg, Jackowic hurried down the ladders to the bottom of the raise to find his partner lying at the foot of the steel-slide, badly injured. Jackowic went for help and soon Temberg was placed in a stretcher and taken to surface.

On admission to hospital, it was found that Temberg sustained a large wound below the shoulder. Four vertebrae and three ribs in this region were broken. Just above the hips, another wound was apparent: here, another vertebra was broken and the spinal cord injured.

On July 23, he developed pneumonia and, ten days later, he died.

### July 21, 1936

## Lake Shore Mine

# Marshall Leslie Stinson

### (1917 - 1936)

Marshall Stinson was born to William and Mary Stinson of Cobalt in 1917. William Stinson was well-known in Cobalt as he was the former fire chief and, in 1936, the town foreman.

Marshall Stinson was a likable lad who in 1934 was a valued member of the rugby team at the Haileybury School of Mines. Stinson, 18 and single, came to Kirkland Lake in September 1935 and got a job at Lake Shore as a drill helper. He lived at 16 Main Street with his sister Flo and her husband, Chick Valentine.

At the time of the accident, Stinson and two drillers, Cooke and O'Hagen, were preparing to take timber into a stope at the 2,200 level. Suddenly, a crash of falling loose occurred. O'Hagen, who was not injured, came to the rescue of the other two. Cooke was in the clear but had an injured foot. Stinson was pinned under the fallen ground. O'Hagen left to get a doctor and more help. With others, he soon returned and Stinson was released but he was dead.

Surviving were his parents of Cobalt, two brothers, Emerson of Lamaque, Quebec and Dale at home, three sisters, Flo, and at home, Doreen and Eunice.

The body was sent by train to Cobalt for burial in Silverland Cemetery. The inscription on the headstone is: CLARENCE MARSHALL LESLIE STINSON – BELOVED SON OF WILLAM AND MARY STINSON – BORN MARCH 28, 1917 – DIED JULY 21, 1936 – 'GONE BUT NOT FORGOTTEN'.

As of 1998, there are surviving relatives in Kirkland Lake: Mrs. Eunice Caldwell, sister; Dennis Caldwell, nephew; Mrs. Emerson Stinson, sister-in-law; and Brenda Ryan, niece. Also presently surviving is Marshall's sister, Doreen Sparling of North Bay.

### August 15, 1936

## Lake Shore Mine

# Three Men Die in a Rock Fall at Lake Shore

Roy Warwick, James Morden and Jack Bottrill were killed when a cave-in occurred in No. 2,702 E stope at the 2,700 level of Lake Shore at about 1:00 a.m. on August 15.

The night-shift crew working in this stope consisted of six men: Vic Swan, Nick McKinnon,

James Morden, Roy Warwick, Clarence Swift and his helper, a recent Polish immigrant known only as 'Johnnie'. Before going on shift that night, Shift Boss Jack Bottrill ordered Swan and McKinnon to pull muck and to scale, Morden and Warwick to build timbered square sets, and Swift and Johnnie to make ladders and clean up muck under the square sets. The men were told to keep watch, as the day-shift had reported that this stope was 'working a little'.

At 9:15 p.m., Bottrill and Night Captain William Allen came to this stope on their round and found all six men not working but seated in the gangway. Queried, they replied that the stope was 'working'. Bottrill and Allen, accompanied by Morden and Warwick, ascended into the stope to inspect.

They observed that the timber in the stope was taking a lot of stress and was "creaking and cracking". Bottrill told the crew to stay out of the centre section of the stope until he came back after lunch at midnight. He gave them work to do in the meantime. Swan and McKinnon were to continue their work in the place they had been working as it was well away from the dangerous section; Morden and Warwick were to work in the gangway, cleaning up and laying a floor; and Swift and Johnnie were to hoist timber for timbering work to be done when the stope quieted down. After lunch, Bottrill returned. He, Morden and Warwick ascended to the stope to inspect its behaviour. While Bottrill and Morden were on the mining floor of the stope and, while Warwick was at the top of the manway and about to enter the stope and Swift was in the gangway, a huge rockburst suddenly occurred.

McKinnon, Swan and Johnnie heard a loud 'crack' followed by a great fall of ground. Swift got out of the gangway in a hurry and returned to level. Swan and McKinnon shouted up to the three men above them but there was no answer so they made their way out through a raise, a gangway, a manway and down to level.

The underground superintendent, a doctor and the district mine inspector were notified. Recovery work started immediately.

The plan was to re-open the gangway. From Swift's information, the approximate position of the trapped men was thought to be at the centre of the stope. Warwick's body was located within eight hours of the cave-in but not removed until about 17 hours after the accident. Morden and Bottrill were not located until the next day and it took a further nine hours to extricate their bodies. The weight of the cave-in had carried them through the mining floor.

The block of ground that fell was 70 feet long, wedge-shaped, 25 feet wide at one end and two feet wide at the other. The block was 30 feet thick and had a total estimated weight of thousands of tons. It is impossible to say what caused the block to fall; there had been no blasting in the vicinity of the stope during the shift.

*Jack Bottrill. Photographer: Esme Hurd*

## Jack John Thomas Bottrill

### (1911 – 1936)

David and Edith Bottrill of Leicester, England were the proud parents of first-born Jack John Thomas on November 25, 1911.

Within months of Jack's birth, his father came to Canada where he worked in Northern Ontario, getting various jobs in Cobalt, Iroquois Falls and Kirkland Lake. The family was re-united in 1914 when three-year-old Jack and his mother left England and sailed to Canada. A daughter, Esme, was born here in 1917.

In 1923, Jack's father suddenly passed away at the age of 35. In 1927, the remaining family of three

*(left to right) step-father George Fraser, mother Edith, sister Esme and Jack Bottrill. (March 18, 1934). Photographer: Unknown*

moved to New Zealand for a short period of time, returning to Canada the next year.

At the young age of 17, Jack Bottrill got a job at Lake Shore, the 'man of the house' now that his father was gone. His mother managed a boarding house at 50 Second Street where she looked after ten boarders.

Younger sister Esme looked to her big brother for friendship and guidance. She remembers that he was very fond of cars in those days. Today, she has in her possession a photograph of her brother standing in front of mining operations with his good friend Ted Adams; Adams was killed at Lake Shore in 1931.

Eventually, Edith remarried. In 1935, her second husband, George Fraser, died of a heart attack just before going down the cage at Sylvanite where he was shift boss. Again there was a family of three living at 50 Second Street.

In 1936, 25-year-old Bottrill was a shift boss at Lake Shore. On August 15, he was doing his regular inspections when a huge fall of ground occurred. He was killed in this mishap. Bottrill's mother and his 19-year-old sister were left to mourn his passing.

The funeral service was held at Trinity United Church where hundreds of people attended and placed floral tributes. The interment took place in Kirkland Lake cemetery. The personal information and photographs for this account were contributed by Esme Hurd, who presently lives in Kirkland Lake (2000). Esme recounts the stories of Ted Adams who died at Lake Shore in 1931 and that of her teenage friend, Edna, who lost a father, Joseph Walton, at the Teck-Hughes in 1933. Jack Bottrill's nephew, Dave Hurd, presently lives in Kenogami, near Kirkland Lake.

## Jim 'Irish' Morden (1888 - 1936)

Jim Morden, born in Ireland in 1888, immigrated to the United States and settled in West Virginia. Leaving the U.S., he came to live in Trenton, Ontario. In 1928, he arrived in Kirkland Lake where he worked for eight years before his accident.

He was married to Ethel Maude Yates of Cardiff, Wales. She was previously married and had five children with the family name Barry: Lillian, Marion, William, Alan and Raymond. The Morden marriage produced four children: Isabelle, George, Patrick and Scotty. The family lived south of Queen Street, in an area adjacent to the town referred to by non-residents as 'Pigtown'. They lived in a log cabin he had constructed himself. Their home was one of many that had sprung up in recent years during the Depression. Their owners were willing to endure the hardships of wilderness life rather than having the amenities and taxes of town life.

At the funeral, he was given a Salvation Army service. All the pallbearers were miners who had worked with him at the mine. Interment occurred in the Kirkland Lake cemetery.

Personal details for this account were provided by Mrs. Judy Hooper, a grandchild of James Morden, who presently (1999) lives in Uxbridge, Ontario.

## Roy Warwick (1909 - 1936)

Twenty-seven-year-old Roy Warwick, Canadian born and married with no children, lived alone at 32 Main Street.

In 1934 he worked in the Noranda Mine in Quebec. He came to Kirkland Lake during the winter of 1936 to work at Lake Shore where he worked for six months before the accident.

Surviving him were his wife and his parents, Mr. and Mrs. T. E. Warwick, all living on a farm at RR #2, Maberley, Ontario.

The body, accompanied by brother Hubert, was taken to Perth, Ontario for funeral services and interment.

## September 15, 1936
## Golden Gate Mine

### Two Miners are Killed in a Blasting Accident at Swastika

Driller Johann Evjent died of injuries about six hours after a blasting accident. Edouard Ouellette, helper, was instantly killed in the same accident that occurred on the 200 level of the Golden Gate Mine (the old Lucky Cross) on September 15.

*Miner loads hole with dynamite, cap and fuse.*
*Photographer: Unknown*

Work was progressing on a drift face 400 feet from the shaft station. The drift was driven by a drilling crew of a driller and helper on one shift, and a mucking crew of two muckers on the opposite shift. The shifts ran from 7:00 a.m. to 4:00 p.m. and from 7:00 p.m. to 4:00 a.m., with an hour off for lunch. The drilling crew started with a fresh face, drilled off 20 holes using one machine and then blasted. The fuse was spit with a carbide lamp. The smoke was blown out with compressed air after the blast. The mucking crew on the next shift mucked out the round and blasted any missed holes, leaving a clean face for the drilling crew on the next shift.

On September 15, Evjent and Ouellette, working on the 7:00 a.m. to 4:00 p.m. shift, drilled off 20 holes and blasted between 3:00 and 3:30 p.m. Hoistman E. Sparling stated that the air compressor was starting to race about this time, indicating that they were blowing air to clear the smoke. At about 4:15, they blasted again. Dan Lalonde, a blacksmith's helper, heard these shots go off about this time. There were no eyewitnesses to the accident.

At 4:00, Mine Captain C. Mathewson, on coming out of the office, noticed that Evjent's car was still parked in the yard, which suggested the drillers were still on shift. He went to the hoistroom to see if the hoistman had any information as to how the work was proceeding underground. Sparling told Mathewson that the drillers were taking a long time. Mathewson waited by returning to his office to discuss some work with E. McCracken, the deckman. At 5:00, Sparling came into the office and told Mathewson that the drillers had not yet come up. Mathewson went underground to see what was causing the delay.

On going into the drift where the men were supposedly working, Mathewson found the environment smoky and gassy. The compressed air was not blowing. Soon he came across Evjent who was lying face down on the track and snoring loudly. Suspecting gassing, Mathewson went back to the station to get help and an inhalator. On his way, he turned on the compressed air. At the surface, he arranged for a doctor and an inhalator from the Fire Hall.

Mathewson returned underground with McCracken at 5:15. They brought Evjent to a crosscut for better air and went to the face of the drift, finding Ouellette at the muck pile, partially covered with muck. They noticed the round had been blasted. Ouellette was brought to the crosscut but he appeared to be dead, so they gave their attention to Evjent and brought him to surface.

At the surface were Dr. R. T. Neelands, Percy Smith of the Mine Rescue Station and two members of the Kirkland Lake Fire Department with an inhalator.

Evjent was given first aid and revived sufficiently to say that he was in bad shape and he asked about his partner. Dr. Neelands put a splint on his broken right leg and sent him by ambulance to the Kirkland District Hospital where he died at about 10:00 that night.

Mathewson returned underground with Dr. Neelands who pronounced Ouellette dead. He had suffered powder burns and bruises along the left side of his body. He had a fractured skull and a serious bruise over his heart.

Evjent, it was discovered during the post-mortem, had the following injuries: fracture of the right leg, powder burns and bruises on his right side, teeth knocked out, a bruise over the eyes and a fractured skull. Also, a very high concentration of carbon monoxide was found in his blood. Death was attributed to shock and cerebral hemorrhage.

An inquest into the deaths was held before Coroner J. F. Edis on September 23. The only two witnesses to the accident were dead. According to evidence, it would appear that Evjent was in a crouching position, engaged in lighting the fuses, and that Ouellette was standing a short distance behind him. Ouellette got the full force of the blast while Evjent got only part of it, the main part going over his shoulders. Evjent got up after the blast and was able to travel a distance of 49 feet before he collapsed or was struck by succeeding shots.

Obviously they were delayed at the face for some reason. Was there a difficulty in lighting the fuses and the men did not realize the time? After considering the evidence, the jury (consisting of J. Lamothe, Robert Pollock, C. W. Mallin, A. Matheson, R. Vigras, E. Orser and Austin Cooper) returned the following verdict: "That the deceased came to their deaths, by blasting, from an indeterminable cause."

According to the Department of Mines Report, the following improvements were to be made at the mine after the fatal accident: "Hot-wire lighters will be used for lighting fuse in the future. This, as well as being a better and quicker method of lighting, will give miners an easy and definite check on the time."

## Johann Evjent (1900 - 1936)

Johann Evjent, single, was a 36-year-old Norwegian. He had nine years experience in mining: four years at Lake Shore, three months at the Beatty and five years with Consolidated Mining Company in British Columbia.

An afternoon funeral service was held at the Kirkland Lake cemetery with Reverend J. E. Graham officiating. Flowers were received from brother Ernest, wife and family, boys and staff of Dominion Hotel, Wilson brothers and staff, Golden Gate Mining Company, Golden Gate Mine employees, Bert and Teddy, Mr. and Mrs. W. Black and Family, Mr. and Mrs. John Brown, Bill Hakala, Hannah and Hugh, Eva and Ivar Lindstrom, and Jean and Joe Kutcher.

## Edouard J. B. Ouellette (1900 - 1936)

Edouard Ouellette was a former employee of Lake Shore. He and his partner, Johann Evjent, drove to work together each morning from Kirkland Lake to the Golden Gate in Swastika. He was 36, married and a Canadian citizen.

The funeral services were held at Astorville, Ontario, located 18 miles from North Bay.

He was survived by his wife, Laura, living in Timmins, and by their three children: Yvette, who was keeping house for him in Kirkland Lake; and Alpha and Lena, who were in a convent in Toronto. Also surviving were a niece in Kirkland Lake, Mrs. R. Kingsley, his mother, Mrs. Joseph Laporte of Astorville, two sisters, Alphonse Point and Mrs. Joseph Gauthier, and a brother living in Astorville.

Wreaths at the funeral were received from Golden Gate Mine, Golden Gate Mine employees and Mr. and Mrs. R. Kingsley.

*Author's note: I found four letters of interest that pertain to the accidental deaths of Johann Evjent and J. B. Ouellette: a transcribed copy is presented here.*

---

*17th September 1936*

*The President*
*Golden Gate Mining Company Ltd.*
*Toronto*

*Dear Sir:*

*I regret to have to tell you that two employees at this mine were killed in a blasting accident at the 200 level about 5:00 p.m. on the 15th inst. It was one of those unfortunate accidents in which the men managed to dynamite themselves and nobody will ever know the exact cause of the accident. The inquest is fixed for the evening of the 23rd inst. Accidental death is the usual verdict arrived at these cases.*

*There will be no liability attached to the Company, for funeral expenses, compensations etc. re. all taken care of by the Compensation Board. The only expenses incurred by us will be hire of the ambulance, two wreathes, and $25 I gave by Company cheque for immediate necessities of the family of one of the men.*

*Diamond drilling was carried on as usual yesterday, and no drifting is being resumed to-day.*

*Yours truly,*
*Signature unclear*

---

*Millar and Hunter*
*Barristers, Solicitors*
*Andrew W. Hunter, K. C. G. Fitz Gerald*

*Telephone Elgin 4495*
*Cable "Rallim, Toronto"*

*59 Yonge St.*
*Toronto, 2*

*F.S.*

*September 19, 1936*

*S.A. Pain, Esq.*
*Golden Gate Mining Co.*
*Swastika, Ont.*

*Dear Sir:*

*I acknowledge your letter of the 17th of September addressed to the President of the Company advising of the accident whereby two of the workmen were killed*

*I note that the inquest is fixed for the evening of the 23rd of the month.*

*If there is any chance that the Coroner's Jury will find that the company was in any way negligent, I would suggest that you engage some good lawyer to look after the interests of the company. As to whether you should take such action or not, I leave entirely up to your judgment.*

*Yours truly,*
*Andrew Hunter*

---

*General Delivery, Timmins, Ont.* *Dec. 18, 1936*

*The Manager*
*Golden Gate Mine*
*Swastika, Ontario.*

*Dear Sir:*

*In regard to the wages and back-time due to my late husband J. B. Ouellette who was killed in your mine on Sept. 15/ 1936. I wish you would either pay this money to me or my daughter. We have not as yet been paid our claim by the Workmen's Compensation Board, although they have written to me several times and are evidently going to pay the claims eventually.*

*During the three months that have elapsed since his death I have exhausted what funds I have to live on and we are absolutely destitute and my daughter is in need of glasses which I cannot get for her.*

*We are not eligible for relief in Timmins and I am becoming desperate for I am faced without food for Christmas.*

*If there is any way in that I can help to get this matter settled, please advise me at once.*

*If you cannot send the money to me then for the love of God send some to my daughter, as we are in poverty now. The furniture company is going to repossess their goods and my rent expires on December 30th with no means of paying the next month in sight so I ask you in the name of decency to do something about this, as it is in your power to offset what is becoming a frightful situation for us.*

*Sincerely hoping you may be moved to act in this matter I am*

*Yours truly,*
*Mrs. Laura Ouellette*

---

*General Delivery* *December 26, 1936*
*Timmins, Ont.*

*Mr. S. Pain*
*Golden Gate Mine*
*Swastika, Ont.*

*Dear sir,*

*Due to the fact that I have no reply to my letter to you regarding my husband's pay and back-time, I am writing to inform you I have been advised by the Workmen's Compensation Board that they have no jurisdiction in the matter of his wages and that you should be made to pay.*

*Now I am destitute and unable to obtain the services of a lawyer to look after my interests but I am going to appeal to the crown prosecutor to handle the case for me as I am at least entitled to an explanation as to why this has not been paid.*

*Yours truly,*
*Mrs. Laura Ouellette*

---

* *Author's note: Did Mrs. Ouellette eventually get her deceased husband's pay and back-time?*
*I don't know. The mine closed in 1942.*

September 19, 1936

## Wright-Hargreaves Mine

### James E. Banks (1906 - 1936)

Thirty-year-old James Banks, married without children and employed as a trammer at Wright-Hargreaves, died in a stope on September 19.

On the night shift, the west rill of No. 501 stope was being prepared for sand filling. Two timbermen, McMillan and Pelletier, constructed structures that acted as vertical barricades to control the sand-filling process. Banks and his partner, Wojnas, were removing a plank covering.

Banks and Wojnas heard a noise above and suddenly a run of sand came down towards them from that area. They jumped towards a ladder but a section of flooring on which they were supported gave way and slid to the bottom of the stope. Wojnas landed against a wall and suffered a bruised ankle. Banks landed at the bottom against a spruce timber. The sand sliding with the floor trapped him there and buried him. Only his feet and part of his legs remained uncovered.

McMillan and Pelletier gave the alarm. George Campbell, the captain on shift, arrived and initiated a rescue effort. Dr. Contway arrived at the scene about the time that Banks' head was uncovered. Banks' heart showed no signs of action and a stimulant was injected with no result. Resuscitation work was continued by the mine rescue team for two-and-a-half hours but had no effect. After the body was taken to surface, a further examination showed no fractures; death was caused by suffocation.

The time of death was established as 1:15 a.m. That afternoon Banks' body was sent by train to Toronto where his wife resided. Funeral services and interment took place there.

October 13, 1936

## Wright-Hargreaves Mine

### Bozo Cop (a.k.a. Chop) (1905 - 1936)

Bozo Cop, born in the town of Hjerlin in Yugoslavia in 1905, came to Kirkland Lake in 1928. He had worked at Wright-Hargreaves from 1930 to the time of the accident. He was married and the mine accident report recorded no children; however, the Northern News reported that he had a wife and son living in Yugoslavia.

On October 12, Shift Boss J. Martine, in charge of the 7:00 p.m. to 3:00 a.m. shift on the 1,800 level, instructed Cop to go the 1,877 stope and level off muck so that timbermen could put in another floor stull. The shift boss came through at 9:00 on his first round and noticed that three men, Cop, Timbermen T. Steele and William Sievert, were working as directed.

Shortly after 1:00 a.m., the timbermen heard a fall of loose and Cop cry out. They rushed to his aid and found him lying on the muck, conscious, telling them that he had found a piece of loose on the wall and, while scaling it, had been struck by it when it fell. Cop died in the Kirkland District Hospital at 9:30 a.m. on October 13. Death was caused by shock attributed to fractures and hemorrhage.

"It is probable that Cop was standing on a plank with his back against a stull, scaling a piece of loose ahead of him. Although he was an experienced miner, he must have misjudged the length of the piece that split over him and, in falling, jammed him against the stull, fracturing his pelvis and sacrum."

The funeral mass was held at Holy Name Church. The pallbearers, selected by the Croatian Fraternal Union, included John Franciskovic, Felix Tuss, Emil Sentil, Mato Regus, Marko Frankovic and Karlo Barisic. Interment took place in the local cemetery. It is presently an unmarked grave.

October 23, 1936

## Moffat-Hall Mine

### Harry Omelchuk (1898 - 1936)

Thirty-eight-year-old Harry Omelchuk, married and Russian, was fatally injured at Moffat-Hall on October 23. The injury occurred during salvage operations in the shaft. The accident happened at 8:40 p.m. and death followed in the Kirkland District Hospital about four hours later.

Mining operations had ceased at Moffat-Hall located east of Bidgood Mine in Lebel township. A salvage crew was on the property and was engaged in the complete removal of all underground equipment and materials.

Seven men were involved in the removal of pipe lines from the shaft. Two were working on surface, pulling the last lengths of pipe up a manway. Two others were in the upper portion of the shaft, stripping the landing platforms and removing pipe clamps. Omelchuk and his partner, Sokoluk, were working their way downwards in the shaft, removing clamps that were holding power cables to the shaft timbers. The seventh man, Dancheski, was working on surface.

Omelchuk and Sokoluk were well aware that men were working in the shaft above them. The two had reached a depth of 250 feet when the accident occurred. Unknown to the men above, a steam line had previously been disconnected and, when a landing platform was dismantled, there was no support left to hold the pipe in place. The two-inch diameter pipe plummeted downward in the shaft and impaled Omelchuk through the abdomen. The long length of pipe continued through his body, penetrated through a two-inch plank on the platform and only came to a halt when it penetrated into an 8x8 end plate of the shaft.

A doctor was called in from Kirkland Lake and, when he arrived, he found Omelchuk alive and conscious. The upper portion of the pipe above the victim was secured in position and then it was cut just above his body. Omelchuk was lifted off the remaining pipe and sent to hospital where he died. Death was attributed to shock and loss of blood.

### November 4, 1936
## Lake Shore Mine

# Triple Fatality Occurs in Lake Shore Shaft

Three men working at the slashing of the new No. 5 shaft at Lake Shore dropped to their deaths on November 4. Killed instantly were Shaft Leader Bert Peterson, Karol Duda and George Browne-Cadman, both shaftmen.

The three were members of a shaft sinking crew of seven men. All were standing on a staging when it suddenly collapsed. The three fell to the bottom of the shaft, a distance of about 320 feet. The other four clung to what they could and saved their lives. The survivors were: Duncan McLeod, Lauchie McLellan, Piotr Sitarz and Paddy Smith.

There were other unfortunate multiple fatalities at Lake Shore in 1936. Two and a half months earlier, another triple fatality had occurred at the mine.

## Bertel 'Bert' Peterson
### (1903 - 1936)

At the funeral, you would hear phrases like: "The finest fellow you could ever meet," "The best leader I ever worked under, "Did more work himself than anyone on shift." These were the words of former fellow workers of Bert Peterson.

He was a 33-year-old married man who was born in Sweden and came to Northern Ontario to work in the mines. He spent 12 years in the north country: five at Silver Centre, two at Matachewan and five-and-a-half in Kirkland Lake.

Ellen Dorothy Jonason from the South Lorraine District became his wife in 1927 when they were wed at Silver Centre. At the time of his death, they had two children, Myrtle (8) and Elda (5). The Peterson family lived at 27 Premier Avenue. According to his wife, Bert was a true family man: "he was never a man to drink; he was never five minutes late coming home; he liked his home and the quiet life." He enjoyed skiing, hunting and fishing, especially in the Montreal River and Elk Lake areas.

Peterson was survived by his immediate family, two younger brothers, Carl and Eric, both working at Lake Shore, and a sister-in-law, Noe. His parents survived him in Sweden.

The funeral service took place at Trinity United Church and the interment occurred at the local cemetery. The inscription on the gravestone reads: BERTEL PETERSON – BORN JAN. 2, 1903 – DIED NOV. 4, 1936 – ELLEN DOROTHY JONASON – BORN MAY 25, 1903 – DIED AUG. 14, 1987 – 'GONE BUT NOT FORGOTTEN'.

## Karol Duda (a.k.a. Charles Dooda)

(1903 - 1936)

Thirty-three-year-old Karol Duda, born in Poland in 1903, was married in 1927 and came to Canada with his wife in 1928. His older brother Fred was already working at Lake Shore for two years when they arrived.

In 1932, Karol Duda went into a partnership to build a large house at 79 McKelvie Avenue. Within three years it was built and his family, now including three children, moved into their new home.

Duda was heavily involved with Polish cultural and community functions. He had become a naturalized British subject and a proud Canadian. He was fluent in both Polish and English. He had many friends and was well liked by his neighbours.

He was survived by his wife and children: Walter (6), Jennie (4) and a week-old baby. He was also survived by five older brothers: Fred, a miner at Lake Shore, John of New York City, and three brothers in Europe and by two sisters living at Juroslaw, Poland. His parents were deceased.

The body rested at the Polish Hall until the funeral mass at Holy Name Church. Interment occurred in the local cemetery. The inscription found on the gravestone is: TU SPOCZYWA – S. P. KAROL DUDA – U. R. 3 SIERP 1903 – U. M. 4 LIST 1936. Translation: Here lies in holy memory Karol Duda – Born August 3, 1903 – Died November 4, 1936.

## George Scott Browne-Cadman (1890 - 1936)

George Browne-Cadman was born in Dalkeith, Scotland in 1890. In his youth he was very athletic and an avid soccer player.

Browne-Cadman was a Northern Ontario pioneer, as he had spent 27 years in this north country including the early days when Cobalt was developed. He and his brother were prospecting in the Kirkland Lake area before the gold camp existed. Little did they realize that beneath their campfire, the Wright-Hargreaves would be developed a few years later. In 1911, they gave up prospecting in the Kirkland Lake bush when the

*George Browne-Cadman (top row, fifth man from left) was an avid soccer player and member of the Lake Shore Football Club 1929. Photographer: A. MacLean*

Great Fire that destroyed the Porcupine had the potential to threaten this area also.

During the World War, Browne-Cadman served overseas in the infantry. He could be found in the trenches in 1915 and a year later he was wounded. He had been a resident of Kirkland Lake for his last 14 years.

Married in Scotland to Greta Laidlaw, they had a daughter, Phyllis. His mother and stepfather were early pioneers of the 1900s; Mr. and Mrs. Albert Cadman farmed at Dawson's Point on Lake Temiskaming.

Browne-Cadman's interest in athletics and soccer never waned as he played soccer in Cobalt and Kirkland Lake; in later years, he became a capable soccer referee. He was a man of clean living and his body was always in remarkable condition. Even at the age of 47, he was in terrific shape.

He was survived by his mother and stepfather of Lakeshore Road in Haileybury, two brothers – Fergy and Alex who was doing field work and working at the Hollinger in Timmins – and a sister, Mrs. Leo Erenhous, the widow of the first president of Bidgood Mine. At the time of George's death, she lived on Lakeshore Road in Haileybury.

The body was taken to Trinity United Church for funeral services and then by train to Haileybury for interment at the Mount Pleasant Cemetery. The inscription on the headstone reads: RIP – LOYAL ORDER OF MOOSE – 47823 PRIVATE GEORGE SCOTT BROWNE – CADMAN – 15TH CANADIAN BATT. CEF. – 1890 - 1936.

### December 9, 1936

## Teck-Hughes Mine

## Two Miners Are Killed Instantly in a Teck-Hughes Rockburst

George Stark and Frank Zalec were killed instantly when they were buried by a rockburst in a Teck-Hughes stope on the 3,800 level on December 9 at about 7:45 p.m.

The day shift did some scaling in the stope but left the place about lunchtime as the ground was cracking and snapping. The shift boss visited the stope in the early afternoon and found it had quieted down. He reported this fact to his opposite partner, the shift boss of the 3 – 11 p.m. shift, when they met in the boss's office at 2:30 that afternoon. The 3 – 11 p.m. shift boss instructed Stark and Zalec, experienced timbermen who were familiar with this stope, to go in and put in the timber. He visited before lunch and again after as is the usual custom. On both visits the stope was quiet and the men reported that everything was all right.

On his second visit, the boss stayed and talked with them for about 10 minutes, then went out and down the manway. He had just reached the level when the rockburst occurred. He immediately returned and found the stope full of muck and the men completely buried. He descended to the level station, sent word to the mine superintendent and returned with muckers who started working to recover the buried men. The superintendent and the assistant superintendent arrived and took charge. Zalec's body was recovered at 12:45 a.m. and Stark's at 2:45 a.m. on December 10.

Doctors examined the bodies underground and later reported that both suffered multiple lacerations and contusions to the head and upper body. Zalec had no fractures. Stark had a crushing of the right side of the chest but no further fractures. Death was caused by general crushing and asphyxiation.

## George Stark (1907 - 1936)

George Stark was born in Yugoslavia in 1907. At the time of his death, he was 29 and single. He was employed at Teck-Hughes from 1927 to 1936. Interment took place in the Kirkland Lake cemetery.

## Frank Zalec (1887 - 1936)

Frank Zalec was born in Yugoslavia in 1887. At the time of the fatal accident, he was 49, married, with a wife and three children. He had been employed at Teck-Hughes from 1931 to 1936. Interment occurred next to George Stark in the local cemetery. Both graves are presently unmarked.

*Employees at Sylvanite Gold Mine 1937. Note the carbide lamps on the hard hats. Photographer; Unknown*

## 1937

- Ten miners are killed accidentally in local mines this year
- Stephen P. Ogryzlo becomes mine superintendent at Toburn and introduces the jackleg drill and the Elmco loader at the operation
- Morris Kirkland Mine in Lebel township opens
- Kirkland Lake has 750 births this year – 60 babies per month
- Larder Lake's population is 1,700
- Wall Street Market declines, signaling a serious economic recession in the U.S.

### February 3, 1937

## Matachewan Consolidated Mine

## Edward Hulm (1897 - 1937)

Forty-year-old Edward Hulm, Finnish and married with a wife and five children living in Finland, died at the Cobalt Municipal Hospital on February 20, 1937. He was employed as a shaftman and working in the No. 3 shaft of Matachewan Consolidated at the time of the accident on February 3.

On this day, the shaft crew descended to the bottom of the shaft to muck the blasted rock. The crew consisted of Leader Pritchard and four shaftmen: Johnson, Sutherland, Burton and Hulm. They removed a bucket from the hoist cable to get a second bucket down. To do this, Pritchard rang one bell. The bucket was raised about one foot. While it was raised, Pritchard and Johnson pushed it into one corner and the one-bell signal was given again to stop its movement. Hulm was in the process of disconnecting the bucket and wanted it lowered. Pritchard thus gave a two-bell signal by pulling the cord twice but the signal system was not working well and the hoistman only received a one-bell signal. Accordingly, he raised the bucket instead of lowering it. Surprised, Hulm could not get clear of the rising bucket that jammed his arm between it and a timber. Noticing the accident, Pritchard rung bell signals to stop and lower the bucket. Hulm dropped to the muck.

He was taken to surface where Dr. Fitch found the left arm fractured above the elbow; there were also lacerations and scratches around the elbow and down to the wrist. The arm was considerably swollen and showed some blue colouration. After plaster-of-Paris splints were applied, Hulm was sent to the hospital in Cobalt.

He arrived there about seven-and-half hours after his accident and was examined by Dr. Case whose diagnosis agreed with Dr. Fitch. The arm was cold, showed no pulse, and the skin was tight.

Two days later, the arm turned dark. Dr. Case wished to amputate but Hulm wanted it saved. Dr. Arnold was called to consult and the two doctors agreed that there might be a chance of saving it. On February 10, they still thought the arm might be saved. Four days later, there was a sudden change for the worse as gangrene set in and the patient's temperature was high. Dr. Case amputated the arm above the elbow. After a day, his body temperature returned to normal but by February 17 it began to rise again. Hulm's condition gradually deteriorated and he died February 20, 17 days after the accident.

The cause of death was his natural desire to save his arm. The cause of the accident was the failure of the signal system at the mine.

Surviving Hulm was his immediate family in Finland. Interment occurred at Cobalt.

An inquest was held by Coroner J. S. McCullough in the office of the Matachewan Consolidated Mines Ltd. The jury's verdict was accidental death but they did make a recommendation: "We recommend that a man be detailed to look after and inspect the signal system at the start of each shift."

### March 11, 1937
## Sylvanite Mine

### Jaska Valitorppa (a.k.a. Jack Kauppa) (1901 - 1937)

Jaska Valitorppa was born in Finland in 1901. He was married there and his family of wife and son came to Canada in 1928. At Kirkland Lake in the same year, Valitorppa got a job at Sylvanite. In 1929, a second child, a daughter, was born. His wife and the two children went back to Finland in 1930 and while there, the mother became ill and died; the children remained in Finland. In 1937, Valitorppa was living at the Northland Café.

On March 11, Valitorppa, a machine runner, was with partner J. Duralia, working in a stope at the 875 level. A fall of rock from the hanging wall fell on him, burying the lower part of his body. Duralia went to get help but Valitorppa had already died from a severe head injury.

The funeral service was held in the Symington Funeral Home chapel with Reverend Koski of the Finnish Evangelical Lutheran Church officiating. Burial took place in the local cemetery. The grave is presently (1999) unmarked.

*Author's Note: So many graves are presently unmarked Fortunately, the Miners' Memorial and this book prevents the complete loss of record of the lives of these men.*

### March 15, 1937
## Lake Shore Mine

### Richard 'Scotty' Daly (1907 - 1937)

Scotty Daly was a 30-year-old motorman at Lake Shore employed at the mine since October 1936. He was killed there on March 15.

The verdict of a Coroner's jury after the fatal accident was "that Richard Daly came to his death through no fault of his own, by meeting an obstruction that was placed across the mine track in direct contravention to the standing rules of Lake Shore Mines." Evidence adduced at the hearing, conducted by Coroner J. F. Edis, indicated that a plank had been placed across the tracks by a machine runner, E. Burley. The plank was placed there to enable J. Kielt, a short man, to get down from a stope that was located above the tracks. The loaded train, with Scotty Daly as motorman on the front end, ran into the plank. The plank jammed against a drift post and hit Daly in the chest, breaking his ribs and puncturing and collapsing a lung. This severe blow and resulting injuries placed undue stress on his heart, causing him to die.

Daly was survived by his wife and two children. He had a brother in Kirkland Lake who was employed at Algar's Jewellery Store in 1937.

The funeral mass was conducted at Holy Name Church with Reverend Father Roney officiating. The pallbearers were: Fred Ames, W. E. Burley, E. G. McCloskey, T. Gibson, W. Zaluski and M. Burke, Daly's partner at work. The body was sent by train to Toronto for additional funeral services and burial.

### March 24, 1937
## Omega Mine

### George 'Scotty' Summers (1907 - 1937)

This was the first accidental fatality at Omega Mine located a few miles northeast of the town of Larder Lake.

George Summers, 30, single, known to his friends as Scotty, was well-liked in the camp and was a very good friend of Sandy Maitland.

On March 24, Summers, a drill runner, was mucking in a stope with his partner, Joe Kokoszka, when unexpectedly a fall of loose crashed down on them. Kokoszka was rescued but badly injured, suffering crushed vertebrae and a flesh wound above the knee of the left leg. Summer's right leg was pinned under a large piece of fallen rock. When he was extricated, he was diagnosed with a fractured and lacerated leg, a crushed pelvis and a crushed back. Both were rushed to the Kirkland Lake District Hospital which was 20 miles away. Shortly after admission to the hospital, Summers died.

Over a period of time, Kokoszka recovered favourably in the hospital.

The funeral service was held in Kirkland Lake. Interment occurred in the local cemetery. The gravesite is presently unmarked.

### July 3, 1937
## Lake Shore Mine

### Archie MacDonald (1916 - 1937)

Twenty-one-year-old Archie MacDonald, British and married with one child, suffered head injuries on July 3 while working underground at Lake Shore. He died from these injuries in a Toronto hospital five months later.

Employed at the mine as an underground mechanic, MacDonald was adjusting the brakes of a locomotive on July 3. He tested them by running the locomotive up and down the main crosscut on the 3,200 level. Then wishing to test the brakes with a load, he ran the locomotive towards two ore cars standing on a curve. Due to the positioning of the locomotive and the cars on the curve, he could not make the coupling without help. W. Ireton, a mucker, was close by and MacDonald told him to get onto the locomotive and move it slowly towards the cars so that he (MacDonald) could guide the draw bar and slip the coupling pin into place. Ireton, explaining that he had no experience in handling a locomotive, got onto it while MacDonald was preparing to couple the cars that were about nine feet away. Ireton moved the locomotive slowly up to the cars and was warned by MacDonald that he was within three feet of them when suddenly the locomotive moved with a jump. In some manner, MacDonald's head was caught between the end of the car and the locomotive.

He was taken to hospital where examination showed fractures of the skull and considerable swelling. There was complete paralysis on the left side of the body. He was conscious although "a little muddled at times." On the day following, his condition became worse as there were signs of increasing pressure on the brain. A brain specialist was brought in from Toronto and an operation was performed to relieve the pressure. MacDonald's condition gradually improved and he was finally released from hospital on September 20. The paralysis had almost disappeared, with the exception of partial paralysis of the left arm.

While in hospital, MacDonald suffered seizures, somewhat like epileptic fits, and these continued after his release with increasing severity. On September 26 he was admitted to Toronto's St. Michael's Hospital where it was discovered that his brain was pulled out of position, to the right, by adhesions. It was thought that this was the cause of his seizures. Several operations were performed over a period of two months but he did not respond well to the treatments and finally died on December 9. An autopsy revealed skull fractures, brain laceration, brain abscess and meningitis.

An inquest was held on January 19, 1938 in the Kirkland Lake Town Hall before Coroner R. H. Armstrong. The jury returned a verdict of accidental death but they did make a recommendation: "We the jury recommend that no inexperienced men be allowed to operate motors without three month's experience in doing so, except they operate them under proper supervision and we wish this to be enforced by the proper authorities."

### July 13, 1937
## Wright-Hargreaves Mine

### Archie J. Rupert (1913 - 1937)

Archie Rupert and his wife, Mabel, arrived in Kirkland Lake in April 1937. The young couple, married for three years and without children, lived on Wishman Street.

Rupert, 24, born in Havelock, Ontario in 1913, had a relative in Kirkland Lake – first cousin Chris McGinnis of 159 Pollock Avenue.

Just after his arrival in town, Rupert got a job working underground at Wright-Hargreaves. He quickly became very popular among his fellow workers. Two-and-a-half months later, he was dead.

He was killed while working in a stope at the 3000 level with his partner, Earl Bittrof. On the night of July 13, he was crushed by a huge piece of loose rock; his partner escaped unscathed. Dr. Contway examined Rupert at the mine and pronounced him dead.

The body was transported to Havelock for funeral services and burial.

*Eddie Miller. Photograph taken just days before the accident. Photographer: Unknown*

## September 8, 1937
## Martin-Bird Mine

### Edward Miller (1918 - 1937)

A very young single man of 19 years of age was killed at the Martin-Bird Mine near Larder Lake on September 8 as a result of an underground drilling accident. His partner was seriously injured in the same accident that occurred on the 375-foot level.

Eddy Miller was born at the Cobalt silver camp to Mr. and Mrs. Harry G. Miller on January 28, 1918. He was raised at Silver Centre along with brothers and sisters: Roy, Ethel, Irene, Gladys and Stanley.

The only work in Silver Centre in the early '30s was hand-cobbing for silver in the mine dumps. Eddy and his older brother Roy worked at this until the fall of 1936 when they left to look for work in the gold mines of Larder Lake. In July of 1937, the rest of the Miller family moved to Cobalt.

Both brothers found employment at the Martin-Bird; Roy got a job as an assayer and Eddy became a helper in the cookery. On September 8, a driller did not show up for work. His helper, A. F. Renton, became the driller for that day and Miller from the cookery was sent underground for the first time to become the driller's helper. They drilled into a face that had been blasted the previous day and hit a missed hole. An experienced driller would have suspected the presence of an unexploded hole by noticing a poor break but Renton and Miller were inexperienced in their new positions. Miller, who was guiding the steel, received the full brunt of the blast and was killed instantly. Renton, who was behind the drilling machine, received serious injuries but survived the blast.

Surviving were Miller's immediate family. His father, Harry Miller, later became well known in the Cobalt area where he developed and became owner and manager of the Silver Miller Mine from 1946 to 1954.

The funeral service for Eddy Miller was held at Symington's Funeral Home and interment occurred in the local cemetery. The inscription on the stone marking his grave reads simply: EDDY MILLER – DIED SEPTEMBER 8, 1937 – AGED 19 YEARS.

The details for this memorial account and the photograph of her brother were supplied by Gladys Wilcox of Cobalt.

## September 18, 1937
## Kerr Addison Mine

### Henry Silvola (a.k.a. Silvala) (1909 - 1937)

An inquest conducted by Coroner J. F. Edis placed blame for the September 18th accident at Kerr Addison on the carelessness of two deckmen. However, Dr. Edis pointed out that some of the blame had to be placed on the deceased, due to his own negligence.

Silvola, 28 and married, was working at the bottom of the shaft with others when a ton of muck was poured down the shaft accidentally. Two deckmen at the top of the shaft neglected to properly close the shaft doors while they were in the process of transporting ore to the ore bin. Some pieces of rock struck Silvola, causing severe brain injuries and he died seven hours later in the Kirkland District Hospital.

Arnes Palken, one of the deckmen, gave a signal to dump the bucket. He thought that the other deckman, N. Nurmi, had closed the doors. Nurmi dumped the bucket but at the inquest said that it was Palken's job to drop the doors and that he had no thoughts of doing so. Both deckmen were experienced miners.

Evidence from witnesses who had been at the bottom of the shaft with Silvola showed they had time to get out of the way and under the shelter of a steel plate placed there for that purpose. The jury mentioned that the deceased was negligent in not seeking shelter as soon as he could.

Interment occurred in the Kirkland Lake cemetery. The gravesite is presently unmarked.

### September 22, 1937

## Kir-Vit Mine

### Charles M. Miller (1890 - 1937)

Organized in 1937, Kir-Vit Mine consisted of 17 claims located in the central part of McVittie township. The camp was reached from Larder Lake Station by a wagon road about three miles long. In 1937, the company did about 7,000 feet of diamond drilling and then set up a mining operation. A shaft was sunk to a distance of 275 feet and levels were opened at depths of 125 and 250 feet. Underground developments were disappointing and the mine closed in the spring of 1938.

Charles Miller, 47 and single, was a victim of an accident at the mine where he began employment on April 12. He worked as a spare man, surface trammer and spare deckman. He was considered "a good and steady workman, although not very fast as a learner."

On September 22, he was to begin his new position as regular deckman. The mine foreman instructed him in his duties and remained with him while five ore cars were hoisted properly. After this short training program, Miller was left in charge on the 250 level. His duty was to place loaded cars on the cage and then to ring the hoistman so that the cars would be hoisted to the surface. He loaded an ore car onto the cage but neglected to put the safety bar of the cage into position. When the cage was raised, the loaded car rolled out and fell on him. He was crushed and killed.

At the inquest the jury's verdict was: "We the jury, empanelled to inquire into the death of Charles M. Miller, find the accident due to failure of the deceased to replace the guard bar on the cage before giving the hoisting signal, thereby allowing the loaded car to leave the cage and to fall on himself."

Miller was survived by three sisters, two of whom lived in New Liskeard and the other in Castleton, Ontario. The funeral service and burial occurred at New Liskeard.

### December 18, 1937

## Matachewan Consolidated Mine

### Turner Barnett ( ? - 1937)

Turner Barnett, single and British, died in the first-aid hospital at Matachewan Consolidated on December 18. He had received injuries earlier in the day at the mill. He was employed as a mechanic and welder.

Shortly after the beginning of the day shift on December 18, a blockage in the crusher at the mill occurred and the switch kicked out. With the help of others, Crusherman Bruce Christie dug out the ore from the jaws and found two five-inch steel balls jamming the crusher. The balls, fastened together with a bolt, had come loose from the control gate of the ore bin, had fallen down with the ore and blocked the crusher.

Attempts were made to release the balls by turning the crusher over with a chain block hooked to the fly-wheel but this was unsuccessful. Other methods to remove the balls were tried but they also failed. The mechanics were called in.

Master Mechanic H. Barnett decided burning out the balls with a torch was the only method that would be successful. He instructed Turner Barnett to do so. Barnett got into the crusher and began to remove one ball by taking off horizontal slices. About halfway through the first ball, the balls suddenly let loose. Turner Barnett was struck in the face, receiving a lacerated chin and eyebrow, a broken nose, split lips, a broken facial bone and knocked out teeth. H. Barnett, standing on a platform at the mouth of the crusher, was struck on the nose and knocked to the floor. He suffered a laceration to the nose but was not seriously injured.

At the mine first aid hospital, Turner was attended by Dr. Fitch shortly after 10:30 in the morning. His general condition was good and his injuries were not considered life threatening. He was conscious and spoke occasionally; there was no hemorrhaging from the eye sockets or ears but there was some bleeding from the nose into the throat. A trained nurse was placed in charge at 3:00 in the afternoon. By 6:00, a deterioration phase had set in as the pulse rate and respiration rate had changed. At 7:00, the doctor was called back to the mine. Artificial respiration and stimulants were used to revive Turner Barnett but there was no reaction and within a half hour he was pronounced dead by the doctor. Death was attributed to shock following the injuries.

### 1938

- March 15 turned out to be a tragic time in the mining community of Kirkland Lake as five miners died while at work on this day; all five were town residents
- A rockburst at Lake Shore kills a miner in July
- Another rockburst at Lake Shore kills two miners in December
- In total, 11 miners are killed while at work in area mines this year
- Kerr Addison Mine and mill at Virginiatown is into heavy production
- Chesterville Mine sinks a shaft and builds a mill
- Anoki Mine in Gauthier township, formerly the Oriole Mine, is incorporated
- Morris Kirkland Mine , a small producer, has a good year

*Mine Rescue Truck destined for Kirkland Lake 1938. Photographer: Unknown*

- Upper Canada Mine, in Gauthier township, begins production
- The 220-foot smokestack at Lake Shore is operating well

### March 15, 1938
## Bidgood Mine

## Alex Maloney (a.k.a. Alex Malinowsky)

### (1901 - 1938)

Alex Maloney, 37, single and a recent Ukrainian immigrant who had changed his surname when he came to Canada, had been in Kirkland Lake for a short period of time and lived at 63 Taylor Avenue.

On the night of March 15, he was instructed by his shift boss, Angus Kennedy, to get a plugger machine and to drill holes in a number of large pieces of ore left by a previous blast. Maloney, however, was unable to locate a machine so he put up a staging and continued with the slashing that he had done the two previous shifts. He was assisted by his partner, Emil Aho.

They had completed one hole and were starting a second when suddenly a fall of ground occurred. Aho, who was working about 15 feet away, heard the crash and immediately came to Maloney's assistance. Finding him unconscious on the muck below the staging, Aho placed him on the staging and went to the above level for help. When help arrived, they realized Maloney had been killed instantly.

The change of name to Maloney and his short residence in Canada made it difficult for authorities to contact his next of kin. He was buried in the Kirkland Lake cemetery. The gravesite is presently unmarked.

### March 15, 1938

## Teck-Hughes Mine

## Rockburst at Teck-Hughes Kills Four

A rockburst on the 4,600 level of Teck-Hughes claimed the lives of four men. There was no prior warning before a mass of rock, estimated at about 500 tons, crashed down on a stope located between the 38th and 39th levels.

Mine Superintendent George Gill* was on a ladder in a raise in the west end of this section when the burst occurred. He then proceeded to remove the crews from the area and went to surface to get Major J. G. McMillan, the general superintendent. Coming back down to this level, they realized the full extent of the damage. Four men mucking and tramming in the affected stope were trapped by the burst. They were: Harvey Peever, Alex Nicholl, Anton Gornik and Bert Lamoureux.

Rescue operations began immediately but slow progress was made because of the amount of drilling and blasting required to get to the area.

** A rock slide at the Teck-Hughes would kill George Gill's son in 1948.*

## Harvey Peever (1895 - 1938)

Harvey Peever, 43, was born and raised in Maynooth, Ontario. During the World War, he served overseas for three years with the Canadian troops, returning after the war to Hastings county in Southern Ontario. In 1931, he came to Kirkland Lake where he got a job at Teck-Hughes.

He and his wife Jessie and their four children lived in a neat, yellow-painted home at 95 Third Street. Thelma (16), Douglas (12), Betty (10), and Jack (7), were very much saddened when they learned of their father's death.

Peever was very popular among his fellow workers and had many friends about town. His sister, Miss Estelle Peever, lived in town and worked as a governess. Three other sisters and two brothers lived near their parents at Maynooth.

Interment occurred in the Kirkland Lake cemetery.

## Alex 'Red' Nicholl (1905 - 1938)

Red Nicholl was born and raised in Guysborough, Argyle County, Nova Scotia. In 1925, 20-year-old Nicholl and his friend Allan Bezanger left Nova Scotia with plans of making a new life in Western Canada. On their way, they stopped at Kirkland Lake and, instead of continuing west, they got a job in the gold mines. Nicholl was hired at Teck-Hughes.

Being single, he did not stay in Kirkland Lake continuously but took numerous and extended trips back east. His last trip to the Maritimes was in the summer of 1937 when he decided to go with a party of friends who were motoring there.

He was a very social person who took an active interest in all sports at the mine. He played inter-department hockey, was a member of the bowling team and liked to box. He had lived at 104 Government Road but a week before his death he moved to 32 Poplar Avenue. On March 15, Nicholl and three others were mucking in a stope. As only two men could conveniently work about the ore car at a time, they adopted a system of working in relays so that each pair filled and trammed two cars while the other pair rested. At the time of the rockburst, he and Bert Lamoureux were sitting on blocks, resting. All four perished in the burst.

Red Nicholl was survived by three sisters: Mrs. James Grant and Miss Alma Nicholl, both of Timmins, and Winona of Boston, Massachusetts. He was survived by four brothers.

The body was taken by train to Guysborough, N.S. for funeral services and burial.

## Anton Gornik (1899 - 1938)

Anton Gornik, 39, born in Yugoslavia on June 16, 1899, left Europe and came to Canada in 1927. He worked in Kapuskasing and Kirkland Lake until 1931 when he returned to Yugoslavia and remained there for five years. In Yugoslavia he married and a daughter was born; soon afterward, he decided to return to Canada with the plan of coming back to his family with money earned in the gold mines.

Back in Kirkland Lake for two years, his home and family overseas were always on his mind. Gornik had a dream the night before his death. He told his friend and roommate at 87 Woods Street, that in his dream, his home in Yugoslavia was burning.

Gornik had been a shaftman at Teck-Hughes and more recently had become a mucker. On March 15, he was killed along with three others when huge blocks of rock collapsed into a stope during a rockburst. It took days for the rescue crews to find and release the bodies. Red Nicholl and Bert Lamoureux were found after five days; Gornik was found on the sixth and Harvey Peever on the seventh day.

The funeral mass was held at Holy Name Church. Yugoslavian friends remarked that Anton was "the most happy-go-lucky person in the world." Interment was in the local cemetery. The inscription on the headstone reads: TUKAJ POCIVA – ANTON GORNIK – IZ MAHAROVA PRIST JEREJU – POLESKO JOGOSLAVIA – ROJ. 16 JUNIJA 1899 – UMRL 15 MARCA 1938 – V TECK-HUGHES MINE – POCIVAJ VMIRO. Translation: Here lies – Anton Gornik – From Maharova Prist Jereju – Polish Yugoslavia – Born 16 June 1899 – Died 15 March 1938 – At the Teck-Hughes Mine – Rest in peace.

## Wilfred 'Bert' Lamoureux

### (1902 - 1938)

Thirty-six-year-old Bert Lamoureax of 19 Woods Street came to Kirkland Lake from Quebec. He had been employed in Kirkland Lake in 1935 and then left. Before the accident, he had returned and had been in town for about three months. Earlier, he had been employed near Cochrane, Ontario. It was evident in town that Lamoureux, who was single, was a great hockey fan "as he had been seen numerous times at the arena during the winter."

On March 15, he and Alex Nicholl were sitting on blocks in a stope, waiting for their turn to muck. Two other miners were mucking ore into an ore car about 10 feet away when suddenly a rockburst occurred. It happened so quickly that they did not have time to even jump to their feet to seek safety. All four miners died in the mishap.

Lamoureux was survived by a brother, Philippe, who was working at Lake Shore at the time.

The funeral mass was held at Assumption Church and interment took place in the local cemetery. The municipal burial records have his name recorded as Joseph Wilbert Lamoureux. The gravesite is presently (1999) unmarked.

### May 25, 1938

# Wright-Hargreaves Mine

## T. B. Booth (1900 - 1938)

Thirty-eight-year-old T. B. Booth lived on Goodfish Road while his wife and children lived in Timmins. He was employed as a mucking machine operator at Wright-Hargreaves and was killed there on May 25.

On the day of the accident, he and his helper, R. Dupalo, scaled and spragged the walls of a drift and did not begin the actual mucking operation until 10:00 a.m. Booth operated the shovel until 11:00, at which time all the men stopped for lunch. During the lunch, the drift was visited by Safety Inspector Morris Hastie and Shift Boss Harry Sauvé, who found the site in good condition and all the sprags in place. After lunch, mucking continued until 12:30 at which time the motor crew came in to pull out the loaded cars. Booth proceeded to do more scaling at the site and took down some ground which caused a sprag to fall out. Harold Mercier, a motorman nearby, asked him if he wanted some help in replacing the timber. Booth made no reply; he did not replace the timber but went to his shovel and began mucking. Soon the crew heard a crash of falling rock and rushed to the area where they found Booth lying beside his shovel with the top of his head removed by a large slab of rock originally supported by the sprag.

At the inquest, the following statement was made: "There is no doubt that Booth's neglect to replace the fallen sprag resulted in his death. His failure to replace it, after having his attention directed towards it by Mercier, is difficult to understand."

## July 19, 1938
# Lake Shore Mine

## Jan Greczek (1903 - 1938)

Thirty-five-year-old Jan Greczek, of Polish descent and married, was employed at Lake Shore as a driller and timberman. He died at about 3:30 p.m. on July 19.

On the morning of the accident, he arrived at the stope before his partner O. Webster, who had gone for nails for the job and was not present at the site when the rockburst occurred. Mine Captain Clarence Whalley*, whose beat included this stope, was at the collar of No. 5 shaft when the burst occurred at 7:15 a.m. He quickly proceeded underground and, on learning that Greczek was unaccounted for, immediately instituted a search for him.

Access to the stope was blocked by the burst but in about an hour the rescue crew had established verbal communication with Greczek. When spoken to, he reported that he was trapped in sand fill but that he was otherwise all right. The crew broke through the debris at 10:00 a.m. and found Greczek. He was given coffee and a stimulant. It took another four hours to free him, as he was trapped about the legs by the fill. During the rescue operation, he freely conversed with the rescue crew and with Dr. Contway who was tending to him. Once he was free, he was fully examined by the doctor and showed no symptoms or serious injuries although he had sustained various slight lacerations. As the surrounding temperature was favourable at the site, the doctor kept Greczek there for about an hour to give him a chance to recover from primary shock before being moved to surface. Frequent checks were made on his condition that remained good until he suddenly developed a definite change for the worse and died.

An autopsy was performed by Dr. M. J. Thompson, who reported finding no evidence of skull fracture or other serious injury and ascribed death as attributed to secondary shock.

The funeral mass was held at Holy Name Church with Reverend Father Roney officiating. There was a memorial service at the Polish Hall where the deceased was a member of the Polish Mutual Benefit Society. Interment was in the local cemetery. The

*Headstone photograph of Jan Greczek. Photographer: Author*

headstone marking the grave has the following inscription: TU SPOCZY WA – S. P. JAN GRECZEK – URODZONY 29 GO KWIETNIA 1903 – UMART 19 GO LIPCA 1938. Translation: Here lies in Holy memory Jan Greczek – Born 29 April 1903 – Died 19 July 1938.

** Clarence Whalley would die in a mining accident at Lake Shore in 1954.*

## August 2, 1938
# Fernland Mine

## Mike Zarcovitch (1901 - 1938)

Mike Zarcovitch came to Canada from Yugoslavia in 1928 at 27 years of age. He began employment as a shaftman on July 7, 1938 at the Fernland Mine, about a mile east of the town of Larder Lake. Within a month, he was dead, killed when he fell down the shaft in which he was working. He was married; his wife and two children were living in Yugoslavia.

The 11:00 p.m. to 7:00 a.m. shift of August 1 included Roy Graham, deckman, and three shaftmen, Nick Bankovitch, Nick Tertany and Zarcovitch. J. Ferguson was the leader in charge. Sinking of the shaft was in progress from a depth of 300 to 550 feet. The work was nearly completed as the shaft was 525 feet deep. The shift mucked out the shaft for four hours and then broke for lunch. After lunch, the shift descended to the bulkhead, 43 feet from the bottom of the shaft.

Within a few minutes, Tertany and Bankovitch returned to surface and reported to Ferguson that

Zarcovitch slipped from the bulkhead and had fallen down the shaft. Ferguson told Tertany to go to the mine office and phone Dr. Richardson in Larder Lake. Ferguson, Bankovitch and Graham got into the bucket and went to the bottom of the shaft where they found Zarcovitch dead. They removed the body to surface where Dr. H. F. Richardson pronounced death to be instantaneous and to have been caused by a compound fracture of the skull with cerebral lacerations. There was a fracture of the right leg.

At the inquest, it was learned that Zarcovitch slipped off the bulkhead while he was moving a timber. The members of the crew testified there was no gas in the shaft, Zarcovitch appeared normal before his fall and he was an experienced shaftman.

The funeral mass was held at Holy Name Church. Interment was conducted in the Kirkland Lake cemetery. It is presently an unmarked grave.

### November 8, 1938

## Wright-Hargreaves Mine

## Edward J. Labine (1907 - 1938)

Edward Labine, born at Trout Lake near Renfrew, Ontario in 1907, was a resident of Kirkland Lake for 10 years, marrying in town and raising three children. Previous to this, he lived in St. Charles, Ontario. He initially spent a short time working at Teck-Hughes but was employed continuously at Wright-Hargreaves since 1931. He was killed instantly on the 1,000 level on November 8.

*Miner prepares fuse for blasting in a stope.*
*Photographer: Unknown*

On the day of the accident, he and his helper, H. A. McGonigal, were drilling, loading holes and cutting fuses in stope 1,007. After 14 holes were loaded and the fuse cut in regular order, Labine started to light the fuse using a hot-wire lighter. He told McGonigal to pick up the lunch pails, go to the entrance of the stope and to wait for him there. McGonigal did so and waited at the entrance about 24 feet from where Labine was lighting the fuse, when he heard a shot. He called to Labine but received no reply. He then called to Jock Brodie, who was passing by, and told him what happened. At this moment, a number of other shots were heard in rapid succession. McGonigal then ran to the shaft station and notified the first-aid office by telephone. Brodie turned on the air into the stope to blow out the smoke.

Upon receiving notice of the accident, Mine Superintendent W. Sampson, Mine Captain J. Martine, Shift Boss L. Doyon and Dr. C. S. Harris arrived at the scene. They found Labine's body lying on the footwall with some broken rocks on his legs. At this point, Dr. Harris pronounced him dead. Labine had suffered a fractured skull, broken collar bone, fractured ribs and a torn aorta.

It seems probable that after lighting about half of the fuses, Labine noticed that an inside fuse was not burning and attempted to light it when an outside charge exploded, throwing him backwards and killing him. The injuries received indicated that he was directly in front of the first explosion.

He was survived by his wife, the former Rose Gauthier, and three sons, Noel (8), Oscar (7), and Morris (6). A fourth son, Eddie, was born eleven days after his father's death.

The funeral mass was conducted at Holy Name Church and burial took place in the local cemetery. Six brothers acted as pallbearers: Hiram of Ansonville, David of Hornepayne, and Cyreel, Benny, Oscar and Jedeon, all of Hagar.

Three sons of the late Edward J. Labine currently (1998) reside in Kirkland Lake. They are Morris, Oscar and Eddie. Noel died in 1973.

*Diagram of rescue attempts to reach Holubec and Vlaisavgevitch. Artist: Globe and Mail Staff, 1938*

## December 27, 1938
## Lake Shore Mine

# Rockburst Shatters the 2700-Foot Level of Lake Shore, Two Miners are Killed

Herbert Breens, a motorman working on the 2,700 level of Lake Shore, heard two or three shots coming from the 2701 E drift; suddenly, a rockburst occurred. The time was 11:50 p.m., December 27.

Recovery operations to find two missing miners began immediately. The rescue crew soon discovered that a large block of ground in the vicinity of 2,701 E drift was shattered, causing immense underground damage. It was estimated that a distance of 70 feet of drift would have to be reopened to reach Vujo Vlaisavgevitch and John Holubec.

After days of round the clock rescue work, Holubec's body was found at 3:00 p.m., January 3, 1939. He was sitting in the driver's seat of a locomotive. Vlaisavgevitch's body was found nearby, an hour later. He had been standing near the blasting switch at the time of the burst. Both bodies were completely buried in the muck and badly crushed. It was evident that death was instantaneous.

# Vujo (Vito) Vlaisavgevitch
## (1901 - 1938)

Vujo Vlaisavgevitch, 38, born in Yugoslavia in 1901, moved to Canada and had lived in Kirkland Lake from 1930. He was married with two sons and three daughters; his family remained in Yugoslavia.

He returned to Yugoslavia in 1937, unknowingly seeing his family for the last time. One of his daughters was now married. He established his family on a farm and promised them that he would return in a few years so that they could finally be re-united. The rockburst would prevent this plan from ever being fulfilled.

*Headstone photograph of Vujo Vlaisavgevitch. Photographer: Author*

On December 27, Vlaisavgevitch left his home at 24 Main Street and walked the short distance to the dry at Lake Shore. Underground, he was working in a drift with his partner. They had just set a blast and were in the process of seeking shelter when suddenly a huge rockburst occurred in this area, shaking the mine and the homes in town. Both were killed instantly and, a week later, their bodies were retrieved from the mine.

Prior to the funeral service at St. Peter's Anglican Church, Vlaisavgevitch's body rested at the home of Mr. and Mrs. Zigich at 43 Hudson Bay Avenue. A large number of people visited the home and attended the funeral service conducted by the Reverend H. A. Sims. The pallbearers were George Reinovich, Uajo Lukich, Joe Derich, Nilan Hinich, Joe Vlaisovich and Dane Mirich*. Interment took place in the local cemetery.

** Dane Mirich would die in a mining accident at Sylvanite in 1941.*

## John Holubec (1905 - 1938)

Thirty-three-year-old John Holubec, born in Ukraine in 1905, came to Canada at 18 years of age in 1923. He was married in Hamilton, Ontario in 1929 to Mary, of Ukrainian descent, who came to Canada in 1924. A year after their marriage, they moved to Kirkland Lake. Their only child, Mary, was born on February 20, 1931.

In December 1938, he was living alone at 17 Poplar Avenue. His wife had been living in Noranda for the past year and their daughter, now seven, was visiting her father at Christmas. Her mother could not be in Kirkland Lake at this time as she was recuperating from a very serious operation.

Two days after Christmas, Holubec was back at work underground at Lake Shore. He and his partner set off a blast, were caught in a large rockburst and killed instantly.

A Ukrainian religious service was held at the Ukrainian National Hall on 54 McCamus Avenue. A large number of "fellow countrymen and fellow workers" attended the service conducted by John Ilcheshen and Joseph Dacek. An address by M. Raschyn was delivered to the crowd of people. Various selections of music were played by the Ukrainian Band directed by leader Sam Mason. A brief service officiated by Reverend Father McMahon and Reverend Father Roney was later held at Holy Name Church. The pallbearers were Art Radul, L. Hrecul, I. Onucki, John Rega, P. Holod and William Susla*. Interment took place in the local cemetery. The gravesite is presently unmarked.

** Another rockburst at Lake Shore would occur a year later. This one would claim the life of pallbearer William Susla.*

### 1939

- Rockbursts at Lake Shore claim seven lives this year
- A run of ore, a fall down a shaft and a pit accident claims three lives at Wright-Hargreaves
- Morris Kirkland loses a miner in a cage accident
- An employee at Toburn is electrocuted
- At Matachewan Consolidated, a fall of loose claims a life
- Statistics for the number of tons of ore milled per day: Lake Shore (2,300), Wright-Hargreaves (1,200), Kerr Addison (1,200), Teck-Hughes (1,000), Sylvanite (525), Omega (500), Chesterville (500), Kirkland Lake Gold (400), Upper Canada (175), Toburn (150), Bidgood (125), Golden Gate (125)
- Morris Kirkland Mine slows down drastically
- In Matachewan, the Young-Davidson and Matachewan Consolidated are still producers
- Work is done at Fernland Mine but later in the year, work stops
- Upper Canada Mine leases the mill at Morris Kirkland and builds five new dwellings on the mine property
- Operations are suspended at Martin-Bird
- Price of gold is pegged at $38.50 U.S.
- Average wage earned by workers in the mining industry is $1,612 per year
- The population of Kirkland Lake peaks at 24,200; of this total, 4,680 are mine employees
- Harry Oakes officially becomes Sir Harry Oakes

- Sir Harry Oakes donates the lot deed to St. Peter's Church; he leaves and never returns to Kirkland Lake
- St. Andrew's Presbyterian Church is built in town
- The recession ends in the U.S.; an economic boom begins due to orders from Europe for arms and war equipment
- World War II begins in Europe on September 3; Canada declares war on Germany on September 10

### January 20, 1939

## Wright-Hargreaves Mine

## Alfred Charles Closs

### (1888 - 1939)

Alfred Closs was a night surface labour foreman working in a Wright-Hargreaves sand and gravel pit about seven miles north of the town in Morrissette Township. On the cold evening of January 20, he and diesel shovel operator Ernie Davis were scaling with long ten-foot poles from the top of an excavation. Noticing that the loosened gravel was starting to move, Closs shouted a warning and both men ran towards the bottom of the slope about 30 feet away. As he reached safety, Davis looked back for Closs but could not see him. Realizing that Closs was probably buried in the recently scaled material, Davis immediately began to shovel in different areas, looking for his partner. Not finding him, he called to truck driver Joe Cahill who had just arrived for a load of fill. Cahill drove back to the mine for help.

A rescue squad from the mine returned and found Closs buried completely in the loose gravel and sand. Dr. T. F. Kelly, who was called to the scene, observed the body when it was removed from the material but detected no signs of life. At the pit tool-house, the doctor applied heart stimulants and an inhalator but after two hours, Closs was pronounced dead due to suffocation.

Fifty-year-old Closs had lived with his wife, the former Sarah Ida Clark (1888 – 1962), at 144 Carter Avenue. They had an 18-year-old daughter.

At the inquest conducted by Coroner J. F. Edis, the jurymen questioned H. Halverson, day foreman of the pit workings. They asked him if he could suggest a safer way to get at the source of the mine back-fill without the men going up to the top and poking the gravel with long poles. He said other methods were tried in the past but this way seemed to be the safest. The jury returned with a verdict of death by suffocation and made two recommendations: 1) a third man should be present during scaling operations at the pit, and 2) stationary flood lights should be installed in addition to the two large lights that were attached to the boom of the diesel shovel.

Funeral services were held locally and interment was conducted in the Kirkland Lake cemetery.

### March 11, 1939

## Lake Shore Mine

## Rockburst Occurs at Lake Shore, Two Men Injured, Two Men Dead

Two men were crushed to death and two others were injured in a rockburst in the vicinity of the 3,400 level at Lake Shore at about 8:00 p.m., March 11.

George Braaten and Michael Harrington died when the burst caused timber and rock to collapse into a drift where they were located. Russell Bates, working in the stope above the cave-in, suffered lacerations of the head, back and arms from spawling rock. He was able to make his way to the safety of the next level. Herb Levon was buried in the cave-in and was fortunately rescued; he suffered minor lacerations to the head and face.

After the burst, Alex Gill and Jim Gannon proceeded to the scene to find the drift timber collapsed and no sign of the men. They immediately started to open up the drift and, within a short period of time, found Levon a few feet inside the collapsed timber. Mine Captain Hector McPhail and Shift Boss George Needham arrived at the scene and immediately organized a larger rescue crew. Braaten's body was recovered four-and-a-half hours after the burst and, within another half hour, Harrington's body was found.

## George Samuel Braaten

(1910 - 1939)

George Braaten, 29, was married in 1935; he and his wife and child lived on Goodfish Road. He was employed at Lake Shore from August 1935.

He was the son of Mr. and Mrs. S. O. Braaten of Gameland, Ontario. He was predeceased by a sister, Mrs. Albert Youngberg of Kirkland Lake.

His interment occurred at Gameland.

## Michael Harrington (1909 - 1939)

Michael Harrington, 30, was from a large family of sixteen sons. He lived with his wife of four years and their six-month-old son, on McCamus Avenue.

The funeral mass was held at Holy Name Church and the body was taken to Killaloe, Ontario for burial.

March 23, 1939

# Wright-Hargreaves Mine

## Run of Ore Kills Two Men at Wright-Hargreaves

Alfred Lafleur and Arthur Bellamy died from crushing injuries received when they were caught in a run of broken ore in the 3101 W stope of Wright-Hargreaves on March 23.

Both had been working in this stope for several months and were familiar with its surroundings. On the day shift, they intended to clean off the bench in the stope and to assist two timbermen, H. Jones and F. Racich.

Coming into their work area, they found muck completely covering the bench. They knew the bench well – its location, its size and perimeter. They knew that muck would be pulled from the lower level by the motor crew sometime that morning. They decided to set up the staging before the muck settled and they were not too concerned about the muck dropping as it always settled slowly in the past.

Lafleur, Bellamy and Jones went onto the muck to do some measurements. They were standing there when it suddenly dropped 50 feet. Lafleur and Bellamy went down with the muck, Jones was saved when Racich, on the timber above, reacted quickly and grabbed the measuring stick Jones was holding. This steadied him just long enough to get a foothold on solid rock and throw his arm around a stull above him. Jones rushed out to the shaft where he gave the alarm, took a cage to the lower level and stopped the motor crew from any further pulling. They had already pulled nine cars of ore that morning. The muck suddenly dropped when they were filling the ninth car.

Jones got a crew of four men to cut a hole through the manway by which they entered the chuteway into the space under the bench. Here they found Lafleur, dead, with about 10 feet of muck over him. Bellamy was found just above Lafleur and he was also dead. Both bodies were badly crushed, indicating they died instantly.

An inquest was held in the town hall before Coroner R. H. Armstrong. The jury returned the following verdict: "We find that the late Alfred Lafleur and Arthur Bellamy met their deaths as a result of muck being held up and suddenly giving way under the 3101 W stope. We find that this accident was attributed to negligence on the part of Wright-Hargreaves Mines Limited. We recommend that greater precaution be taken when muck is being pulled from the chute in stopes where men are working on broken rock above. These men should be informed definitely of the exact time that rock is to be removed."

## Arthur Walker Bellamy

(1911 - 1939)

Arthur Bellamy, 28, his wife, the former Catherine Enright, and a daughter, Doris, lived in King Kirkland. He was a machine runner's helper at Wright-Hargreaves where he was employed since June 17, 1937.

He was buried in the Kirkland Lake cemetery. The epitaph on the headstone over his grave reads: ARTHUR BELLAMY – BELOVED HUSBAND AND FATHER – 'NOT LOST BUT GONE BEFORE'.

*Alfred Lafleur. Photographer: Unknown*

## Alfred Lafleur (1894 - 1939)

Forty-five-year-old Fred Lafleur was married with ten children: Albert, Marcelle, Pierrette, Fernand, Patricia, Jacqueline, George, Girard, Aline and Francine. The family lived at the end of Third Street near Toburn Mine. He worked as a machine runner at Wright-Hargreaves, employed there since February 16, 1937.

The funeral mass was held at Assumption Church and interment occurred in the local cemetery. The inscription over his grave is: A LA MEMOIRE – ALFRED LAFLEUR – NÉ LE 1 DECEMBRE 1894 – DÉCÉDE ACCIDENTELLEMENT A LA MINE – WRIGHT-HARGREAVES – LE 23 MARS 1939.

Information and photograph were provided by great-grandson Jason Kilbride of Sault Ste. Marie, Ontario.

### April 11, 1939

# Wright-Hargreaves Mine

## Arne Harmala (1902 - 1939)

Arne Harmala, 38, Finnish and married with one child, was an experienced miner employed at Wright-Hargreaves from 1933.

Shortly before 7:00 p.m. on the day of the accident, loose rock developed on the wall of the No. 4 service and supply shaft. A repair crew organized by Night Mine Captain J. Newman included five men: John Myelde, Adolphe Proulx, Lennis McKay, Napoleon Gauthier and Harmala.

A platform of planks was installed across the top of the cage that was lowered to just below the 2,250 level. The work of scaling and loading the loose rock onto the top of the cage was started. On the level, Newman was looking down the shaft as the work progressed and saw Harmala standing on the platform on the cage. While scaling the loose rock from the wall, Harmala stepped back out of the way in case the loose came with a rush. He lost his footing and McKay managed to get a grasp of him with one hand but he could not hold on. Harmala fell backwards through an opening in the timber into the counterbalance compartment of the shaft. He dropped a distance of 1,700 feet and landed in the flooded sump at the bottom of the shaft.

Newman immediately sent word to surface for a doctor and then proceeded to the 3,900 level by means of the No. 3 shaft. There he met Shift Boss J. Elliott and together they descended the manway to the sump where they found Harmala's mutilated body floating in the water. Part of the head and most of the arms and legs were missing. The body was sent to surface.

Interment took place in the Kirkland Lake cemetery. The inscription on the gravestone reads: ARNE HARMALA – BORN JANUARY 13, 1902 – DIED APRIL 11, 1939 – BELOVED HUSBAND OF HILDA HARJU – BORN MAY 2, 1902 – DIED JANUARY 29, 1973 – 'LEPA A RAUHASSA'.

### April 27, 1939

# Morris Kirkland Mine

## John Young (1908 - 1939)

Thirty-one-year-old John Young of King Kirkland died at the Kirkland District Hospital following an accident at the Morris Kirkland Mine. He was hired as a surface workman on May 29, 1936 and by November became a deckman. A year later, he was severely injured when he was caught between the shaft timbers and the cage near the top deck. It wasn't until February 9, 1939 that he returned and was hired as a cagetender. Two-and-a-half months later, he was involved in another serious accident – this one claimed his life.

He came on shift at 7:00 p.m. on the day of the accident. As a cagetender, he took miners down to the new levels below 1,250 feet and returned to surface. Here, Mine Captain H. Van de Kamer and J. D. Cumming, a consulting engineer, entered the cage and were taken down to the 1,625. From there, Young took the cage up to the 1,500 where he loaded 12 pieces of steel. He failed to make use of the available chain to secure the steel and did not close the cage doors; he rang the cage to surface. On the way up, two pieces of steel slid out of the cage, striking the shaft timbers, causing damage to the cage and injury to Young.

Richard Wyper was standing beside the shaft gate on the surface deck at the time and heard the noise of the cage catching the timbers. He immediately rang a one-bell signal to stop the cage. He descended the manway and found the cage stopped at about 65 feet below. Young was sitting on the cage floor, conscious but bleeding. Again using the manway, Wyper returned to surface where he told H. W. Biddulph to go to the office and phone a doctor. Wyper returned to the cage with a hammer, made quick repairs to the cage so that it would not engage the shaft timbers and he called to the hoistman, H. Teeple, to raise the cage slowly.

When Wyper and the cage arrived at the collar, Dr. Wickett and others removed Young and placed him in a basket-stretcher. He was taken to the mine office where a tourniquet and bandages were applied to stop the bleeding. Dr. W. E. Bryant and an ambulance arrived from Kirkland Lake and Young was rushed to hospital.

There it was discovered that Young had suffered a compound fracture of the right leg and swelling and contusion of the left foot; he was in deep shock. For two days, he was given blood transfusions but gangrene was setting into his broken leg. On May 1, his right leg was amputated and another transfusion was given but he died that morning.

Born in Glasgow, Scotland and formerly a resident of Ottawa, he was survived by his wife, the former Ina Hodgins, a sister, Christina (Mrs. Milton Beattie) of Ottawa, and two brothers, Harry of Rainy River and Bruce of Glasgow.

At the funeral, many floral tributes were received from friends in King Kirkland and Kirkland Lake. As reported by the Northern News, floral tributes were received from many families, including the widow and daughter of Arthur Bellamy of King Kirkland who was killed at Wright-Hargreaves four weeks prior. An inquest was held before Coroner J. F. Edis. The jury's verdict was as follows: "John Young met his death through negligence while riding the cage with steel bars, in not having closed the cage doors, in not having chained the steel bars so that they could not fall out of the cage."

*Miners leave the cage as cagetender stands nearby.*
*Photographer: Unknown*

## July 12, 1939
## Toburn Mine

### Onide Tremblay (1908 - 1939)

Onide Tremblay lived at 25 Queen Street with his wife and five children. He was 31 years of age, Canadian-born and employed as a carpenter at Toburn since October 13, 1936. He was killed on surface at the mine on July 12.

On the day before the accident, lightning struck the substation on the Toburn property, destroying the building by fire and causing the complete shutdown of the mine. That afternoon, the mine employees erected a temporary switchhouse and a fence around the transformers and lightning arresters. The power was restored by 9:00 that evening.

Louis Rouillard, the carpenter foreman, sent Tremblay to the sub-station the next morning; he instructed Tremblay to build a support for two posts to shoulder new electrical work that would be installed later that day. These posts were away from the new switch-house and transformers. He reminded Tremblay that everything was 'live' in the enclosure and there was no need for him to go into this location to complete his work. Tremblay confirmed his orders and indicated that he understood that everything was 'live' in the enclosure.

In the area of the sub-station, Jack O'Connor and Archie Gillies were spreading crushed rock and levelling the ground. O'Connor talked to Tremblay and noticed he had removed two boards from the enclosure fence. He reminded him of the live enclosure and asked if he knew about the orders to stay out. Tremblay replied that he had to go in to measure a post. O'Connor returned to his work and was talking to Gillies when suddenly their attention was attracted to a crackling sound from the switch-house. They ran over to the area and saw Tremblay crouching between a lightning arrester and a transformer. Their cries for help attracted workers in the vicinity. The hydro company was told to turn off the power and Dr. Kelly arrived to observe Tremblay and to pronounce him dead.

It would seem that Tremblay entered the enclosure though the opening he made. He apparently walked around a transformer, trying to pass between it and an arrester. It was speculated that the hammer he was carrying made contact with the electrical apparatus, causing 12,000 volts to pass through his body, electrocuting him.

The body was taken by train to Buckingham, Quebec for funeral and burial services.

## September 14, 1939
## Lake Shore Mine

### Malcolm MacLeod (1897 - 1939)

Malcolm MacLeod, born in Lewis, Scotland on January 5, 1897, was a veteran of World War I, having served in the British Navy.

His family of wife and two children, Katie and Kenneth, lived at 35 Taylor Avenue. MacLeod was a well-known member of the Scots community in town and a staunch member of the Sons of Scotland Society. He had been employed at Lake Shore since 1937.

The day shift of shaftmen on September 14 included Shaft Leader G. Dickenson and Shaftsmen P. Miletovich, M. MacDonald, A. Williamson, M. Ragus and MacLeod. They scaled and mucked and by 9:20, only picking at the bottom of the shaft remained to be done. The shaft, now 4,775 feet deep, was subject to strain bursts at the bottom. However, none were reported this September.

A strain burst did occur on this day, affecting an area about 6x6 feet on the shaft wall. MacLeod, standing within three feet of this area, was knocked over and the lower part of his body was partially covered with muck. He was uncovered and taken to surface. He was suffering from bruises on the left ankle, the left forearm and the back. Medical opinion suggested the injuries were not life threatening and would only incapacitate him for about two weeks.

He received medical attention at his home each day but on September 16 suffered an attack of severe abdomen pain and a rise in temperature. He was admitted to hospital and appeared to recover. He suffered another similar attack on September 24 and recovered again. On the evening of October 5, another attack occurred and he suddenly died in the early morning of the next day.

An autopsy was performed revealing that death was attributed to a pulmonary embolism derived from the injuries received.

Surviving MacLeod were his immediate family in town, a mother in Stornoway, Scotland, three brothers, Angus and Murdoch of Kirkland Lake, and Kenneth of Timmins, two sisters, Mrs. R. MacLeod of Vancouver and Mrs. A. MacLeod in Scotland. There were four cousins present at the funeral: John and William MacLeod of Guelph, Alec MacLeod of Prescott and Angus MacLeod of Kingston.

Prior to the funeral service, an impressive lodge service was conducted by the Sons and Daughters of Scotland at the Symington Funeral Home chapel. Chaplain and Worthy Chief Lang officiated in the presence of about 40 lodge members.

A graveside ceremony at the local cemetery was held; many flowers accompanied the casket. The service was officiated by Reverend J. E. Graham of Trinity United Church. The pallbearers were: William King representing the Canadian Legion, Jack Price, Duncan McLeod, Gordon Laverre, George Dickinson, William Johnston, and Charles MacDonald representing the Sons of Scotland.

### October 26, 1939

## Lake Shore Mine

# Extensive Rockburst at Lake Shore Claims Four Lives

At about 12:55 p.m. on October 26, an extensive rockburst occurred, centered at about the 3,700 level at Lake Shore. It affected an area on the north vein of approximately 200 feet in vertical extent over a horizontal distance of 160 feet. Drifts and stopes in this area were damaged.

When the burst occurred, two producing stopes, 3501 W and 3801 W, were seriously affected. Josef Kovac, Stanley Slack, William Susla, Griffith Jones, Melbourne Crowe and Pat Harrington were all working in 3501 W when it collapsed. All six were buried in timber and loose rock.

Alex Turus, Allan Gray, Fred Kulchynski and Shift Boss A. Nelson were working in 3801 W when it partially collapsed during the burst. Nelson, who was uninjured, released Kulchynski who was caught between timber and broken rock in a manway. Gray was struck by falling rock and made it to the manway on his own. Turus was completely buried by fallen rock.

Rescue operations were instituted immediately under the supervision of mine captains, Hector McPhail and Walter Selnes. Crowe was released at 4:15 p.m. with a serious contusion on his forehead. Forty-five minutes later, Jones was found and released, having numerous lacerations, multiple bruises and exhibiting symptoms of severe shock. At 5:45, Harrington was released after he was found buried in the rubble up to his head. He suffered serious injuries: a fractured left leg, cuts on his head and abrasions on his back. Susla was confined under a heap of rock and broken timber until he was released at 11:00 p.m. He was conscious, had multiple bruises and was suffering from severe nervous shock from his extended confinement. At 4:45 a.m., about 16 hours after the burst, Slack's dead body was recovered. Judging by the injuries, death was instantaneous. Kovac's body was not extricated until 2:10 that afternoon.

In stope 3801 W, A. Nelson and Shift Boss Tom Ramsay located the body of Turus who was examined by Doctor Lynch and pronounced dead.

Susla was taken to hospital, placed in an oxygen tent and given blood transfusions. Blood was donated by fellow miners but within two days, he died. Doctors said that the cause of death was severe shock.

The rockburst had claimed the lives of Josef Kovac, Stanley Slack, Alex Turus and William Susla.

### October 26, 1939

## Lake Shore Mine

# Alex Turus (1899 - 1939)

Alex Turus, 40, born in Western Ukraine in 1899, died along with three others in a rockburst at Lake Shore on October 26.

He was a World War I veteran who had served as a sergeant in the Ukrainian army in 1918. He came to Canada in 1928, settling in Saskatoon. In 1934, he left and came to Kirkland Lake. By 1939, he and his wife Rosie and their three-year-old daughter Vera were living at 14 Woods Street.

*Alex Turus. Sergeant in the Ukrainian Army 1918. Photographer: Lawrence Chrismas*

He was highly regarded in both the Ukrainian and English communities of Kirkland Lake. He was a prominent member of the Ukrainian National Federation and was secretary of the Ukrainian War Veterans' Association. When the call was issued through branches of the Canadian Legion all across Canada early in 1939 for war veterans to register for service in the event of an emergency, Turus was among those who registered at the local branch.

An impressive funeral was held in the Ukrainian Hall at 54 McCamus with about 1,000 people attending. During the funeral cortege, music was played by a band in which Turus had been a drummer. A large wreath was placed on his drum carried by a fellow band member. Present in the cortege was Brigadier-General Swift, a great friend of the Ukrainian community.

Turus was a member of Trinity United Church where the funeral service conducted by Reverend J. E. Graham was held. A Ukrainian memorial eulogy was given by John Lichesyn. The pallbearers were Mike Antoniuch, Mike Wasney, Metro Manchulenko, Joe Galeida, Pete Mischie, Harry Sowka, John Onutsky and Steve Holynka.

Following the church service, the casket was carried as far as the Lake Shore Mine gates and then by car to the cemetery where interment ceremonies occurred. At the grave, John Datzyk gave a farewell as many mourners came to the funeral from the Nickel Camp. A second farewell was given by Mike Paschyn on behalf of Turus' father, wife, daughter and sister.

## Stanley Dorian Slack
### (1915 - 1939)

Stanley Slack, born at North Cobalt on December 27, 1915, was educated at Cobalt and Nipissing before coming to Kirkland Lake in 1935 at the age of 20. He gained employment at Lake Shore in 1936. On May 20, 1939, he married Adrienne Lauzon and the couple lived at 50 Prospect Avenue, the residence of Mrs. G. E. Ferguson, Stanley's mother. Five months later he was killed instantly by a rockburst at the mine.

When interviewed by a newspaper reporter, Mrs. Adrienne Slack sobbed, "I was always telling him to get out of mining but he loved it. He was never afraid and he always went to work happy."

The funeral service at Assumption Church was well attended. Present were many relatives: Mr. and Mrs. Hoyle of Nipissing, Joseph Busch of Nipissing, Mr. and Mrs. Donat Coueillard of Haileybury, Miss J. Lavigne of Val d'Or, Mrs. A. Lauzon of Timmins, Mrs. J. Judge of Timmins, and Mrs. O. Matte of Rouyn. Many other uncles, aunts and cousins attended the service. Five priests took part in the mass: Reverend Fathers Deshaies, Jodoin, McManus, Cavanagh and Champagne. The pallbearers were Hreve, Lawrence, Ray, Edgar, and Donat Lauzon, all brothers of the widow, and W. T. Scharfe.

According to the Northern News, spiritual bouquets, floral tributes and messages of sympathy were received from many individuals and families.

Interment occurred in the Kirkland Lake cemetery.

## Josef Kovac (1900 - 1939)

Josef Kovac, born in Czechoslovakia on February 2, 1900, arrived in Kirkland Lake in 1926. A year later, he married Maria. In 1939, they lived at 62 Woods Street.

*Headstone photograph of Josef Kovac.*
*Photographer: Author*

He was an employee at Lake Shore for three-and-a-half years. He died along with three others when they were buried and crushed in a rockburst at the mine.

"Joe left home for work at about 6:00," Mrs. Kovac said. "I was having lunch when I heard the burst. A cold chill ran up my spine. I thought of Joe. I often wondered whether he would be killed. I am lonely now that Joe is gone; we have no children."

Elaborate funeral rites were held by members of the Czechoslovakian community. The funeral cortege led by a group of women carrying floral tributes, proceeded to the gates of the Lake Shore Mine. The cortege continued on to the cemetery by means of automobiles and a bus.

Josef and Maria are presently buried side by side. The inscription on the gravestone is: TU ODPOCIVA – JOSEF KOVAC – ROD 2 FEB. 1900 – U OBCI VELKE UHERCE NITRIANSKA, CECHOSLOVAKIA – BOLUSMRTENY 26 OCT. 1939 – LAKE SHORE MINE K. L. – POMNIK VENOVALA MU NA – PROCEST JEHO MANZELKA – MARIA KOVAC – 1906 - 1971.

## William Susla (1904 - 1939)

Thirty-five-year-old William Susla was born in Ukraine on March 12, 1904 and died on October 30, 1939 as a result of injuries sustained in a rockburst at Lake Shore four days earlier.

Susla came to Canada in 1926, moved to Kirkland Lake in 1932 and got a job at Lake Shore. After a period of time, he left to become employed at Wright-Hargreaves and subsequently Bidgood; he returned to Lake Shore in 1936.

He was single and boarding at 46 Lebel Avenue; all his relatives including his parents were living in Ukraine.

His landlady stated, "Bill got up as usual but he cleaned up his room which wasn't normal for him. He said he wasn't feeling well and that he had no ambition to go to work. He was later than usual when he left the house."

A Ukrainian funeral service arranged by his friends was held at the Ukrainian Hall on McCamus Avenue. From there, the funeral procession led to Holy Name Church for a funeral mass. Interment was in the local cemetery. A lengthy inscription in Ukrainian is found on his headstone and is translated generally as: Here lies William Susla – Born March 12, 1904 – In the town of Radihev, Ukraine – Died of injuries sustained at Lake Shore Mine – October 30, 1939 – Always remembered Rest in Peace.

*Headstone photograph of William Susla.*
*Photographer: Author*

## November 1, 1939

# Matachewan Consolidated Mine

## Ernest Jones (1902 - 1939)

Thirty-seven, single and Canadian, Ernest Jones had been employed at Matachewan Consolidated from December 1936 as a mucker, timberman, machineman and scaler. He died of injuries received in a stope above the 267-foot level.

While mucking, he uncovered on the footwall a piece of rock that suddenly fell. In stepping clear, he lost his balance and rolled down the 40-degree sloped footwall. The rock broke into pieces, rolled down the slope and struck Jones. Another mucker, J. Pichenuk, working above in a sub-drift, noticed the accident and, using a safety rope, dropped down to aid Jones. He immediately removed the rocks from the injured miner and called for more help. E. Desmarais and G. Christianson, who were working in the stope, came to his assistance. Christianson went to the station and informed cagetender T. Guilbeault and R. Crete of the accident.

Realizing that Jones was seriously injured, Pichenuk went to the station, took the cage to surface and went to the office to phone for a doctor; he then returned to the scene with a basket-stretcher. Still alive, Jones was carried from the stope through a manway to the first level where he was placed in the stretcher and taken to surface. Dr. George Crann and Dr. F. Fitch were waiting and immediately attended to him. Observation showed that he had 10 broken ribs, lacerations of the head, internal hemorrhage and a fractured spine. With Dr. Crann in attendance, Jones was rushed to the New Liskeard hospital but he died in transit.

*Part of the Ukrainian Federation Orchestra 1937. William Susla (front row 3rd from left) and Alex Turus (drummer, front row left) were killed in the same mining accident. Photographer: Unknown*

Chapter IV

# THE FORTIES

## 1940

- Wright-Hargreaves, Sylvanite, Toburn and Kirkland Lake Gold attain their highest level of production this year
- Lake Shore and Teck-Hughes are now past their peak
- A miner dies after being gassed at the Golden Gate three years previous
- Two men are killed in rockbursts at Lake Shore in separate incidents
- A Wright-Hargreaves miner is killed while scaling
- A Chesterville miner is killed while tramming
- A Kerr Addison miner falls to his death underground
- Jack Bland, Bob Carlin, and 45 others are fired at Teck-Hughes for their involvement in union activities
- Federal government needs gold to pay for the war effort; Kirkland Lake camp is pushed to higher production

## 1940

## Golden Gate Mine

### John M. Brown (1899 - 1940)

John M. Brown is one of the many names inscribed on the Miners' Memorial Monument in Kirkland Lake. Apparently, he died in 1940, three years after being gassed at Golden Gate Mine in Swastika. Burial took place at the Haileybury Roman Catholic Cemetery.

The inscription on the headstone reads: IN LOVING MEMORY OF A DEAR HUSBAND – JOHN MURRAY BROWN 1899 - 1940 – MAY HE REST IN PEACE – GONE BUT NOT FORGOTTEN.

## February 14, 1940

## Lake Shore Mine

### Joseph B. McDonald

(1898 - 1940)

Joseph McDonald, 42, British and married, was killed in No. 3209 E drift at Lake Shore. He was hired on at the mine in June 1934. He left its employ in November 1935 and returned in June 1936. He worked as a driller.

On the night shift of February 14, he and Peter Selbie scaled and began to muck in the drift. At about 9:00 p.m., Joseph Marinich was there with a locomotive, engaged in taking out a loaded ore car. He was seated on the locomotive and McDonald was standing nearby with Selbie when a rockburst suddenly occurred. About five tons of rock came off the wall, causing injuries to McDonald and partially burying Selbie. Escaping with slight injuries, Marinich was able to make his way to the shaft station and call for help.

Night Captain William Allen was on a lower level when he was told of the burst. At the 3,200 station, he met the first-aid man and together they entered the drift. There, they met Ray Belec and R. Thompson who had taken a stretcher to the scene and were now coming out for more help. Other miners soon arrived and Allen organized a rescue operation. Assistant Underground Superintendent Walter Selnes arrived and took control of the operation at 9:30. By 11:00, Selbie was extricated. Dr. Lynch was now present at the scene and administered medical aid to him. A half hour later, McDonald's body was recovered. He had sustained crushing injuries to the chest, multiple fractures to the ribs, rupture of the left lung, multiple contusions of the scalp, severe contusions of the right arm, abrasions of the right leg and contusions of the right side of the face. The chest and lung injuries alone would have caused instant death. Dr. Lynch pronounced him dead.

May 4, 1940

## Lake Shore Mine

### Eli Kordic (1905 - 1940)

Born in Yugoslavia in 1905, Eli Kordic was married there and had three children. He immigrated to Canada, leaving his family in Europe. In 1931, he was hired on at Lake Shore where he worked as a driller. After residing in this country for nine years, he became a naturalized Canadian.

During the day shift of May 4, Kordic and J. Kopek were working in a raise located about 2,301 feet below surface. Scaling at the time, they found it necessary to pull muck from the raise. Kopec went down to the level and pulled three cars. He was preparing to fill a fourth car when he heard a signal from the raise. He was in the manway approaching the raise when he heard Kordic call for help. He continued up the manway and found Kordic injured on the second ladder from the top. Kopec helped him to the next landing and went for help. Carrying timber at the bottom of the raise, J. Steele was sent up to look after Kordic while Kopec made a call for first aid.

Kordic, conscious, while he waited for Dr. T. E. Kelly to arrive told the men that his injuries were due to a rockburst. Dr. Kelly arrived and attended to him. Kordic was removed to the surface first aid room and subsequently to the Kirkland District Hospital. He suffered compound fractures of the lower left leg, contusions on the right side and back, laceration of the right hip, multiple fractures of the pelvis and five fractured vertebrae in the lower back. He died in hospital that day. Doctors said death was due to internal hemorrhage and shock.

Interment occurred in the Kirkland Lake cemetery. The gravesite is presently unmarked.

*Timberman and partner working in a raise.*
*Photographer: Peak Studios*

June 3, 1940

## Kerr Addison Mine

### David Niskanen (1909 - 1940)

David Niskanen, 31 and married, was a naturalized Canadian of Finnish descent.

The accident occurred on June 3 but the inquest conducted by R. W. McBain, assisted by Crown Attorney Dalton Dean and Mining Inspector James B. Taylor, was not held until months later.

According to evidence given during the hearing, Niskanen, a timberman in the employ of Kerr Addison, was injured at the mine when he fell 40 feet from a ladder in a raise. At the time, he was working with his partner, John Fleming. Niskanen was not wearing a safety rope but survived the fall and was taken to hospital. According to the medical evidence given by Dr. D. H. Pollock of Virginiatown, Niskanen suffered from a fractured neck and a severed nerve cord; as a result of this injury, paralysis ensued in the lower part of his body. Over a period of several months, ulceration developed over his body, resulting in death on December 11, 1940. The jurymen selected for this hearing were: Carl Hedman, Lloyd H. Thompson, Irvine Kokotow, Allan Mitchell and Glen MacDonald. They returned the following verdict: "We, the jury, find that David Niskanen came to his death from a fall, resulting in a fractured neck and a severed spinal cord, resulting in paralysis of the lower part of the body, that caused ulcers, that resulted in death."

## July 5, 1940
## Chesterville Mine

### Percy Michael Robinson
(1916 - 1940)

Twenty-four-years-old and single, Percy Robinson died in the early morning of July 5 when he was crushed between a locomotive and an ore car on the third level of the Chesterville Mine. He had been a motorman at the mine for about a year. On April 3, 1940, he was caught in a fall of loose and buried alive for several hours but later he was rescued by fellow miners. He terminated his employment at the end of the month, only to return a few weeks later to his original position. A couple of months later, he was involved in another accident at the mine – this one claimed his life.

He was survived by his parents, Jim and Theresa Robinson of Kearns, Ontario and by a brother, Louis, and two sisters, Geraldine and Della. Percy's cousin, Blaine Vallier, presently lives in Kirkland Lake and supplied some personal information for this account. The funeral service was held locally and interment occurred in the Kirkland Lake cemetery. The headstone over the grave has the following inscription: IN MEMORY OF OUR DEAR SON MICHAEL PERCY ROBINSON – BORN JUNE 8, 1916 – DIED JULY 5, 1940 – 'TO MEMORY EVER DEAR'.

## November 12, 1940
## Lake Shore Mine

### Johan Galinoc (1903 - 1940)

Born in Yugoslavia in 1903, 37-year-old Johan Galinoc was married and had two children. He immigrated to Canada, leaving his family in Europe, and began employment at Lake Shore in 1931. During his nine-year residency in this country, he became a naturalized Canadian. He resided at 58 Prospect Avenue.

On the night shift of November 12, he and his partner, William Connelly, entered a stope down through a manway from the 3,575 level. They washed the face and began to scale. As Galinoc scaled a slab off the brow, Connelly noticed that a timber block had become loose and had fallen out. Immediately, timber and the fill fence collapsed and the two men were trapped in timber, muck and wet sandy fill.

At level, Charles Welsh heard the collapse below and noticed that the timber in the stope had given way, thus blocking any entrance into it from this level. He immediately informed Shift Boss George Needham who subsequently gave the alarm at the shaft station. Needham climbed down to the next level and entered the stope from the bottom; he found Connelly trapped but alive. Galinoc was found caught in a mass of timber and muck; it was evident that he was dead. Jack Redfern, the night shift first-aid man, administered to Connelly who was still trapped. Night Captain William Allen immediately launched a rescue operation that was taken over later by Day Captain Hector McPhail and by Assistant Underground Superintendent Walter Selnes. Connelly was extricated alive and Galinoc's body was released three hours later. Galinoc had suffered a crushed chest, fractured neck, fractured leg and multiple contusions of the body.

Interment occurred in the Kirkland Lake cemetery. It is presently an unmarked grave.

## December 10, 1940
## Wright-Hargreaves Mine

### Albert H. Bourassa (1909 - 1940)

Albert Bourassa, 31, Canadian and married with two children, lived at 3 Taylor Avenue; he was fatally injured at Wright-Hargreaves on December 10.

On the night shift of December 9, P. Vale and G. Tripp mucked eight cars in a shrinkage stope. Dan McDonald, the shift boss, conducted an inspection while the men worked and found conditions to be satisfactory. They did notice, however, an apparently loose slab on the hanging wall, about 18 inches above the muck. McDonald instructed them to pay particular attention to this and to scale it down as soon as sufficient muck was removed. At the end of the shift, the slab was not yet removed, so the crew left a scaling bar wedged in the crack to attract the attention of the next shift.

*Miner scaling. Photographer: Unknown*

The day shift crew of Bourassa and W. S. 'Biff' Loach entered the stope, began to scale the wall and remove some loose. They noticed the large slab and tried to take it down but it was supported by the muck. They removed five cars of muck. Bourassa, with a scaling bar, tried this piece of loose again. While doing so, another piece fell from the hanging wall and knocked him down. Loach found him in a doubled-up position with loose muck covering his body and a 50-pound piece of rock resting on his head. Within ten minutes, he was extricated and sent to hospital where death occurred later that day. He had received the following injuries: a fractured skull, fractured thoracic vertebrae, fractured jaw and nasal bones, and multiple contusions and abrasions. Death was attributed to the fracture at the base of his skull.

A funeral mass was held at Assumption Church and the body was taken by train to Sturgeon Falls for further funeral services and burial there.

### 1941

- Falls of ground, a cage accident, a blasting accident, a powder magazine explosion, a collapsed bulkhead, a run of fill and a compressed air explosion claim 12 lives at various local mines this year
- Kerr Addison treats 2,000 tons of ore daily
- Kirkland Lake area men are killed overseas
- A strike vote is held at various local mines. Those voting for a strike are the majority of miners from Teck-Hughes, Wright-Hargreaves, Kirkland Lake Gold, Lake Shore, Macassa, Toburn, Sylvanite and Bidgood. The majority of miners at Morris Kirkland, Golden Gate, Upper Canada and Brock vote to stay on the job. Other local mines do not participate in the vote
- The miners' strike starts on November 18 at Teck-Hughes and spreads to other mines; 183 provincial police officers are sent to Kirkland Lake to keep the peace. During the three-month work stoppage, about 9,000 people move away. Although the mines continue to function by hiring scabs, there are no fatal accidents at the picketed mines during the strike. There is a story about striking miners at Lake Shore crossing the picketline to help rescue scabs trapped underground by a rockburst
- 1,600 men from Kirkland Lake are in the military service
- Before the strike, Kirkland Lake's population is 21,500

### January 31, 1941
## Sylvanite Mine

## Dane Mirich (1901 - 1941)

Forty-year-old Dane Mirich, born in Yugoslavia in 1901, immigrated to Canada and later became a naturalized Canadian. He was married; his wife and three children remained in Yugoslavia. He was employed at Sylvanite since January 1929, a period of 12 years.

Timberman Mirich was retimbering a drift with partner A. Dukacz on January 31. They were removing a 10x10 B.C. fir support. Dukacz

*Headstone photograph of Dane Mirich.*
*Photographer: Author*

commenced to drive out the head block by striking it with a hammer but it only moved slightly. Mirich told him to step aside as he would give it a try. Dukacz told him to be careful and reminded him that the hanging wall was in poor shape and hadn't been previously sounded. Mirich replied that it would be fine because there was a stull seven feet above the head block. In the process of knocking out the block, a slab of rock came off the hanging wall, striking Mirich and knocking him to the floor. The impact smashed his hat and lamp and broke the handle of his hammer. After removing the 60-pound slab from Mirich, Dukacz moved him to a safer place and went to the shaft to call for help. C. Booth and M. Brajkovic responded to his call and came into the drift. Dukacz was in an extremely nervous condition and did not return to the workplace. Soon, Dr. T. F. Kelly arrived at the scene and attended to the injured miner. Mirich was taken to the mine first-aid room and sent to hospital.

The next day, an operation was performed to repair his foot tendons. After surgery, Mirich was taken to the x-ray room and, while preparations were being made, he stopped breathing. Stimulants and oxygen were given; artificial respiration was conducted but he died.

An autopsy performed by Dr. G. A. Cowie revealed fractures of the right leg, ankle, foot and right arm. There was a puncture wound in the right hip and the right knee and numerous contusions and lacerations over the body. Death was due mainly to a ruptured liver and internal hemorrhage.

Interment occurred at the Kirkland Lake cemetery. The headstone has the following inscription: OVIE POCIVA U MIRU BOZJEM DANE MIRICH – RODJEN 1901 U MESTRA KRBAVICA, LIKA, YUGOSLAVIA – PROKOPAN 3 FEMBRUARA 1941 – II GLAN ORUSTVA BROJ. Translation: Here lies in blessed peace Dane Mirich – Born 1901 in the town of Krbavica, Lika, Yugoslavia – Buried 3 February 1941.

### March 1, 1941

## Morris Kirkland Mine

## Simon M. Joseph Thibault

### (1916 - 1941)

Newlyweds Mr. and Mrs. Simon Thibault lived at 26 Dixon Avenue. Twenty-four-year-old Thibault was employed as a cagetender at Morris Kirkland from September 1940 and had previous experience at this position.

On March 1, he loaded drill steels onto the cage at the collar of the shaft. The large seven- and nine-foot pieces were chained and secured but the shorter ones were left loose and unchained. Shift Boss M. McEwen and Thibault descended to the 250 and 750 levels where some steels were unloaded, and then to the 1,125 where McEwen got off and went into the mine workings. Thibault unloaded some steels and then descended to the 1,500 where he unloaded steel for a machine there. The steels for the 1,375 still remained on the cage. Jack Perry got on at this time and noticed that the steels were not chained. The cage was ascending towards the 1,375 when suddenly some long steels slid along the cage and ended up projecting over the door. Some of them caught onto the timbers lining the shaft. The cage travelled a short distance before a steel jammed between a shaft timber and the cage floor. By this time, Thibault had been struck by some of the flying steels, was thrown out of the cage and fell to the chain chairs on the 1,500 level.

Perry miraculously escaped uninjured and subsequently climbed down to the 1,500. Rescue workers removed Thibault's body from the anchorage bolts but he had a serious skull fracture and was likely killed instantly. An inquest was held before Coroner R. W. McBain. The jury consisting of James Robb, Harman Lehrman, Mandel Gurevitch, William Haviland and Angus McDougall returned the following verdict: "We find that Simon Joseph Thibault came to his death due to a compound fracture of the skull caused by being thrown from a cage by dislodged steel and falling down the shaft, being caught by the chain hooks at the 1,500-foot level. We recommend that provisions of the Mining Act regarding handling of steel be more rigidly adhered to."

Towards the end of 1941, operations at Morris Kirkland ceased; the machinery and equipment were sold and the mine closed.

### March 29, 1941
## Brock Mine
# William Harry Lalonde
### (1920 - 1941)

It was 1941 and the world was embroiled in a world war. Twenty-year-old William Lalonde expected a call for military service but the call hadn't yet come. So the eldest son of Peter Lalonde of Larder Lake got a job at Brock Mines Limited, a small new mine incorporated in 1938 and located near Dobie.

A shaft was started in 1940 and, by 1941, reached a depth of 630 feet with levels at 200, 325, 450 and 575 feet with crosscuts on the 325 and 575 levels. On March 29, Lalonde and a helper worked with three others on the 575. Mine Manager J. G. Tatham and J. Starevih were loading holes for blasting and L. Hansman and Lalonde were loading other holes nearby. The blasting signal was given and answered. Tatham and Hansman lit their fuses simultaneously. When Tatham finished lighting , he went to see Hansman and found that he still had several holes to light. Tatham helped light some holes and then told everyone to go to the cage. Hansman delayed for an instant to light another fuse when suddenly a shot fired. He was thrown against the wall and injured his head. He picked himself up and in the darkness staggered to find the cage. He managed to find the manway instead and he climbed to the first platform seeking safety.

Tatham and Starevih were about eight feet from the cage when the shot fired. They were thrown down to the ground but recovered and crawled to the cage. On hearing Hansman calling from the manway that he was safe, they rang the cage up. It took 75 seconds to reach surface and during this time they heard five or six more shots. The cage was sent down to pick up Hansman and Lalonde but neither was located.

Hansman finally appeared at surface after climbing the manways. Where was the young inexperienced Lalonde? Men later found his dismembered body at the blasting site.

The funeral mass for the single 20-year-old was held at Holy Name Church and interment occurred in the local cemetery. The gravesite is presently unmarked.

At the inquest held before Coroner R. W. McBain, the jury returned the following verdict: "We find that W. H. Lalonde came to his death by a premature blast of undetermined origin."

The Brock Mine produced no gold ore and, in 1941, it closed. The property was purchased by Upper Canada Mine in 1946.

### April 14, 1941
## Lake Shore Mine
# Leo Wilmot 'Busher' Murphy
### (1920 - 1941)

Twenty-one-year-old Busher Murphy, born in Haileybury in 1920, was adopted by Mr. and Mrs. John McCauley and raised in Kirkland Lake. As a boy, he was a Northern News seller. He had a brother, Everett, who also lived in town and a sister, Helen, who lived in Timmins. Busher married the former Ann Olsen. The couple lived at 71 McCamus Avenue.

Murphy was killed instantly at 7:40 a.m. April 14, when the powder magazine on the 3,700 level at Lake Shore exploded. Many theories were advanced to account for this explosion. Several have some supporting evidence but none satisfactorily explain the cause. The fatality was the first of its kind in this mining camp.

At Lake Shore, there is a powder magazine at each level. On the day of the accident, several men got off the cage at the 3,700 station at the beginning of the shift. Murphy, who was responsible for checking the powder inventory on this shift, went directly to the magazine. G. Varichuk, J. Toth and N. Motrich passed him on their way to their work site and later testified that they saw him in the magazine, sitting at his desk. Motrich, the last man to see him, was already 800 feet west of the magazine when the explosion occurred. Varichuk, who was entering his manway, was slightly injured by the blast. Another miner, Steve Bodnar, who was almost at his work place, was also slightly injured.

Immediately after the explosion, a rescue crew was organized and Murphy's body was found on the tracks of the drift. The body was partially dismembered and completely disemboweled. All bones appeared to be broken and there were severe head mutilations and fractures. At a point opposite the magazine, on the wall of the drift, one could see where his body had struck. Other than the complete destruction of the magazine, there was very little material damage done elsewhere.

At the inquest held before Chief Coroner J. S. McCullough, the jury returned the following verdict: "Leo Wilmot Murphy met his death on April 14, 1941 and we the jury believe from the evidence given, that the cause of death was the result of an explosion, cause of said explosion we find was an accident."

The funeral mass was held at Holy Name Church and interment occurred in the local cemetery. Presently, the grave is unmarked.

*William Luoma. Photographer: Unknown*

### April 30, 1941

## Bidgood Mine

## William Luoma (1901 - 1941)

Born in Finland in 1901, William Luoma immigrated to Canada, arriving in Kirkland Lake to work in the gold mines. He worked at Wright-Hargreaves, Teck-Hughes and Macassa before becoming employed at Bidgood in June 1935. At the time of his death, he lived on Federal Street with his wife, Maija, and their one son, Kalevi (Bill).

The 11 p.m. to 7 a.m. shift, consisting of Leader Bert Fallis, William Luoma, Neil Armstrong and Hugh O'Malley, proceeded into a raise. Armstrong and O'Malley were sent down to the level for timber to construct a drilling platform. Using the manway, Fallis and Luoma climbed the raise by means of ladders and platforms, a distance of 48 feet. At the top of the raise was a bulkhead, damaged by the previous shift while blasting. Fallis and Luoma examined it. Considering the damage slight, they entered by means of a crawl-through, intending to repair the damage later. On entering, they found about 18 inches of muck on the bulkhead including a slab of rock of about 1,200 pounds. The two, standing on the bulkhead, began to clear the muck. Suddenly, it gave way. Fallis was able to grab a stull and hang on but Luoma fell 36 feet to the bottom platform. Fallis climbed down and found him unconscious. He notified the surface. Then he, Armstrong and O'Malley removed Luoma from the raise and secured a stretcher and blankets.

Dr. Contway was called in from Kirkland Lake and went underground, finding Luoma still unconscious with little pulse and severe head injuries. He administered stimulants and had Luoma moved to surface and by ambulance to Kirkland Lake where he hoped to operate. Luoma, however, died on the way to hospital.

Surviving Luoma were his wife and son, and brother Jack who later died in a mining accident at Yellowknife in 1953.

Interment occurred in Kirkland Lake cemetery. The inscription on the headstone over his grave reads: WILLIAM LUOMA – b. JULY 25, 1901 d. APRIL 30, 1941 – RAKKAUDELLA MUISTELLEE/KALEVI JA MAIJA. Translation: Lovingly Remembered By Bill and Maija.

William Luoma's son Bill presently (1999) lives in Mississauga and comes north on a regular basis. He was at one time a miner at Teck-Hughes. He was present at the underground rescue of Maxwell Parker and Jean Paul Morissette in 1957 and he gave aid on surface to the dying Morissette.

### May 3, 1941

## Macassa Mine

## Leonel Roy (1919 - 1941)

Twenty-two-year-old Leo Roy, single and a worker at Macassa, was killed on the 1,625 level on May 3. This was the first fatal accident at this mine since the death of Thomas Cuthbertson in 1934.

On the night shift, Roy and his partner, Dyer, were scaling and mucking in a stope. The shift boss, Roy Sipprell*, visited them at 10:45 and noticed that all was well. After their midnight lunch, the two men continued to scale and muck. At 1:45 a.m., without warning, 12 tons of rock fell from the wall of the stope. Roy was buried and his partner was not injured. Dyer went up immediately to the next level to get help. He found two diamond drillers and ran to the shaft to phone for more help, a stretcher and a doctor.

Shortly after his return to the scene, Night Foreman C. Green arrived with five more men. Dr. Contway arrived at 2:10 and administered stimulants to Roy who wasn't extricated until 3:45. He was being moved to surface when he died. He had suffered lacerations of the scalp, numerous contusions, extreme shock and fractured ribs, vertebrae, skull and ankle. After the accident, Roy's distraught father made the following statement to a newspaper reporter concerning his young son: "This was his first mine job. He had worked here for only eight months." (According to the accident report, Roy began his employment at Macassa on September 12, 1940.)

He was survived by his parents, Louis and Valleda Roy. He had three brothers: Eugene, Joe and Robert, and three sisters: Annie, Marie and Mrs. P. Desgagné. All of his siblings resided in Kirkland Lake.

Interment took place in the local cemetery.

** Roy Sipprell died in a 1966 accident at the same mine.*

# Two Miners Die in Town on the Same Day in Separate Accidents

At about 10:00 in the morning on June 21, 1941, John Wesley Coulson was killed accidentally at Teck-Hughes. At about 8:00 that night, a miner at Lake Shore, Christopher R. Conroy, was killed underground.

### June 21,1941

## Teck-Hughes Mine

## John Wesley Coulson (1905 - 1941)

Canadian-born John Coulson came to town in 1929 and got a job at Teck-Hughes that year.

In 1941, he was living at 24 Queen Street with his wife, the former Adeline Perry of Angus, Ontario, and three children: Bernice (15), Elmer (13) and Evelyn (8).

On the morning of June 21, Coulson and partner Albert Saunders were putting in additional posts in a stope on the 41st level. Mine Captain Jas Ainslee and Shift Boss George Gill Jr.* were in the stope before 8:00 a.m. and gave instructions to the two men to proceed with the timbering. The back and wall was solid and there was no sound of the wall 'working'. At around 10:00, Coulson and Saunders were seen at the 41st station getting some timber and returning to their work place. The two put in a post with Saunders holding it and Coulson pushing the head block on top of it when suddenly there was a loud noise and a section of the stope fell on them. Saunders was knocked down parallel to the post and pinned under a piece of rock weighing about two tons; however, the felled post sustained most of the weight. While trapped, he could hear more rock falling in the stope and in the manway. He felt Coulson next to him, noticing that there was a lot of blood. He spoke to Coulson, who answered, "I am done for." There was no further conversation.

At noon, Wilfred Lively and other trammers were having lunch at the 41st station. Realizing Coulson and Saunders should have joined them, they went towards the stope to investigate. Going up the manway, Lively noticed muck and dislodged ladders. He called into the stope and Saunders answered. Lively immediately returned to the station and notified Gill and Ainslee who went with others into the stope by means of another entrance. Dr. Messenger was called.

Not seeing Coulson at first, the men rescued Saunders and freed him by 1:15. Saunders then showed them to Coulson. He was covered with about 18 inches of rock and later pronounced dead by the doctor. Death was caused by the pressure of the muck on his chest that prevented breathing. He also suffered internal injuries. There was a severe laceration on the right knee with consequent loss of blood. Saunders was taken to hospital where x-rays showed compression fractures of two lumbar vertebrae and a fractured rib.

It was later determined that a rockburst had occurred over the work place where the two were installing timber. It was a small local burst that was not detected anywhere else in the mine. The rock was still working during rescue operations.

The funeral service with Reverend Wilfred L. Cullis officiating was held at the All People's United Church in Barrie, Ontario. Interment took place at a cemetery there. Coulson was survived by his immediate family in Kirkland Lake and by his mother and brothers in Angus.

** George Gill's brother would die in an accident at Teck-Hughes on January 13, 1948.*

## June 21, 1941
## Lake Shore Mine

# Christopher R. Conroy

## (1915 - 1941)

Twenty-five and married, Christopher Conroy came to town from Cobalt. He was the fifth of six children born to George Conroy and Rose May Buckland. Both of his parents and a brother Howard and sister Abigail were born in Buckingham, Quebec. His other brothers and sisters, William, Margaret and Gladys, were born in Cobalt. While working at Lake Shore, he lived with a brother at 78 A Prospect Avenue. He had been working at the mine for about three weeks prior to the accident. He had no previous mining experience and was classified as a miner's helper.

On the night shift of June 21, Shift Boss William Jones instructed Toivo Huito and Conroy to proceed to the 4,200 level to work in the rill stope. They climbed up the rill floor and started lowering the drilling gear with a rope, Huito above and Conroy below. As they were doing this, Huito noticed the boards of the floor rising and realized that wet fill was starting to run down the rill stope. He yelled down to his partner but he probably wasn't heard due to the noise of the running fill. Twelve tons of wet sandy fill flooded the stope. Huito escaped by climbing up the manway to the 4,000 level where he met many miners who responded to his calls for help. They rushed down to the stope and found Conroy lying on his back with his feet in the air and his head and body covered with fill. He was quickly freed and artificial respiration was applied. Shift Boss Jones had called mine officials and a doctor. Dr. C. S. Harris arrived, cleared sand from Conroy's mouth, injected stimulants and continued artificial respiration but he noted no pulse. At no time were there any signs of life and, after two hours of artificial respiration the doctor pronounced the miner dead.

It's likely that at the time of the fill run, Conroy was carrying the drill along the last two sets of timbers. In this position and location he was caught by the run and was unable to clear himself. If he had been standing at the bottom of the stope when the run occurred, he possibly could have jumped clear when he heard the fill coming.

*Christopher Conroy. Photographer: Unknown*

At the inquest held before Coroner R. W. McBain, the jury returned a verdict as follows: "We the jury, after giving careful consideration to all the evidence, find that Christopher Conroy came to his death by suffocation from being buried in wet sand. We feel that there should be no blame attached to anyone in the cause of this accidental death."

The funeral was held at Cobalt and interment occurred at the Mount Pleasant Cemetery in Haileybury. The inscription on the stone over his grave reads: IN MEMORY OF OUR DEAR SON CHRISTOPHER R. CONROY – BORN NOV. 9, 1915 – DIED JUNE 21, 1941.

Some details concerning the family and the photograph were supplied by Isabel Murphy of Kirkland Lake and Jim Murphy and Angela Hopper, both of Cobalt.

### July 9, 1941
## Kirkland Lake Gold Mine

## Samuel Napoleon Vachon
### (1899 – 1941)

Sam Vachon, a miner with 13 years of employment at Kirkland Lake Gold, died on July 13 as a result of injuries sustained in a rockfall near the 4,450 level.

He was 42 and a widower. His wife, the former Rose Gougeon, predeceased him on November 25, 1928. At 26, she had died of pneumonia following childbirth complications.

Vachon, a machineman, was working in a stope with helper William Lockhart. "I heard some rock fall," stated Lockhart, "I turned around and I could not see Sam. I turned off the machine and I looked down the raise. I saw him about 20 feet down. I pulled him out and called for a stretcher."

Lockhart pulled Vachon out of the raise and placed him in the safety of a manway; he then left for help. With the aid of two others, A. Dumais carried Vachon a distance of 90 feet down the ladders to the drift, at which point he was placed in a stretcher and taken to surface.

At the Kirkland District Hospital, he was diagnosed with numerous contusions and abrasions, fractured vertebrae, and multiple fractures of the hip. Dr. C. S. Harris attributed Vachon's death to thrombosis of a mesenteric artery that subsequently caused peritonitis. These combined injuries undoubtedly contributed to his death.

William Smith, a shift boss at the mine, stated, "Sam Vachon was one of the most careful men in the mine. He was one of the best workers."

Surviving Vachon were his four children: Leo, Manuel, Valois and Thérèse. Also surviving were his father, a brother and a sister.

Funeral arrangements were handled by Symington's Funeral Home on Prospect Avenue. The funeral mass was held at Assumption Church and interment occurred in the local cemetery.

*Sam Vachon. Photographer: Unknown*

Personal information and the photograph concerning this account were supplied by son Valois and grandson, Denis Vachon and his wife, Jennifer who presently (1999) live in Larder Lake. Son Manuel lives in Powassan, Ontario and daughter Thérèse lives in Sudbury. Son Leo is deceased.

### August 5, 1941
## Morris Kirkland Mine
## John Baptiste Landry
### (1923 - 1941)

Funeral and burial services for John Baptiste Landry were held at Rockmore, Quebec; the Reverend Father Couture officiated. Landry, single and just 18, was the son of Mr. and Mrs. Adelard Landry of Rockmore. He was survived by his parents and 11 other members of his family.

He was employed at Morris Kirkland for only a month, being hired as a mucker on July 4.

On August 5, he was working in a stope with his partner, Dominic Pecaski. Shift Boss M. A. McEwen visited while they were mucking. Shortly after he left, about two tons of loose fell into the stope. A chunk weighing about 200 pounds hit Landry, killing him almost instantly. He suffered a fractured skull, partially severed arm, broken leg and internal injuries. Pecaski was not injured. Dr. Messenger from Kirkland Lake arrived and officially pronounced Landry dead.

The body was taken from the mine to Symington's Funeral Home; it was sent from there the next day to Noranda, and from there to Rockmore.

The Morris Kirkland Mine closed later in the year.

### September 23, 1941
## Macassa Mine
## John Tomljenovich (1900 - 1941)

Forty-one-year-old John (Ivan) Tomljenovich was born in Yugoslavia on March 24, 1900. He came to Canada without his wife and family. He was employed continuously at Macassa from April 13, 1934.

A 45-gallon gasoline drum, used to store gas for an auxiliary engine that was used during electric power shutdowns, was located outside the building housing the engine. A leak in the drum was suspected. On September 23, Master Mechanic Karl Helmer instructed Pipefitter Tomljenovich to remove the ground around the drum, to fill the drum to the top with gasoline and to note any possible leaks. Tomljenovich dug away the ground, disconnected two fittings and hand pumped the remaining gasoline into another drum and took the suspect drum into the machine shop.

Two mechanics in the shop at that time, Jas Fowler and Henry Klawitter, saw him bring in the drum; they assumed he was going to clean it out. Tomljenovich and a young shop helper, Mendel Thrasher, made air pressure connections to the drum and then Tomljenovich instructed Thrasher to turn on the air valve slightly. He did so and returned to the side of Tomljenovich, who was listening for any escaping air from the drum. Thrasher saw the sides of the drum beginning to bulge but he said nothing. About two minutes later, Tomljenovich told him to turn on more air. Thrasher was walking towards the valve when the explosion occurred. Seeing that Tomljenovich was knocked down and the drum broken, he turned off the valve in the air

*Headstone photograph of John Tomljenovich. Photographer: Author.*

line. Tomljenovich apparently had his head over the drum listening for air leakage when the head of the drum blew out, striking his face. Dr. M. J. Thompson was called. Tomljenovich did not regain consciousness and died a few minutes after the doctor's arrival. He had received multiple lacerations and fractures of the face and skull. A fracture at the base of the skull was the probable cause of death. Helmer later checked out the recording gauge in the foreman's shop, finding out that the pressure in the air line at the time of the accident was 87 pounds per square inch.

Tomljenovich was buried in the Kirkland Lake cemetery. The inscription on the headstone is in Yugoslavian and is translated as follows: Here rests in blessed peace, our brother Ivan Tomljenovich, born 24 March, year 1900 in Gospic, Yugoslavia – Brother died 23rd September, year 1941, leaving in the old country wife Majkukatui, son Franuiluku and many friends who praise you brother.

### November 12, 1941

## Toburn Mine

### Michael Malenoski (1908 - 1941)

Thirty-three-year-old Michael Malenoski was born in 1908 in the Ukrainian Province of Bukovina; he was married with one child. He was employed at Toburn continuously from April 30, 1931, a period of almost 11 years. During his residence in this country, he became a naturalized Canadian. At the time of the accident, he was employed as a driller.

A stope was started in a section of a drift on the 13th level and the first cut was drilled and blasted. On the morning of November 12, when the day shift began at 7:00 a.m., Malenoski was told by Shift Boss A. Moreno to proceed to the drift, scale down loose and start drilling beyond the last section blasted. He went to the work place, set up his stoper drill and prepared to make the area suitable for his drilling operations.

In the meantime, Armand Belmore, the cagetender, came into this area to ascertain the amount of muck in the drift as this was to be loaded with a mucking machine and hoisted on the day's shift. He noticed a considerable amount of loose where the blasting was taking place and told Guido Trento, a trammer, not to go under this section as the place was too dangerous.

Mucking Machine Operator Alex Ellis also came in to inspect the muck pile in this area. He noted the considerable amount of loose and mentioned this to Malenoski before he went to another level to get a mucking machine.

Joseph Oreck, a trammer, was sent to the drift. He procured a water hose and began to wet down the muck pile to lay the dust. He also noticed the large amount of loose in this area. He was joined by Pipefitter Frank Modola who said he would continue with the wash down if Oreck would clean up the track ahead of the muck pile so the mucking machine could get at it.

Malenoski had his stoper set up at the face. He told Modola he would complete the washing job if Modola would get the pipe lines fixed so the drilling could commence.

Oreck cleaned up the track and picked up a scaling bar to bring down some of the loose. Malenoski was in the area still washing down the muck pile. Oreck touched the back with the bar, causing a 15-foot piece of loose to fall. The large piece fell from a point directly over Malenoski and not more than one foot above his head.

Oreck and Modola immediately went for help and notified mine officials of the accident. When the slab of rock was lifted off, the rescuers found

*Headstone photograph of Michael Malenoski. Photographer: Author*

Malenoski doubled up under it. Dr. C. S. Harris was present at the site and pronounced the miner dead. He had suffered a crushed chest, a broken back and a fractured leg. Death was probably instantaneous.

A mine spokesman, M. W. Hotchkin, expressed regret at losing the miner. "Mike Malenoski was one of the original workers here. He worked for Toburn for eleven years and I am sorry this had to happen."

At the inquest, it was learned that Malenoski, widely considered a careful miner, did not do any scaling in his work area this day. At least two tons of rock were ready to fall at the slightest touch and 10 minutes of scaling would have made the working place quite safe. Cagetender Belmore stated that Malenoski told him that he was going to hurry and get his drilling finished early so he could 'talk.'

Interment occurred in the Kirkland Lake cemetery. The inscription on the headstone is in the Ukrainian language and its translation reads as follows: Here rests forever Michael Malenoski of the family from the village Sinivchi. Born 1908. Killed instantly on the 12th day of November 1941.

** Author's Note: The miners' strike vote took place on November 8; 63 per cent of the men voted for action. This fatal accident occurred on November 12 and the strike began November 18. Mike Malenoski was an 11-year employee at Toburn but strangely on this day he did not scale his workplace. He was in a hurry to get his work done so he could "talk." Did he have something else on his mind that affected his work?*

### 1942

- During the year, seven men are killed in local mines
- Kirkland Lake Miners' Strike ends on February 12 with approximately 1,000 fewer miners in the camp
- Kerr Addison loses about 50 men to the armed forces
- Mining companies make a contribution to the country by using their highly skilled mechanics and machinists to manufacture items for the war effort
- Gold mines in Canada are declared a non-essential war industry by the government; a decline in local gold mining begins
- Fernland Mine closes
- Kirkland Lake-area men are injured and killed overseas

### January 3, 1942

## Omega Mine

### Peter Furlan (1904 - 1942)

Born in Croatia in 1904, 38-year-old Peter Furlan was married with no children. He was employed as a drill runner at the Larder Lake Omega Mine with only seven months of experience working underground.

On January 3, he was doing routine scaling work in a stope on the 400 level when suddenly a 20-ton rock struck and hurled him a distance of 30 feet. Two others were working in the same area at the time but did not witness the actual accident as they were busy working with their machines.

Injured at 11:50 in the morning and taken to the Kirkland District Hospital, Furlan went into surgery to repair multiple injuries including a dislocated hip, a fractured pelvis, a fractured leg, ten fractured ribs and fractured vertebrae. About five hours later, he died. Death was attributed to multiple fractures and shock.

The funeral mass was held at Holy Name Church. Interment occurred in the local cemetery and the site is presently unmarked.

Although a huge miners' strike was taking place in the Kirkland Lake gold camp at the time, the Omega Mine was not directly involved in the strike. (Kirkland Lake Miner's Strike: November 18, 1941 – February 12, 1942.)

### March 26, 1942

## Toburn Mine

### James Dubreuil (1904 - 1942)

James Dubreuil lived at 8 Green Street with his wife and eight children. He was a 38-year-old Canadian who hired on at Toburn four days before the Kirkland Lake Miners' strike ended. Six weeks later, he was killed at the mine.

On March 26, Timbermen John Moras and Dubreuil were constructing a staging in a stope located between the 400 and 500 levels. Dubreuil sent Moras up to the 400 for a measuring stick. Moras was at the uppermost rung of the ladder when a piece of loose became dislodged and crashed down on the staging below. The falling rock demolished a portion of the staging and caused Dubreuil to lose his balance. He had a safety rope in his hand but did not have it attached to his body. He fell to the 500 level and died instantly.

An inquest was held and the jury made the following statement: "We the jury, empanelled to enquire into the death of James Dubreuil, find that he came to his death at 12:50 a.m., March 26, 1942, at the Toburn Mine by accidental death caused by a fall as a result of being knocked off a staging by the falling of loose rock. This jury recommends that each man should be supplied with a rope or a safety belt while working in dangerous places."

Interment occurred in the Kirkland Lake cemetery. The gravesite is presently unmarked.

### May 27, 1942
## Kirkland Lake Gold Mine

### Stive Rubanets (1906 - 1942)

The coroner's jury returned a verdict stating that Stive Rubanets' death was accidental; it also noted that better instructions and training should be provided to inexperienced workers.

Rubanets had only been working at Kirkland Lake Gold for four days. His partner underground, Thomas Bahun, also had only four days experience in the mine. The third man working with them, Jason Metherall, was working his first shift underground. Apparently, Rubanets had stated that he had worked at the Frood Mine in Sudbury for 18 months, but a witness at the hearing stated, "he was working in a local grocery store only a few days ago."

On the day of the accident, the three men were tramming muck out of a drift on the 375 level. Rubanets stepped under a chute to let a loaded 3,300-pound ore car pass by. He was jammed between the chute post and the moving car and suffered a fractured skull. He died within a few minutes.

Born in the Ukrainian Province of Bukovina, Rubanets was married and lived with his wife and child at 49A Taylor Avenue. Employed as a mucker and trammer, he was 36 at the time of his death.

He was buried in the Kirkland Lake cemetery. His headstone inscription in Ukrainian translates as follows: Here rests Stive Rubanets – Born 1906 – Suddenly killed at Kirkland Lake Gold Mine on May 27, 1942 – A man with a kind heart.

### May 30, 1942
## Lake Shore Mine

## Rockburst at Lake Shore Kills Two Miners

On May 30, a rockburst of medium intensity occurred in a stope at the 4,000 level of Lake Shore. This was the first serious burst to occur in the camp in ten months. There were four men working in the stope at the time: Wilfred Chartrand, William Smarz, C. Rox and Victor Young.

Rox was thrown clear and escaped with only a cut chin. Young, a healthy and strong 24-year-old, was buried and trapped for a period of one-and-a-half hours but was rescued. He survived, no doubt due to his physical conditioning. Chartrand and Smarz were buried and killed instantly.

### Wilfred C. Chartrand (1922 - 1942)

Twenty and single, Wilfred Chartrand was a man of French-Canadian descent living on Young Avenue. He was previously employed at the White-Guyatt Mine and the Ross Mine in Holtyre, Ontario. He was hired on at Lake Shore on May 29, one day before the burst.

### William Smarz (1900 - 1942)

Forty-two-year-old William Smarz of 24 Woods Street was married; his wife and three children lived in Poland. The Polish man was employed intermittently in Canadian mines from 1927 and was most recently employed in the mines at Noranda, Quebec. His first day of employment at Lake Shore was May 28, two days before the burst. Interment was conducted in the Kirkland Lake cemetery; the gravesite is unmarked.

## July 1, 1942
## Chesterville Mine

### Fernand Trudeau (1922 - 1942)

Single and 20 years old, Fernand Trudeau was instantly killed at Chesterville. The Canadian worker was hired on May 18, 1942 as a mucker and trammer; he'd had previous experience as an underground diamond driller.

On July 1, a tramming shift was working underground under the supervision of Trammer Boss R. Webber. The fifth level crew included W. Westbrook, Donat Lessard and Trudeau.

At the time of the accident, a car of muck was dumped through a pocket into the skip. Levi Payne, the cagetender's helper, held onto the shaft gate while signalling the skip up. Using a backwards movement, he scraped his foot across the plate on which the pocket gate was hinged, to wipe off muck; this apparently prevented the door from closing completely.

Cagetender John Stos and Trudeau were pushing the empty ore car away from the shaft while the skip was up. Lessard and Stos saw Trudeau fall through the improperly closed pocket gate into the shaft. After falling a distance of 535 feet, he was found dead on a bulkhead below the ninth level. He sustained fractures to an arm, both legs, pelvis and neck; he'd also suffered lacerations and internal injuries to his lower abdomen.

Interment took place in the Kirkland Lake cemetery. The gravesite is unmarked.

## October 31, 1942
## Kirkland Lake Gold Mine

### Fred (Theodorus) Melnyk (1903 - 1942)

Polish and married with three children, 39-year-old Fred Melnyk died on October 31 while working near the 4,450 level of Kirkland Lake Gold.

He was standing on a plank flooring in a stope and receiving timber that was being hoisted upwards into the stope through an ore chute. On the level, 49 feet below, his partner, N. Mihailuk, was sending up the timber. Melnyk was in the act of removing the hoisting cable from a post when a portion of loose from the stope fell on him and knocked him down. A timber came to rest diagonally across his chest and abdomen and a rock weighing about 150 pounds came to rest on this timber. Lloyd Bates, Joe Duchene and George McCann were working nearby and came to his aid and freed him immediately. He died a few minutes later, before he was removed from the stope.

Death was caused by the blow to his chest that ruptured internal organs. His left hip and a bone in his left hand were also fractured.

Interment occurred in the Kirkland Lake cemetery; the grave is unmarked.

## 1943

- Antti Latvala is killed at Toburn; Vital Vincent is killed at Kerr Addison; a fall of loose kills Adolphe David at Wright-Hargreaves; a run of ore kills John Parrott at Kerr Addison; a scaling accident kills Fred Lafond at Lake Shore; a fall of ground kills Juho Huhtala at Bidgood; a tramming accident kills George Ojala at Wright-Hargreaves; and a fall of loose kills William Demko at Chesterville
- Sir Harry Oakes is murdered in his home in the Bahamas
- Kirkland Lake's population drops to 15,800
- Women in the Kirkland Lake area knit socks and sew clothing for soldiers overseas
- Kirkland Lake area men continue to die at war
- Butter rationing is acute in Kirkland Lake homes; it is being saved for the sandwiches placed in the lunch boxes of the miner who requires the extra nutrition

## March 24, 1943
## Toburn Mine

### Antti Latvala (1898 - 1943)

Born in Finland in 1898, Antti Latvala had two brothers and a sister. In 1919, he married the former Aino Taipalus. Within a few years, the family grew to five; the three children were named Oliver, Ina and Anne.

In 1922, at the age of 24, Latvala left his family in Finland and came to Canada, seeking employment. While in Northern Ontario, he worked in various mines including the Tough-Oakes. In 1930, he was re-united with his wife and children when they came to Canada to live in Cobalt. Two years later, the family moved to Kirkland Lake where Latvala worked in the gold mines of the area. He also spent some time working in Labrador. He was a hardworking man whose main objective in life was to support his family.

In 1943, he was a steady employee at Toburn. Aino and Antti had decided that he would quit mining and move to a farm, already purchased in the Sutton Bay area near New Liskeard. Their plan was to move there within weeks.

On March 24 at 8:30 p.m., Chief Timberman Latvala was caught by a falling rock while inspecting a chute. The rock, held up in the chute, became dislodged and fell, pinning him to the side of the chute. Dr. T. F. Kelly was called and assistance was speedily procured but he was killed instantly.

Funeral arrangements were organized by the Symington Funeral Home where the service, officiated by Reverend Hjelt, was held. Interment took place in the local cemetery. The inscription on the headstone reads: IN LOVING MEMORY OF ANTTI J. LATVALA 1898 - 1943 – LEPAA RAUHASSA-AINOS 1896 - 1972.

Latvala's daughter, Mrs. Ina Pulkinen, presently (1998) lives in Kirkland Lake. She provided much of the information for this memorial to her father. His sister, Tyyne Janisoja, presently lives in Finland. Ina communicates with her on a regular basis.

*Antti Latvala. Photographer: Unknown*

## May 21, 1943
## Kerr Addison Mine

### Vital Joseph Vincent
### (1919 - 1943)

Vital Joseph Vincent was 24, Canadian and married with one child. He was killed instantly on May 21 on the 300 level at Kerr Addison. He had been employed underground at the mine since January 28, 1942. At the time of his death, he was classified as a chute puller.

The accident occurred at the No. 10 chute below the 309 stope. This stope was mined through to surface and was being pulled. The chutes drew from grizzlies which in turn drew from the stope. In this instance, No. 10 chute was pulling ore from No. 10 grizzly that was getting ore from the stope.

At about 5:30 a.m. on the graveyard shift of May 21, Mucker Boss John Beverly Parrott* visited No. 10 grizzly and found J. W. Couture and A. Matthieu washing down the raise, trying to bring down the muck from the stope. As the muck remained blocked and did not come down, the men decided to blast it. A bomb of 10 sticks was prepared, tied with cord, fixed to a 16-foot blasting pole and attached to a five-foot fuse.

Couture called down the chute-raise to Vincent twice, telling him that they were setting up a pole blast. Vincent replied, "Okay, go ahead and blast." The chute was empty and it was Vincent's duty to guard it. The fuse was lit. Parrott was walking to another area of the mine when the shot went off. Since the shot sounded unexpectedly heavy, he went back to the No. 10 grizzly where he found Couture and Matthieu awaiting the clearing of the smoke. They said the muck still hadn't come down. Parrott then went down to the No. 10 chute where he found Vincent lying face down. Parrott at once called up the chute-raise to Matthieu and Couture not to do any more blasting and said Vincent was dead. Parrott summoned aid at the shaft and waited there until Mine Superintendent Ken Dewar and the doctor arrived.

What had happened? When the bomb of 10 sticks of powder was placed in the raise above the grizzly and lit, it became dislodged and fell into the chute-raise where it exploded. Vincent, who was cleaning up under the chute, was told to guard the chute but he had not moved away from it. The blast in the chute-raise near the chute created a strong air concussion and particles of flying rock.

Vincent suffered the following injuries: the back of the skull at the base of his brain was blown off; all the vertebrae in the neck were broken; there was evidence of a long laceration on the left side of the neck; there was powdered rock in the brain, a small puncture wound on the right forearm and several abrasions on the right shoulder. Death would have been instantaneous.

Interment occurred in the Kirkland Lake cemetery. The gravesite is unmarked.

** John Beverly Parrott would die in a mining accident about three weeks later in this area of the Kerr Addison mine (see June 11, 1943).*

## June 8, 1943
## Wright-Hargreaves Mine

### Adolphe David (1890 - 1943)

Fifty-three-year-old Adolphe David died at Wright-Hargreaves where he had been employed since 1930. A timberman, he was killed instantly June 8 by a fall of loose in a stope. His partner, J. A. Comptois, escaped injury.

When the accident occurred, Comptois obtained the assistance of two miners. By the time the doctor arrived, they had succeeded in prying up the block of rock sufficiently to release David, who was dead. The injuries included a broken neck, a broken back, a crushed chest, a crushed pelvis and a fractured left arm.

David, of French Canadian descent, was married with no children. He and his wife, the former Augustine Delay (1911 – 2000), resided in Kirkland Lake on Second Street. A brother of the deceased also lived in town and was employed at Sylvanite.

Funeral services took place at Assumption Church with Reverend Father Jodoin officiating. Interment occurred in the local cemetery.

## June 11, 1943
## Kerr Addison Mine

### John Beverly Parrott
(1911 - 1943)

Thirty-two-year-old and Canadian, John Beverly Parrott was married with four children; he was killed at Kerr Addison on June 11. He was employed there from March 1941 to July 1941 and from March, 1942 to June 11, 1943. His numerous duties at the mine were machineman, timberman, stope leader and grizzly leader. He had been a mucker boss since May 20, 1942.

The 309 stope was a large shrinkage stope that was completed to surface and being pulled from the 300 level. The ore from this stope would pass through raises in the floor, through grizzlies, chute-raises, chutes and into ore cars.

The crew pulling this stope on the afternoon shift of June 11 included Chute Pullers D. Paquette and John Wnuk, Motorman J. MacKenzie, Grizzly Leader Philip Hamel, and Grizzly Men Ernie Rainville and Tom Denner. As Hamel's partner was absent this shift, Parrott took his place.

The No. 3 grizzly was damaged earlier that day. The centre member of the grizzly was completely severed and was allowing some large chunks of rock into the chute-raise. The crew decided a blast was required to hang up the ore above the damaged grizzly so it could be repaired. Hamel set up a pole blast while Parrott went down to the chute area to instruct the motor crew to guard it during the blast.

Arrangements were also being made to blast the No. 7 grizzly after this blast was completed. After the No. 3 blast, Hamel went to the No. 7 grizzly to set up the blast there and Parrott instructed Denner to tell the motor crew to move to and pull ore from the No. 3 chute, then to guard the No. 7 chute during the blast.

Parrott, Hamel and Rainville went to the No. 3 grizzly to wait out the shot in No. 7. The No. 3 grizzly was now full of ore. Parrott was inspecting the raise with his light and standing on the ore at the edge of the grizzly. The men below, instructed to draw ore from the No. 3 chute, did so. Suddenly, the muck on which Parrott was standing dropped four feet. Simultaneously, two large blocks of ore began to slide down the muck pile towards him. Parrott tried to pass across the grizzly, apparently forgetting the centre part of the damaged grizzly was missing. He did not succeed in grabbing the safety chain and fell through the grizzly. As he fell through, he was struck by one of the sliding blocks of ore and impaled on a piece of iron. The other sliding block slid onto the first one.

Death was instantaneous. Parrott suffered the following principal injuries: a crushed heart, punctured lungs, fractured ribs, fractured clavicle, torn muscles in the chest and various abrasions.

A doctor in attendance pronounced Parrott dead and the body was removed to surface.

### July 27, 1943

## Lake Shore Mine

### Fred Joseph Lafond (1911 - 1943)

The French-Canadian family of Mr. and Mrs. Fred Lafond and their five children lived at 139 Duncan Avenue. Lafond, 31, was a mucking machine operator at Lake Shore where he had worked since August 1936. He was killed instantly at the mine on July 27.

He and his helper, J. Grant, came on shift at 7:00 a.m., went down to the 5,325 level and began to make preparations for a shift's work. Grant brought in the mucking machine, one car and a motor while Lafond attempted to scale down some loose from the back of the drift about 20 feet from the face. He was having difficulty removing a piece so Grant helped out but they failed to remove it. Grant put down the scaling bar and began to clean up the track so that he could move the equipment closer to the face. Meanwhile, Lafond was still trying to remove this stubborn piece. Suddenly, an eight-foot wedge-shaped block of ground fell from the wall and back, crushing Lafond. Grant was not injured but saw it happen. He rushed to Lafond's aid but he could not free the body. He went to another part of the mine and contacted some drillers who soon removed the body from under the fall of ground. Lafond had already died of multiple fractures.

### October 18, 1943

## Wright-Hargreaves Mine

### George Ojala (1901 - 1943)

Forty-two-year-old George Ojala and his wife Hilda lived at 32 Rand Avenue. He had been employed at Wright-Hargreaves for several years. On October 18, he and his partner, Curtis Johnson of Swastika, were tramming from a chute and dumping ore into a dump hole. After dumping, they returned but this time the sand fill beneath the track collapsed and dropped, causing them and the equipment to fall a distance of about 20 feet. Ojala was completely covered with debris and it took about an hour to rescue him. By the time they got to him, he was dead. Johnson was also buried but to

*Headstone photograph of George Ojala. Photographer: Author*

a lesser degree; he was rescued earlier. He had no serious injuries but suffered from shock and spent one night in hospital.

An inquiry into the death was held. The jury came in with the following statement: "We the jury, empanelled to enquire into the death of George Ojala, find that he came to his death in the 1,304 substope of the Wright-Hargreaves Mine at about 11:00 a.m. on October 18, 1943 from suffocation caused by being buried by a run of back fill that occurred shortly after 10:00 a.m. through failure of a bulkhead. We are unable to determine the cause of the failure of the bulkhead and find death to be accidental."

He was buried in the Kirkland Lake cemetery. The inscription on the headstone is:

IN LOVING MEMORY OF GEORGE OJALA – BORN JULY 19, 1901 – DIED OCT. 18, 1943 – HIS BELOVED WIFE HILDA OJALA – BORN MARCH 4, 1884 – DIED AUG. 23, 1967.

*Aino and Juho Huhtala. Photographer: Author*

### October 6, 1943

## Bidgood Mine

## John (Juho) Fredrik Huhtala

### (1896 - 1943)

Juho Fredrik Huhtala was born to Fredrik and Susan Huhtala on June 1, 1896 and was raised in the village of Kurikka in the province of Vaasa, Finland. He was a veteran of the First World War.

At the beginning of the Depression, he left Finland in 1929 to find a better living in the new world. He came to Kirkland Lake looking for employment in the gold mines but the Depression had had its negative effects in Canada as well.

At first, he did not get a job in the mines so he worked where he could, doing odd jobs and "trying to make a buck a day." His family – a wife and four children – were still in Finland and he wished to have them in Canada. He borrowed $1,000 from a Finnish woman in town and sent for them.

On November 23, 1937, his wife, Aino (née Laitanaki), and the children – twins Olavi and Oiva (17), Anne (12) and Pentti (10) – arrived in Montreal by passenger liner. After a long train trip north, they were finally and happily reunited.

Huhtala had been working at Bidgood since August 27, 1935. He lived in one of the half-dozen company homes located along the Bidgood Road near Mud Lake. The newly reunited family moved into this home, lived there for part of the year, and then moved to King Kirkland.

At the time of the accident, the 46-year-old Huhtala was one of the older and more experienced employees at the mine (he worked there eight years.) He was working in a stope at the 1,775 level with partner Art Quinn when suddenly, 30 tons of loose came down and instantly killed him. The loose narrowly missed Quinn who escaped with minor injuries.

Huhtala left behind his immediate family; the twins were 23 and the daughters 18 and 16. The funeral service was held at Symington's Funeral Home and interment was in the Kirkland Lake cemetery.

Personal details of this story were given by Huhtala's son, Olavi, who presently (1999) lives in Kirkland Lake. He has many descendants and proudly notes that he has many great-grandchildren.

### October 20, 1943

## Chesterville Mine

## William Demko (1900 - 1943)

Machine Runner William Demko was killed in a stope at Chesterville on October 20. He was Ukrainian, 43 years of age, married with no children, and employed at Chesterville from 1939.

On the day of the accident, he and J. Rouleau were to drill in No. 7C1 stope. They scaled until about 10:30 a.m. at which time Shift Boss Val Pedersen and Safety Inspector H. Tough visited the stope. About 10 minutes later, Demko went to an area of the stope to do more scaling. He sounded the back over his head and began to pry down some broken muck. Rouleau was standing about 10 feet away when suddenly a slab of rock fell from the back onto Demko, crushing and killing him instantly.

His body was released within an hour and was found to have suffered the following injuries: pulmonary hemorrhage, a fractured skull, a jaw fracture and numerous bruises and lacerations. Death was ascribed to pulmonary hemorrhage.

Interment occurred in the Kirkland Lake cemetery. The grave is unmarked.

### 1944

- Three employees at Wright-Hargreaves are accidentally killed; an underground employee is killed at Kerr Addison; a run of sand buries and kills a miner at Lake Shore
- Many local mines reduce production due to a manpower shortage and the government's decision to label the gold mining industry as non-essential to the war effort
- Many mine volunteers supply skilled labour to build the Kirkland Lake swimming pool on Fifth Street
- More Kirkland Lake and area men are killed and wounded overseas

### January 31, 1944

## Wright-Hargreaves Mine

## Joseph Paul Emile Samson

### (1908 – 1944)

Paul Emile Samson of French-Canadian descent came to Kirkland Lake in 1936 and began his employment at Wright-Hargreaves on July 9 of that year. Together with his wife, the former Annette Tremblay, and their five children, he lived at 30 Rand Avenue next door to Mr. and Mrs. George Ojala. Ojala had been killed at Wright-Hargreaves just four months earlier.

For an extended period of time, Samson and his mining partner, Frank Kruzich, were working in No. 3005 stope, repairing the manway, installing a timber slide, scaling and retimbering. During the morning of January 31, they scaled and hoisted timbers until lunch at 11:05. Later, Samson returned to the stope, and Kruzich went to the shaft station to get a fresh lamp. Returning, and about 750 feet from the stope, he felt a rockburst. Entering the stope, he noticed fresh muck on the floor but was unable to see Samson. He left to find the motor crew (Dolan and McConnell), then found J. K. Smith, A. Rychel and Shift Boss C. Anderson. Some of these men had also felt the burst; in fact, Anderson was already looking for its location.

They entered the stope and found Samson almost completely buried in the muck; about seven tons had come down in the burst. Anderson immediately went to the station, called for more aid and then returned. The crew spragged the walls and lagged the back, both of which were in a dangerously broken condition. Mine Superintendent W. Sampson arrived, the body was recovered, and Dr. G. M. Cameron, who was also in attendance underground, pronounced the miner dead.

Samson had suffered the following injuries: a broken neck, a fracture of the left femur and right tibia, and multiple bruises and lacerations. Death was probably instantaneous.

Wilfred Tremblay, Samson's brother-in-law, assisted the family with funeral arrangements. The funeral mass was held at Assumption Church with Reverend Father Archambault officiating. Interment took place in the local cemetery.

### March 29, 1944

## Kerr Addison Mine

## James Edward Demoline

### (1909 – 1944)

Thirty-five-year-old James Edward Demoline was fatally injured in a drift on the 2,500 level of Kerr Addison on March 29. The British miner was married with five children and had been an employee at the mine since September 17, 1941.

On this day shift, he and Alfred Sedore scaled, cleaned up and mucked six ore cars before Mine Captain Jack Sagin came into the drift at 10:15. Sagin instructed them to gad a large slab of loose extending across the back of the drift. They tried but could not get the slab down so Sedore was sent to get some fuse and three sticks of powder. Sagin left at 11:20 after instructing them to blast at lunch hour.

They blasted, took their train to the dump and then had lunch. By 12:35, they were scaling in the drift again. Suddenly, a long slab dropped out of the back, striking Demoline and pinning him to the muck pile on which he stood. A track crew (Leo Samuel and Nick Wediach, who were working in the drift) came to his aid and attempted to release him. Sedore went to the station to notify the cagetender and to obtain the help of John Musial and Henry Gauthier who were working nearby.

Demoline was released at 1:10. Death was considered to have occurred at the time of his release from under the slab. He suffered the following injuries: fractures of the pelvis, lower spine, iliac and ribs. His left foot was torn off at the ankle and he had internal injuries. The body was removed to surface where Dr. D. H. Pollock pronounced him dead.

### August 9, 1944

## Wright-Hargreaves Mine

## Hugh McMaster (1897 - 1944)

Hugh McMaster, 47, lived with his wife and two children at 103A Carter Avenue. On October 6, 1943, he began his employment at Wright-Hargreaves where he became a mucker helper and trammer.

On the day of the accident, he was working on the 400 level with Shift Boss Eugene Jodoin and another man. They were engaged in passing waste fill through a control chute into a pass below to fill a lower stope. During the process, the timber lining of the chute gave way above them, allowing the fill to sweep across the sill timber on this level. When the sill timbers gave way, Jodoin shouted a warning. He and the other man ran one way to safety and McMaster ran the other way. The rush of fill caught McMaster and carried him to the stope below. It is believed that several hundred tons of fill covered the body.

### December 18, 1944

## Wright-Hargreaves Mine

## Peter John Towtriwsky

### (1923 - 1944)

Peter John Towtriwsky was born in Cobalt to Michael and Maria Towtriwsky on November 12, 1923. He would be their only child.

The family moved to Kirkland Lake and lived at 26 Woods Street. Peter was 21 at the time of his death.

He was well known in local music circles, especially at KLCVI where he took a leading part in musical affairs. He spent a year at Queen's University but in 1944, decided to earn money to further his education. He found work at the Wright-Hargreaves assay shop, one of the best-equipped shops in Ontario and noted for its safety appliances.

Employed as an assay helper on December 18, he was engaged in his usual duties of crushing assay samples prior to them being weighed. He had a number of samples ready in pans; for some reason,

*Headstone photograph of Peter Towtriwsky. Photographer: Author*

he went behind the pulping and crushing equipment and became entangled in the main line revolving shaft.

Loyal Leahy, a pulp weigher working in another part of the shop, was passing at the time and looked in when he heard a noise. He saw what was happening and at once halted the machinery by the use of the push button control located near the door. Towtriwsky's body was removed and Dr. C. S. Harris and Dr. T. F. Kelly were called but the young man was found to be 'beyond human aid.' Later in the day, an investigation was conducted and attended by Mine Inspector Mr. Taylor. Findings of this investigation would be used in the inquest.

The inquest into the death was held and the jury returned the following verdict: "We, the jury, find that the deceased, Peter Towtriwsky, came to his death on December 18, 1944, at approximately 10:00 am at the Wright-Hargreaves assay office by becoming entangled in a moving shaft that resulted in him having sustained a fractured skull and brain injuries, also multiple fractures of the body that caused his death."

Mr. and Mrs. Towtriwsky were devastated by the loss of their only son. Funeral services were held and interment occurred in the Kirkland Lake cemetery. Peter Towtriwsky is presently buried with his parents. His father passed away in 1972 and his mother in 1991.

### December 30, 1944

## Lake Shore Mine

## James Breakwell (1893 - 1944)

Fifty-one-year-old James Breakwell had been employed at Lake Shore from March 11, 1930. He was killed at the mine on December 30. He and his wife Hanna (1890 – 1986) had one son.

The fill-pass system at Lake Shore was filled periodically by dumping railroad cars of sand transported from the McNamara sand pit located near Crystal Lake. The sand was dumped into the system at the surface and the last of 19,074 tons was dumped on December 28. The sand being passed at the inactive 2,325-foot level was of exceedingly fine texture with very poor flowing qualities.

At 11:00 am on December 30, Mine Captain John Conlon took Breakwell and his partner, William Fortey*, to 2,325 to clean up a spill of sand on the track. When Breakwell had cleaned up his side of the track, he said to Fortey, "I'll clean off the slide." He began to shovel sand from the door of the chute into the pass that was full. The time was 1:50 p.m.

John Craig and Walter Draker were at a lower level at 11:20 a.m. when Conlon met them and told them to bypass enough sand from the fill-pass to drop the sand level on the 2,325 about ten feet to give Breakwell and Fortey more room to dispose of the sand they were cleaning up. Sand was pulled from this level but Craig and Draker noticed the sand was hung up 50 feet above. They blasted three sticks of powder to loosen it. The sand at the 2,325 suddenly dropped and Breakwell, losing his balance, fell headlong into the sand pass.

Shift Boss T. Ramsay arrived at the scene and was lowered on a rope about 25 feet down into the fill-pass; he could see no trace of Breakwell. The chute on the lower level was then pulled and, at 4:10 p.m., Breakwell's body was recovered. Death was ascribed to suffocation as he had no other injuries.

He was survived by his wife and son, and nephew Bill and his family. Interment occurred in the Kirkland Lake cemetery.

** William Fortey would die in a similar accident 12 years later, on October 15, 1956.*

*Miner using a shovel near a chute. Photographer: Unknown*

## 1945

- Mining accidents at Wright-Hargreaves, Upper Canada and Young-Davidson claim three lives; six children are left fatherless
- Bidgood Mines Ltd. employs 106; Chesterville Larder Lake Gold Mining Company Ltd., employs 173; Golden Gate employs 17; Kerr Addison employs 285; Kirkland Lake Gold 210, Lake Shore 584, Macassa 189, Matachewan Consolidated 108, Omega 135, Sylvanite 262, Teck-Hughes 207, Toburn 121, Upper Canada 124, Wright-Hargreaves 415, Young Davidson 99.
- Kirkland Lake and area men continue to be wounded and killed overseas

*Miner drawing ore into an ore car. Photographer: Unknown*

### April 16, 1945

## Wright-Hargreaves Mine

### Eric William Kurze (1903 - 1945)

Eric Kurze, a naturalized Canadian of German birth, came to Kirkland Lake in 1937 and began work at Wright-Hargreaves on May 14 of that year.

He was well known in town and lived with his wife and their 16-year-old son, Leroy, at 36 Kirkland Street. He had three brothers: William of Shaughnessy, Alberta; Claude of Bluffton, Alberta; and Max with the Royal Canadian Artillery in Italy. There were three sisters: Mary (Mrs. Howard Lyon) of Diamond City, Alberta; Louise (Mrs. Peter Anderson) of Bluffton, Alberta; and Hilda, (Mrs. Ulrich Luginbuhl) of Kirkland Lake.

On the morning of April 16, Motorman Kurze and his partner, Nick Allick Jr., were drawing ore from a raise into ore cars. Kurze was standing near the chute when they heard a slab of loose drop on the lagging over Kurze's head. Almost immediately, the timber above collapsed and a rush of waste rock and fill carried broken timber, debris and Kurze into a fill raise. He was unable to escape, owing to the lack of sufficient warning and was buried in the waste and debris. Allick escaped injury.

Recovery operations began at once and continued steadily until Kurze's body was recovered two days later. He sustained the following injuries: a fractured skull, fractured neck, fractured pelvis and multiple contusions and abrasions. Death was instantaneous.

The body rested at Symington Funeral Home until funeral services officiated by Reverend John K. Moffat were held at Trinity United Church. The body was interred in the Kirkland Lake cemetery. The inscriptions over and near his grave are thus: IN LOVING MEMORY OF ERIC WILLIAM KURZE – BORN JAN. 12, 1903 – DIED APRIL 16, 1945 – SON ERIC LEROY 1928 - 1972 – GRANDSON RICHARD ERIC 1952 - 1972 – 'ALL IS WELL'.

### May 30, 1945

## Upper Canada Mine

### Frank Huncar (1902 - 1945)

Forty-three-year-old Czechoslovakian immigrant Frank Huncar lived at 115 Wishman Street with his wife and five children who ranged in age from six to 16.

An experienced shaftman, he was employed at Upper Canada in Dobie from October 16, 1943. The No. 2 shaft was being deepened from 750 to 1,250 feet. On May 30, the shaft was 1,145 feet deep.

On this day, the afternoon shift of shaftmen went to the shaft bottom to muck it out. Within three hours, the mucking was done and Leader T. Dodd and Huncar went up 37 feet to the blasting

*No. 2 Headframe at Upper Canada (very few remain in the area). This one is a monument to Frank Huncar who was killed in its shaft in 1945. Photographer: Lawrence Chrismas*

set, using the bucket. The bucket stopped even with the blasting set. Dodd was going up to a higher level, so only Huncar got off at this point. In stepping off, he lost his footing; the bucket swung away from him and he fell the 37 feet to the bottom.

He suffered a fractured arm and skull. He was rescued and brought to surface where he died.

Interment occurred at the Kirkland Lake cemetery. The gravesite is unmarked.

### December 12, 1945

## Young-Davidson Mine

## Reino Jussila (1897 - 1945)

Reino Jussila, 48, Finnish and married, was a timberman at Young-Davidson near Matachewan.

The No. 3 chute in the 205 drift had been pulled during the graveyard shift of December 12. At the end of this shift, the chute was left hung up about 27 feet above the throat.

Mine Captain H. Byberg and Shift Boss R. Stewart, who both knew that the chute was hung up, decided to discontinue pulling the chute temporarily and to install a partition of planks and canvas across the throat of the chute to prevent the cold downdraft from entering the level. Jussila and partner J. A. Forbes, another timberman, were instructed to install this brattice.

Byberg and Stewart examined the hung-up chute independently and were satisfied it was safe. Jussila and Forbes procured the materials required and proceeded with the installation; Jussila stood in the chute. Forbes was in the drift, cutting planks to the required length and handing them up to Jussila as needed. Suddenly, a chunk of rock fell from the area above the chute and struck Jussila on the head. Hearing a dull thud, Forbes turned and caught Jussila as he fell from the chute. His hard hat had an irregular puncture hole in it and blood was flowing from a wound in his head.

Forbes called for assistance and Stewart, F. Desrochers and A. Lefevre, who were nearby, arrived. They took the unconscious Jussila to the first-aid room. Within a few minutes, Dr. F. Fitch arrived from Matachewan and attended to the injured miner. He found that Jussila had sustained a fractured lower jaw and a complete fracture of the crown of the skull. The depth of the wound indicated probable brain damage. Within about 95 minutes of the accident, the injured man died.

### 1946

- Many Kirkland Lake and area men return home from war service
- Felix Mutka returns after serving in the Canadian Army but dies after an accident at Wright-Hargreaves
- John Kuros dies in hospital after an accident at Omega
- An elderly employee at Upper Canada, Fred Harrop, is injured at the mine and later dies in hospital
- Anthime Soucie is killed in an accident at Macassa
- Plans are made to build 100 wartime houses in Kirkland Lake to make up for the housing shortage after the war
- Population of Kirkland Lake is about 20,000

## June 24, 1946
# Wright-Hargreaves Mine
## Felix J. Mutka (1912 - 1946)

Born in Finland on November 30, 1912, Felix Mutka immigrated to Canada and was hired at Wright-Hargreaves in 1936. He began his employment as a mucker and by 1939 had become a motorman. He became a naturalized Canadian and from 1941 to 1946 served in the Canadian Army. He was back at Wright-Hargreaves three-and-a-half months after his discharge. He was single and 33 at the time of the accident.

This tragedy happened at the very beginning of the day shift. The previous shift had sent down a steel-truck loaded with a variety of drill steels. This truck was placed on the track at the 4,350 station so the afternoon shift could unload the steels and place them in the steel-rack. The truck was a standard type commonly used in the mining industry at this time; the steels were placed in the truck vertically. Designed to be pushed by hand, the truck had two hand grips on each side. When loaded with steels, it had a total weight of 3,080 pounds.

Motorman Mutka had to move the truck out of the way to switch his locomotive into a position where it could push a mucking machine into the mine workings. S. E. Neuman, who was standing at the station, assisted him in moving the truck. With Neuman pushing and Mutka pulling, it was moved over to the switch. In some manner, Mutka's left foot caught and the truck flipped over its front wheels, pinning him to the track. His partner, F. Wilkinson, and another miner, A. Charron, were standing at the station when the accident occurred; they assisted Neuman by removing the weight from Mutka and calling the first aid room. Dr. C. S. Harris arrived within minutes and attended to the injured miner. Mutka was removed to the Kirkland and District Hospital where he died the next morning, on June 25.

He had sustained a series of crushing injuries: fractures of the tibia, fibula, femur and pelvis of his left side. His bladder was ruptured and parts of his intestine were torn. The cause of death was shock and internal hemorrhage.

At the inquest, the following statement was made: "Mutka acted contrary to the company's safety regulations in that he was pulling the truck instead of pushing. With his mining experience, he must have been fully aware of this regulation."

The funeral service took place at the Symington Funeral Home. Interment occurred in the Kirkland Lake cemetery.

## August 26, 1946
# Omega Mine
## John Kuros (1903 - 1946)

John Kuros, 43 and a Polish immigrant, worked at Omega and lived in Larder Lake. He was married: his wife and children lived in Poland. He was an experienced miner who had worked in the Larder Lake and Sudbury areas. He was first employed in 1937 at Omega, where he worked until November, 1942 and then worked in the Sudbury district until October 1945, when he returned to Omega.

He had worked for several weeks before the accident on a fill-pass system that ran from the 1,050 level to the 1,400. On August 26, he and partner Roland Tremblay commenced the running of fill into a stope at the 1,400. When they inspected a stope on the 1,175, they found that the sand fill was being held up. A deflector was preventing the fill from flowing properly; Kuros decided to knock it out with a sledge hammer. He took out one post, which permitted some of the sand to flow. He knocked out a second post and this caused a huge rush of sand, as the whole slide was now collapsed. Tremblay ran clear as the sand spread out on the stope floor but Kuros hesitated, then turned and backed away. The spreading sand knocked him down and buried him before he got clear.

Tremblay immediately tried to shovel the sand off Kuros but was unable to make headway as the sand flowed in as fast as he could remove it. In his valiant effort, he broke his shovel. He ran down the manway and contacted the cagetender who brought a couple of men to assist. After about 15 minutes of shovelling, they uncovered Kuros' head and shoulders. Dr. T. J. Elliot of Larder Lake arrived at the scene; resuscitation was attempted but without success. In a post-mortem examination of the body, the doctor found a bruise on the forehead and a fractured lower left jaw. Neither of these injuries would have been fatal in themselves and death was attributed to asphyxiation.

### September 23, 1946

## Upper Canada Mine

## Frederick Harrop (1880 - 1946)

Frederick Harrop was born in Georgetown, Ontario in 1880. He was married and had nine children: four sons: Jim, Wilfred, Edgar, Stanley and five daughters: Lila, Rita, Vera, Dorothy and Enid.

The family moved to Dobie from Iroquois Falls when the 58-year-old Harrop was hired at the new Upper Canada on November 15, 1938 as a surface labourer and classified as a boiler fireman. Here he worked in the mechanical department, tending the automatic stoker in the heating plant and servicing the miners' lamps.

On September 22, 1946 at approximately 8:30 a.m., Harrop, now 66, reported to Master Mechanic Hartley West that he had tripped over the feed screw of the boiler room's mechanical stoker and fallen, injuring his side. West called a taxi and sent him to Kirkland District Hospital where Dr. C. N. Brebner diagnosed his injury as fractured ribs.

The doctor strapped his chest with adhesive tape. On clinical examination, his temperature was normal. There was no indication of internal injuries and he was sent home.

That evening, Dr. Brebner received a call that Harrop was suffering pain. He was brought to hospital by ambulance and again examined. His temperature was still normal but he complained of considerable pain in the lower right side of the chest where the ribs were cracked. Dr. Brebner removed the adhesive strapping, gave him a sedative and kept him in hospital.

On the morning of September 23, Harrop's temperature was higher and Dr. Brebner examined him again, finding pneumonia in the left lung. He did not respond to treatment but became progressively worse and died 31 hours after his accident.

Coroner J. F. Edis ordered an autopsy; it was performed by Dr. J. M. Leblond. He found the lower lobe of the left lung showed evidence of acute

*On right Fred Harrop at camp while collecting firewood for Upper Canada Mine. Photographer: Unknown*

lobar pneumonia. There was no injury to the lung or internal organs from the fractured ribs. In the opinion of both doctors, the injury was a contributing cause to the rapid development of acute pneumonia, from which the patient died.

Frederick Harrop was buried at Iroquois Falls. His son Edgar provided some of this information and the photograph of his father. Edgar, now retired, was also an employee at Upper Canada. He and his wife Cecile presently (1999) live a retired life in Dobie. The other members of the Harrop family live in various parts of Canada.

### December 30, 1946

## Macassa Mine

## Anthime Leo Soucie (1904 - 1946)

Mr. and Mrs. Clafase Soucie were pioneer farmers settled near Fort Coulonge, Quebec. The Soucie family became a large one with 16 children. A son, Anthime, was born in 1904; he grew up on the farm at Fort Coulonge and, when an adult, became a barber. Rose Graveline, who also came from the area, became his wife. In time, they had a daughter, Nola, and a son, Laurier.

The family later moved to the Quebec hamlet of Mud Lake where Anthime and Rose became restauranteurs while Anthime still did some barbering. Friend Bob Freitag enticed them to come to Kirkland Lake as opportunities for a new barber looked promising.

Thus the family of four (Nola 12 and Laurier 10) moved to the area, living in Chaput Hughes where a barber shop was established. Business was slow at first, so Soucie found employment at Macassa on January 2, 1942. He remained there until May 22, 1946, when he again returned to full-time barbering. A friendly barber, he was well liked and, if a customer enjoyed talking about hockey, all the better because Soucie was an ardent hockey fan.

He was rehired at Macassa on December 23, 1946. Why he was working at the mine for a week through Christmas is uncertain. He may have offered to be a Christmas replacement for another miner. At the time of the accident, he was employed as a chute blaster.

On December 30, he was working on the 23rd sublevel with a crew of three inexperienced men: A. Gilroy, D. Wilson and D. Kroeker. The three had all been hired in the month of December and had no previous mining experience. As senior man, Soucie supervised the work of his helpers. He did the blasting when necessary and performed other jobs that required the knowledge and experience his partners lacked. Several chutes were pulled in stope 23G prior to the accident and the ore hand-trammed to the ore pass.

The four men were at the ore pass dumping cars when Soucie left without saying where he was going, walking back up the stope. There, he apparently picked up a bar and started to work at some muck. When he did not return, the others went into the stope to look for him. Gilroy saw Soucie's light gleaming underneath a five-ton slab of rock. Wilson and Kroeker ran down to the 24th level, notified a crew of the accident and arranged for help.

Recovery operations began under the supervision of Mine Superintendent C. Chisholm and Mine Captain C. M. O'Shaughnessy.

*Anthime Soucie. Photographer: Unknown*

Dr. Neelands and Dr. T. W. Armstrong arrived to give medical aid. The body was removed at 11:45 a.m. after the slab was jacked up. Dr. Armstrong pronounced Soucie dead. The cause of death was a fractured neck and a crushed chest. He had died instantly.

His family learned of the accident while home for lunch. On hearing the news, 17-year-old son Laurier ran from his home at 737 Government Road East in Chaput Hughes to the dry of the nearby Macassa. He reached the mine just as they were bringing his father to the surface.

Soucie's death devastated his family. Laurier was a student at Kirkland Lake CVI and the 19-year-old Nola was working as a clerk at Baker's Jewellery Store. The fatal accident left the family facing a financially troubled future. Laurier had made plans to attend university and possibly become a dentist. There was no chance of such a career now.

The funeral mass occurred at Assumption Church; the body was taken by train to Fort Coulonge for further funeral services and interment. Wife Rose is now buried beside him.

Parts of this history were provided by Laurier Soucie who was the long-time manager of Seymour's Men's Wear in Kirkland Lake. Laurier and his wife Shirley presently (2000) live in Kirkland Lake.

## 1947

- Leonard Bennett, father of four, loses his life at Teck-Hughes; Michael Drul, father of a three-year-old, dies at Sylvanite; William Alexander, father of three daughters, is killed at Teck-Hughes; Lucien Chouinard, single, is killed at Lake Shore; Ernest Porter, father of a two-year-old daughter, is killed at Kerr Addison; Ernest More is accidentally killed at Teck-Hughes

### April 7, 1947

## Teck-Hughes Mine

# Leonard Joseph Bennett

### (1898 - 1947)

Leonard Joseph Bennett was born in Newfoundland in 1898. He married the former Mabel Hayes of Englehart in 1929. In 1947, Mr. and Mrs. Bennett resided at 87 Woods Street with their four children: Gloria (17), Marcia (14), John (13) and Earl (10).

Forty-nine-year-old Bennett, an employee at Teck-Hughes since September 1946, had considerable previous experience in mining. On the day of the accident he and partner Kjell Haraldson were visited by Shift Boss Archie McDonald who instructed them to continue cleaning, and to retimber a stope where necessary as they went along. Some time afterward, Bennett went down to the bottom of the stope to get the muck moving into the box-hole. He was scraping and poking around with a shovel when suddenly a large piece of rock fell from the hanging wall and pinned him to the footwall. Haraldson heard the fall of ground and noticed that his partner was not around. He immediately went down to level and got the shift boss who organized a rescue crew and took them into the stope. The body was found within an hour. Dr. Cole was called and the miner was pronounced dead.

Bennett sustained a fractured skull and a severely crushed chest with perforation of the lungs. Death was probably instantaneous.

Surviving him were his immediate family in town and two brothers and a sister living in Newfoundland. Interment was conducted in the Kirkland Lake cemetery. The gravesite is presently (1999) unmarked.

### June 11, 1947

## Sylvanite Mine

# Michael A. Drul (1916 - 1947)

Thirty-one-year-old Michael Drul, a Canadian of Ukrainian parentage, came to Kirkland Lake from Manitoba. A family man, he lived with his wife and their three-year-old at 93 Third Street.

He was hired on at Sylvanite during the miners' strike on December 9, 1941 as a surface labourer. In January 1942, he went underground where he worked as a mucker, helper and driller. He was killed at the mine on June 11, 1947. He was considered an experienced miner.

Before going underground on the day shift of the accident, he and helper Ross Hildred were told on surface by Shift Boss J. Marni to muck and prepare to blast in a stope on the 625 level. Marni visited them about three hours into the shift and found conditions good. After lunch, Hildred went to muck and Drul went to work on the staging. Drul moved some planks and started barring down the muck pile to make a flat surface as an approach step to the staging. As the muck rolled down the pile, a large slab of loose on the wall 15 feet above him rolled out as the muck around it was released. This slab was sufficiently rounded to roll down edgeways, striking the staging, taking it, the drilling gear and Drul down the slope of the muck pile.

When everything came to rest, Drul was covered by this slab. Hildred rushed to his aid but was unable to move the rock and immediately went for help. A crew of rescue workers under the direction of Underground Superintendent H. Waller freed Drul using chain blocks and tackle. At the scene, Dr. Kelly pronounced the miner dead. Drul had suffered the following injuries: a fractured skull, a crushed chest and a fractured right arm. These injuries were sufficient to cause almost instant death.

## July 25, 1947
## Teck-Hughes Mine

### William Alexander (1889 - 1947)

Two employees at Teck-Hughes, Kjell Haraldsen and Archie McDonald, were witnesses to the Leonard Bennett accident at the mine on April 7. Three-and-a-half months later, the same two men witnessed a similar accident that would claim the life of William Alexander.

Alexander, 59 and Austrian, had been employed continuously at the mine from November 1920, a period of almost 27 years. He lived at 50 Main Street with his three daughters, Helen, Doris (Mrs. Earl Watt), and Mary (Mrs. S. Lapierre).

Alexander and partner Haraldsen went on shift at 7:00 am on July 25 and proceeded to their place of work in a stope at the 2,500 level. They were visited by Shift Boss McDonald who instructed Alexander to continue drilling. Alexander drilled six more holes when a large piece of loose fell from the back, broke the staging and pinned him to the floor. Haraldsen, who was also on the staging, fell but was not injured.

Haraldsen tried to move the large rock but could not. He went out to the shaft, reported the accident and returned with five men. The rock was pried up with a long piece of timber and the body freed and taken to the surface. Dr. R.H. Armstrong examined the body at the collar of the south shaft and pronounced Alexander dead. He had sustained multiple fractures to his skull, face and ribs. Death was probably instantaneous.

The funeral service was held at St. Peter's Anglican Church. Interment occurred in the Kirkland Lake cemetery. The inscription on his gravestone reads: IN MEMORY OF WILLIAM ALEXANDER – BORN APRIL 26, 1889 – KILLED JULY 25, 1947 – TECK-HUGHES MINE.

## July 29, 1947
## Teck-Hughes Mine

### Ernest Gordon More

(1903 - 1947)

Another fatality at Teck-Hughes occurred five days later when 45-year-old Ernest More of 55 Rand Avenue West was fatally injured.

Born in Ontario in 1903, he came to Kirkland Lake in 1924. He was employed at Teck-Hughes from 1927. In 1941, he left the mine to serve in the Canadian Armed Forces for four years. He was rehired on November 28, 1945, his military service complete. One-and-a-half years later, he was killed at the mine. He was survived by his wife, the former Anne Brown.

On July 9, Motorman More and helper H. McLeod were pulling muck from the ore pass and dumping it into an ore pocket at the 3500 station. While there, they noticed Skiptender Helge Byberg was trying to dislodge a small piece of rock jammed between the finger gate and the side of the ore pocket. He was using two diamond drill rods but these also jammed during the process. More offered help by climbing into the upper part of the pocket and trying to get the rods out. Byberg got into the skip and lowered it until he could work on the gate from the shaft side. The two worked on the jammed gate from opposite sides until the rock and rods were finally dislodged. Byberg asked McLeod to ring the skip back to level.

Byberg turned the control lever for the finger gate to the down position and immediately heard More yell, "Whoa – it's coming down again." More had slipped under the gate and, when Byberg turned the control lever, the gate came down, striking More across the abdomen. Byberg raised the gate again. More extricated himself and, when a ladder was placed in the pocket, climbed up unaided.

Dr. W. E. Bryant was called. He examined More at the scene and suspected internal injuries. More was taken to surface and to Kirkland District Hospital where he was given blood plasma. He was taken to the operating room about seven hours after the accident; within a half hour, he died. The time of death was listed as 12:15 a.m., July 30. An autopsy revealed ruptures in his liver, intestines and pancreas.

The funeral service was held in Kirkland Lake and interment was conducted in the local cemetery.

## August 9, 1947
## Lake Shore Mine

### Lucien J. G. Chouinard
(1924 - 1947)

Lucien Chouinard was born to Mr. and Mrs. Conrad Chouinard on August 21, 1924; he was the eldest of a family of 12 children.

Like other 18-year-olds living in a country at war, Lucien joined the Canadian Armed Forces. He rose to Petty Officer in the Royal Canadian Navy and served four years. At the age of 22, he was back in Kirkland Lake.

He worked at Lake Shore but his future plans did not include underground mining. In fact, he had mentioned to his friends that he did not like mining and would not stay in it for long. Chouinard had other plans. His plan was to work in the mine to make money so he could return to Queen's University and become an engineer. He also had marriage plans, as he was engaged to Mary Ramsay of Kirkland Lake.

*Lucien Chouinard. Photographer: Unknown*

On August 9, Chouinard and partner Alvin Culpan were sawing a piece of blocking when a rockburst occurred directly over their heads. The burst in this 4,900-level stope displaced about 100 tons of rock. Chouinard was almost completely buried in fallen rock. Culpan was partially covered and was unable to extricate himself until Shift Boss Peter West and Night Captain William Allen came to his assistance. Culpan received minor cuts and abrasions and a severe bruising but was not seriously injured.

Dr. G. M. Cameron was called to the scene and remained there until the body of Chouinard was removed six hours after the burst. Death was caused by traumatic asphyxia resulting from a crushing weight of broken rock on his chest and abdomen.

The funeral mass for the Navy veteran was held at the Church of the Assumption. Pall bearers were brothers Leo and Ernie, Fernand St. Cyr, Jack Kellaway, Lucien Desjardins and Roger Benoit.

One brother, Leonel, had predeceased him. Surviving were his parents and five sisters, Theresa (18), Laurette (17), Rita (16), Lucille (10), and Vivian (7), and five brothers, Leo (20), Ernie (14), Emile (12), Real (9) and Conrad (5).

Interment occurred in the Kirkland Lake cemetery. The epitaph on his gravestone reads: 'PARTIE MAIS NON OUBLIEE'.

The photograph and a confirmation of the accuracy of the story was given by Lucien's sister, Lucille (Mrs. Arnold Merrick), presently (1999) living a retired life with her husband Arnold in King Kirkland.

## December 21, 1947
## Kerr Addison Mine

### Ernest M. Porter (1915 - 1947)

Thirty-two-year-old Ernest Porter, an electrician's helper at Kerr Addison, was instantly killed on December 21. The Canadian miner was married to the former Terry Theriault and they had a two-year-old daughter.

Porter was first hired at Kerr Addison in May 1941. He served three-and-a-half years in the Armed Forces and was rehired at the mine on March 1, 1946. He worked as an underground sampler and on surface in the steel shop. He became an electrician's helper only a week before the fatal accident.

He and partner George Durham, also an electrician's helper, were servicing underground locomotive batteries on the day of the accident. The two got off the cage at the 500, looked after the batteries there, and started down the shaft manway to the 700. Durham was climbing down ahead, carrying a bag of tools over his shoulder. Porter was following, carrying a wooden box that contained a Winchester of distilled water packed in sawdust. The 10-pound box was provided with a rope handle that he slung over his right arm.

Progressing down the ladder, Durham was about halfway between levels when he heard a bumping noise behind him and saw Porter fall down the pipe compartment of the manway. He had not heard Porter say anything nor did he see him slip from the ladder.

Durham immediately continued down to the 700 where he phoned Skiptender Robert Little. Within minutes, Little arrived with Captain Bill Hargrave and Shift Bosses George Rathwell and Ralph Buntin. Hargrave went back to surface to call a doctor and obtain a stretcher while the others climbed down the manway, looking for Porter.

He was found one landing below the 1,000. He had fallen about 400 feet and his body was arrested in its fall by steel bearers supporting pipes across the pipe compartment. Dr. D. H. Pollock arrived at the scene, observed the severely injured body and pronounced Porter dead.

An inquest was held at the Virginiatown Public School before Coroner H. F. Richardson. It appeared that Porter in some manner lost his grip on the ladder and fell backwards over the guardrail into the pipeway. The ladders, landings and guardrails were examined immediately after the accident and, with the exception of the guardrails broken by his fall, they were all found in good condition. The jury determined an accidental death but a recommendation was made: a safer means of separating the pipe compartment from the manway should be constructed.

Interment was conducted in the Kirkland Lake cemetery.

## 1948

- Fall of rock kills two at Teck-Hughes; a surveyor is killed underground at Matachewan Consolidated; a falling slab of rock severely injures a miner at Young-Davidson and he dies in hospital; a miner is killed at Teck-Hughes; a worker is electrocuted at Kirkland Lake Gold; a miner is killed while scaling at Chesterville
- Omega Mine closes
- Many men and their families come to this area to work in the gold mines; they come from displaced persons' camps in Poland, Germany and Italy
- Lake Shore, Wright-Hargreaves, Sylvanite, Kirkland Lake Gold, Toburn, Bidgood, Kerr Addison and Chesterville introduce a cost of living bonus of $4.25 per week per miner
- Miners are paid from the time they arrive at their place of work underground until the time they leave their underground place of work; unions want a miner to be paid from the time he passes the collar into his work place until the time he leaves the collar
- Miners are allowed to work alone; unions want changes made so that a miner never works alone

## January 13, 1948
## Teck-Hughes Mine

# Fall of Rock Kills Two at Teck-Hughes

George Gill was a long-time underground superintendent at Teck-Hughes. On January 13, his son Kenneth was killed at the mine in a fall of ground. Harry Brady, Kenneth's working partner, was also killed in the mishap.

Mining had been discontinued in stope No. 38L30W in 1941. New plans were made in 1947 to revitalize this stope by clearing out the broken ore, retimbering and laying new floor. Kenneth Gill and Brady had worked in this stope for two months before the accident, retimbering from top down and cleaning the foot wall as they progressed.

On January 13, Marcel Chenier and Alfred Paul were working as a motor crew on the level below

*Upper Canada Miners "hamming it up" while waiting for the cage after shift in 1948 (3,600-foot station, shaft No. 1)*
*Photographer: Unknown*

the stope, pulling chutes and hauling ore to the south shaft extension. In the stope above, Gill and Brady were at work. At 1:30 p.m., Brady called down from the stope to Chenier, telling him to pick up a stick of powder and fuse on his next trip to the station. Chenier and Paul returned from the station on their locomotive at 1:50. They called up to the stope but got no reply from either Gill or Brady. Chenier went up into the stope to investigate and found both men dead at the edge of a heap of rock that had fallen from the hanging wall. Chenier left and immediately called Shift Boss Maxwell Parker*. The bodies were soon brought out. Dr. A. W. Souter was summoned and the two were pronounced dead.

An estimated 25 tons of rock had fallen. The stope had been examined by Safety Engineer Ivan Sloan three hours before the accident and no unsafe conditions had been noticed. As there were no eye witnesses to the accident, it is not known what the men were doing when the loose fell. Since they had ordered powder and fuse, it was suspected that they were planning to blast a stull or a piece of loose.

** Maxwell Parker would die in a 1957 mining accident at this mine.*

## Kenneth E. Gill (1918 - 1948)

Kenneth Gill was born in Cobalt on January 14, 1918. His father, George Gill, was born in Nova Scotia in 1887 and moved to Cobalt in 1905. There he married the former Emma L. Lively and in time they had three sons: George Jr., Emanuel and Kenneth. The family moved to Kirkland Lake in 1922 when George Sr. became employed at Teck-Hughes. In 1938, 20-year-old Kenneth also became an employee at the mine.

Gill was married to the former Margaret M. Morrison; they had no children. By 1948, he was an experienced miner with about 10 years of service at the mine. On January 13, he and partner Harry Brady were killed by a fall of rock in a stope. Gill suffered an extensive fracture of the skull with laceration of the brain and partial decapitation; his right forearm was also mangled. Surviving him were his immediate family, an uncle Arthur Gill and three aunts.

Funeral services were held at the home of his parents at 33 Prospect Avenue. Reverend R. J. Cochran of the First Baptist Church officiated. The pall bearers were Stuart McLean, Irving Angus,

Neil McDonald, Jack Moodie, Ken Griffin, Robert Ash, A. McKinnon and E. Trembley. Interment took place in the Kirkland Lake cemetery.

Following Kenneth Gill's death, his father retired from the mine the same year and later that year, his mother passed away on October 18. Brother Emanuel died on August 28, 1953. Father George died on May 19, 1960 and brother George Jr. died in the 1990s. All are buried in the Kirkland Lake cemetery.

## Harry R. Brady (1895 - 1948)

Harry Brady, originally of Winnipeg, came to Kirkland Lake in 1933 with his family. In 1948, they lived at 44 Taylor Avenue. The family included his wife, the former Emily A. Edgar, and three children: Jeanette, Eva and Thomas. Brady was a Lieutenant in the local reserve company of the Algonquin Regiment and a World War I veteran who had enlisted in May 1915. He was an employee of Teck-Hughes from 1933.

He and partner Kenneth Gill were buried and killed by a fall of rock in a stope. Brady sustained compressive fractures of the thorax and spine and probably died quickly of traumatic asphyxia.

The funeral service was held at Trinity United Church with Reverend J. F. Moffat officiating. The pall bearers were George James, J. Hogg, R. Drolet, W. E. Blewett, Hugh Boyd and C. H. Telfer.

The body was laid to rest in the Kirkland Lake cemetery vault awaiting spring interment.

### March 25, 1948

# Matachewan Consolidated Mine

## James Harvey Ellis (1920 - 1948)

Twenty-eight-year-old Canadian James Harvey Ellis was married with two children. Employed as a surveyor at Matachewan Consolidated, he died on March 25 as a result of injuries sustained in a stope at the mine. He was hired on January 21, 1941; he then served four years in the RCAF and returned to the mine after his discharge.

On the morning of the accident, Garnet Durling installed two posts in the stope, joining them at the bottom with three or four pieces of round lagging. Machinemen E. Demerais and A. McInnes and helper N. Al-Moky were working in the stope at the time. Shift Boss W. T. Copping was also in the stope that morning and found conditions satisfactory.

After Durling completed his work and left, Ellis and helper A. F. Everest came into the stope to make a survey. In the line of sight of a survey was a post that Durling had just erected; they removed it. When they completed their survey work, the post was put up and wedged back into place.

Ellis was replacing the lagging at the bottom when suddenly a two-ton slab fell from the back.

*Underground surveyor. Photographer: Unknown*

He was in a stooping position at the time and the slab pinned him to the muck. With the assistance of others, the men in the stope removed the slab of rock. Dr. F. Fitch arrived within 20 minutes. He had Ellis placed in a stretcher and taken to surface. Ellis died in an ambulance en route to the New Liskeard Hospital. He had sustained crushing injuries to his chest, back and hips. Death was attributed to shock and internal and external hemorrhage.

An inquest was held in the presence of Coroner J. S. McCullough in the Legion Hall at Matachewan. It is likely that in replacing the post and wedging it into place, a strain had been set up in the rock in the back of the stope causing a piece to break away suddenly. The jury's verdict was 'accidental death'.

*Burial site at Matachewan Cemetery. Photographer: Author*

## April 28, 1948
## Young-Davidson Mine

### Jonas Vytautus Miciunas
(1929 - 1948)

In the Matachewan cemetery, headstones identify some of the graves but a large number of gravesites are marked only with white wooden crosses. Among these white crosses is an impressive black headstone. The perimeter of the burial plot is traced with six black cement pillars joined to each other with heavy black chain. Inside the burial site, within the confines of its pillars and chain, is a black cement crypt with a white cross lying upon it. The headstone is inscribed with the following epitaph: JONAS VYTAUTUS MICIUNAS – 1929. V. 15 ~ 1948.IV. 28 – TEBUNA LENGVA TAU SVETIMA ZEMELE NULIUDE TEVAI SESUO IR BROLIS.

Twenty years old and single, Vytautus Miciunas was a recent Lithuanian immigrant. He was injured on April 27 at the Young-Davidson near Matachewan and died a day later as a result of the injuries.

Miciunas was hired at the mine on January 20, 1948; a brother also worked there. Their parents lived in Germany.

On the morning of the accident, Miciunas and J. Petrauskas were instructed to blast a breast located in a stope on the third level. They did some scaling and were loading holes when the accident occurred. Petrauskas was on a ladder loading a hole and Miciunas was behind him handing him the powder when a slab of rock fell. Miciunas was knocked down under the rock but it was bridged in such a manner that he did not appear crushed. Petrauskas called for help and four men came to his aid. They lifted the slab and freed the miner who was removed to the surface in a stretcher. Miciunas was conscious and complained of pain in his arm and side; x-rays later showed he had sustained a fractured arm and fractured ribs. On the surface, he was treated by Dr. F. Fitch of Matachewan. Suffering from shock, he was given intravenous glucose and a grain of morphine.

*Attached to the headstone is a photograph of a young man. Photographer: Author*

Within hours, he was taken to the Red Cross Hospital in New Liskeard where he was examined by Dr. J. S. McCullough. He had a high temperature and his breathing was rapid and shallow. Throughout the night, he was watched closely; by morning, his condition had improved. He was given penicillin to combat any infection that may have developed.

At 1:00 p.m., April 28, his temperature rose dramatically; he became semi-conscious and his general condition deteriorated. At 8:30 p.m., he died. The doctor stated that the cause of death was traumatic pneumonia.

Presently (1999), the gravesite of Jonas Vytautus Miciunas is neat and well-kept. People in Matachewan state that his brother, now living in Chicago, comes to the cemetery on a regular annual basis.

### May 3, 1948

## Teck-Hughes Mine

## Richard Greenall (1889 - 1948)

"He was the workingest miner you would ever know." This is how a Northern News reporter quoted Harry Greenall, the 34-year-old son of the deceased Richard Greenall, killed at Teck-Hughes on May 3.

Greenall was a miner for 46 years. Born in England in 1889, the 13-year-old was already working in the coal mines of Lancashire. By 1926, he was in Canada, working underground at Lake Shore where he laboured for eight years. In 1934, he left Lake Shore to work at Teck-Hughes where he was employed a period of 14 years up until the time of the accident.

Fifty-nine-year-old Greenall was married and lived with his wife Sarah at their residence at 109 Main Street. They had two adult sons: Harry (34) and Harold (26).

On May 3, he and partner Harry Walker were drilling a face at the 30th level when suddenly about 580 tons of fill fell out of the bottom of the stope. Greenall was standing on a plank platform with his drilling machine set up when the vortex of the sudden subsidence swallowed him, the platform and the drilling gear. Walker caught an air hose and pulled himself to the safety of a manway.

Recovery operations began on the level by working in the fill and debris. When this dangerous operation got no results, it was abandoned (May 7). As an alternative, a 123-foot-long sublevel crosscut was driven from an adjacent stope to the stope where Greenall was located. Three more days were required to draw the muck and debris from the stope. On May 18, 15 days after the mishap, Greenall's body was finally found. He had dropped a total distance of 62 feet from the working floor of the stope at the time of the accident.

Dr. A. W. Souter made an examination, finding that Greenall died of extensive crushing injuries; death would have been instantaneous.

An inquest was held before Coroner J. F. Edis. Company officials were unable to explain the cause of the subsidence since the area below the accident was last mined and filled 17 years previously. There must have been a void in the fill below the stope and, on May 3, an unfortunate breakage occurred, causing the fill to run into this void and create a subsidence. The jury believed the death was accidental.

Interment occurred in the Kirkland Lake cemetery. The inscriptions on the gravestone are: GREENALL – FATHER RICHARD 1889 - 1948 – MOTHER SARAH E. 1887 - 1952 – 'IN THEIR DEATH THEY WERE NOT DIVIDED'.

### June 16, 1948

## Kirkland Lake Gold Mine

## Johnston Brown (1918 - 1948)

Born in Canada on March 16, 1918, Johnny Brown was a Canadian Army veteran and a one-time resident of Englehart. His parents were Mr. and Mrs. Stewart Brown of Englehart.

Brown, 30, was married to the former May Johns (1918 – 1989), the daughter of Alfred Johns, the mine captain at Kirkland Lake Gold. The childless couple lived at 682 Government Road West in Chaput Hughes.

Brown, an electrician's helper at Kirkland Lake Gold, died while working in the transformer house

*Electrician and electrician's helper at work underground. Photographer: Unknown*

on the 2,475 level. He was employed from November 23, 1938 until June 1942 as a cap-lamp repairman. Joining the Canadian Armed Forces in June 1942, he returned to the mine in February 1946 after being discharged.

On the day of the accident, Brown and Electrician Armand Lapierre were installing an oil circuit breaker panel to control a 550-volt feeder to a new crusher being constructed on the same level. During the work, Brown had to leave to repair a locomotive at a lower level. He returned and continued to work side by side with Lapierre, completing the last connections. Lapierre was tightening a clamp on a horizontal pipe and Brown was tightening another clamp on the same pipe when Brown suddenly collapsed.

Lapierre administered artificial respiration within three minutes; it was continued for three hours without success by Fred Giddings, Joseph Pulchinski, B. Strange, Claude Goodman, William Smith and Dr. C. S. Harris.

Following the accident, Chief Electrician A. Prangley cut the source of power to investigate the cause of the accident. Tests revealed the insulation had deteriorated on one of the 2,200-volt leads. The metal clamp that Brown was tightening broke through the weakened insulation, making the clamp live. A circuit was completed through his body as indicated by second degree burns on his left hand and third degree burns on the right side of his abdomen where he was leaning against an insulator fitting. Both men had been warned by Prangley previously to treat all 2,200-volt leads, although insulated, as live. Brown was always regarded as a very careful workman.

The funeral service was held at the home of his father-in-law in Kirkland Lake and interment took place in the local cemetery. The epitaph on the headstone reads: 'AT REST'.

## December 9, 1948
## Chesterville Mine

### Albert Sliva (1889 - 1948)

Albert Sliva, 59 and Czechoslovakian, was married with a wife and grown family in Czechoslovakia. He was killed in a stope in Chesterville on December 9. He had eight years of experience as a driller at this mine and seven years of mining experience elsewhere.

At 8:00 a.m. on the day of the accident, he and partner S. Dutzak were sent to No. 8D2 stope by Shift Boss B. Bowes to scale and set up the drill at the breast. They went to their work place and replaced a ladder. Two pipefitters, J. Selbie and A. Girard, replaced the air and water lines that had been disconnected. Sliva and Dutzak scaled the area extensively prior to any drilling. At 11:30, they blasted a large slab of loose at the breast and went to the seventh level station for lunch while the smoke cleared.

At 12:30, they returned. The smoke had now cleared and they continued to scale the work area. Dutzak was using a scaling bar and Sliva used a gad and hammer. After about 15 minutes of scaling, Dutzak heard the sound of Sliva's hard hat hitting the rock. He turned around and saw him lying on the muck. His hat was off but the light was still on and hanging over his shoulder. He was lying in a doubled up position with his chest against his knees. There was no rock on top of him but there was a small slab to his right and left. It appeared that the two pieces were originally one but had broken across his body. Dutzak tried to assist Sliva, who gasped a few times and then went limp. He was placed in a more comfortable position and Dutzak went up to the seventh for help.

There, he located J. Luchian and N. Dumais; Luchian returned to the stope with Dutzak and Dumais called to the surface for a doctor and a stretcher.

Sliva was brought to surface where Dr. H. E. Richardson examined him, pronounced him dead and sent him to Kirkland District Hospital for further examination. This revealed extensive abrasions on the back and hips, fractures of a tibia and femur, two fractured lumbar vertebrae, eleven fractured ribs and extensive internal organ injuries. The injuries were sufficient to cause instant death.

An inquest was held before Coroner J. F. Edis in Virginiatown. The jury brought in an accidental death verdict and they made a recommendation: "That during scaling operations, loose rock be taken down with scaling bars where possible. If it is necessary to use gads, that these gads be used to loosen the rock only and that the rocks be taken down by scaling bars."

Interment occurred in the Kirkland Lake cemetery. The gravesite is presently unmarked.

### 1949

- Switchman Elwood Skinkle, is killed instantly at Kerr Addison
- Mucker Wasyl Horbyk, dies in hospital two-and-a-half hours after an accident at Macassa
- Timberman Valentino Bellon is killed at Toburn
- Machineman's helper Frank Stupica Jr. is killed at Teck-Hughes
- Chute Blaster James Larmond is asphyxiated at Kerr Addison
- Timberman Stefan Szumelda dies at Teck-Hughes
- Sawyer's helper Joseph Sancartier dies a day after an accident at Upper Canada
- All activities are suspended at Bidgood after 16 years of operations and the workings allowed to flood
- Kerr Addison: 764 men employed, Kirkland Lake Gold: 281 men employed, Lake Shore: 774, Macassa: 294, Matachewan Consolidated: 125, Sylvanite: 340, Teck-Hughes: 275 , Upper Canada: 203, Wright-Hargreaves: 464, Young-Davidson: 138

*Sampler using a gad and hammer to get an ore sample.*
*Photographer: Unknown*

### March 8, 1949

## Kerr Addison Mine

## Elwood E. Skinkle (1926 - 1949)

Twenty-three and single, Elwood Skinkle was from Warkworth, Ontario. He and his brother Lorne were employed at Kerr Addison in 1949.

A tramming crew consisting of Leader John Campbell, Motorman J. Wanat and Skinkle were engaged in chute-pulling operations on March 8. The train that transported the ore from the chute to the ore pass included a locomotive and nine ore cars. Before the accident, it was loaded with ore and taken to the ore pass and dumped. Wanat was on the motor and Skinkle acted as switchman.

Campbell remained at the chute and cleaned up the spillage on the track. A one-ton Hudson ore car was kept for this purpose as a clean-up car. When full, it was attached to the tail end of the train and taken to the ore pass and dumped.

When the empty train returned after being dumped, Skinkle found the partially loaded Hudson car parked under the chute; Campbell had meantime gone up to a drift. Skinkle gave one blast on his whistle to signal the train to stop. On receipt of another signal from Skinkle, Wanat started the motor and backed the train further into the drift. The train moved 10 feet, then stalled.

Hearing the train return, Campbell came down from the drift. When no further signals were heard from Skinkle, the two men went to the end of the train to investigate. They found the four end cars derailed and the train jack-knifed. The Hudson car was upset, lying crosswise on the track. Skinkle was found slumped against the wall.

Dr. D. H. Pollock was summoned to the 1,000-foot level and pronounced Skinkle dead. The miner's head was apparently crushed between one of the derailed cars and the wall of the drift.

It was assumed that Skinkle had improperly coupled the Hudson car to the end of the train. When the train backed up, the Hudson car derailed, overturned and caused other cars to derail. At the spot where the body was found, there was only 18 inches between the outside of the cars and the wall. Once the cars derailed, there was even less room and a car crushed Skinkle's head against the wall. He suffered lacerations and multiple fractures including a fractured skull and neck. He was killed instantly.

The body was taken to Warkworth for funeral services and burial.

### March 23, 1949

## Macassa Mine

### Wasyl Horbyk (1925 - 1949)

Twenty-four and single, Wasyl Horbyk was born in Kiev, Ukraine in 1925. He immigrated to Canada in 1948 and began work at Macassa on April 6 of that year.

Only six days before his accident, his sister, Mrs. Ivan Radkiewycz, had arrived in Kirkland Lake from Germany to visit her brother and to make plans for

*Motorman operating a locomotive attached to a train of ore cars. Photographer: Canadian Government M. P. B.*

settling in town. She had travelled thousands of miles to see him but ended up mourning his death.

On March 23, Horbyk and partner O. Markussen were mucking into a millhole when a minor pressure burst occurred in the back of the stope, causing about a ton of loose to fall. A slab of rock struck Horbyk a glancing blow. Markussen, who was standing a few feet away, was not injured. Horbyk was removed from the stope where he was examined by Dr. C. N. Brebner. He was brought to surface and sent to Kirkland and District Hospital.

Failing to respond to blood transfusions and other medical aid, Horbyk died two-and-a-half hours after the accident. He had suffered a fracture of the left femur, internal injuries of the chest and abdomen, and various lacerations. Death was attributed to primary shock.

Funeral services were held at St. Peter's Anglican Church with Reverend Prof. Nicholas Pacholkwi of Sudbury officiating in a Greek Orthodox ceremony. Interment occurred in the Kirkland Lake cemetery. The inscription on his headstone is: HORBYK WASYL (12 3 - 1925) (23 3 - 1949) KIRKLAND LAKE – SISTER – BROTHER-IN-LAW – FRIENDS.

## June 16, 1949
## Toburn Mine

### Valentino Bellon (1900 - 1949)

*Valentino Bellon. Photographer: Unknown*

On Saturday June 11, Valentino Bellon was one of three men chosen to receive the Canadian Institute of Mining and Metallurgy (CIM) John T. Ryan trophy from Toburn Manager Mr. Hotchkin at a gathering of employees at the Legion Hall.

The John T. Ryan Trophy is an annual safety award presented to metalliferous mines which in the previous year experienced the lowest reportable injury frequency per 200,000 hours worked in Canada. Bellon was selected from Toburn because he had worked there for 18 years without a compensable accident. The Toburn Mine in 1948 was the only mine in Canada without a single major compensable accident.

That Saturday night, Bellon had listened to a speech by Angus Campbell, President of the Mines Accident Prevention of Ontario. "In 18 months," he said, "Toburn has not had one compensable accident with over six days of lost time. Even we of the mining industry wonder at such a record. Many people wonder if there is such a record. Mining can be a dangerous business. Too often in the past, accidents have taken their grim toll of broken limbs and life itself."

Five days later Bellon was dead, killed instantly by a three-ton rock fall at Toburn. The rock had crushed him against the side of a stope.

A 48-year-old naturalized Canadian of Italian birth, Bellon lived at 96 First Street with his wife Adele and daughters Loretta (11) and Mary (9). He was a timberman and a shift boss at Toburn.

Bellon and partner John Ramsay had been working for several months replacing old timber in the 561 west stope. On June 16, Bellon decided to remove some loose below the staging. Two ladders were hung below the staging and the two men worked from the ladders, sounding the back for loose as they went down. Bellon had a safety rope that he held for added security. Using a short scaling bar, he reached over to pry off some loose from the hanging wall. In doing so, he caused a large slab to fall from directly overhead. The rock struck and squeezed him against the footwall before sliding down and lodging on the muck below. Ramsay went immediately to his partner's assistance and found him unconscious and apparently dying. He placed him in a more comfortable position and went for help.

Dr. T. F. Kelly arrived and pronounced Bellon dead. Death was caused by crushing injuries to the chest and pelvis and was almost instantaneous.

Interment occurred in the Kirkland Lake cemetery. The inscriptions on his headstone read: IN LOVING MEMORY OF VALENTINO BELLON – BORN AUGUST 29, 1900 – DIED JUNE 16, 1949 – 'REST IN PEACE'.

The photograph was given by the Bellon family. They wish to add that Valentino is sadly missed by his loving wife Adele and daughters Loretta and Mary.

## June 21, 1949
## Teck-Hughes Mine

### Frank Stupica Jr. (1923 - 1949)

Twenty-five, single and living at 44 Main Street, Frank Stupica was crushed to death at Teck-Hughes by a rockfall on June 21.

This was the third accident – two of them fatal – within a week in the mines of Kirkland Lake. Ironically, Valentino Bellon, who had received a safety award a few days earlier, was killed at Toburn on June 16. Harry Rushton survived an accident at Kirkland Lake Gold on June 21 – at the expense of his left arm. Also on that day, Stupica died.

Stupica and partner Nick Prentovich were completing slusher operations in a stope located between the 2,500 and 2,600 levels. A slusher hoist was used to scrape broken ore into an opening through which the ore went directly into ore cars.

At the beginning of the afternoon shift, Shift Boss Perry Burke visited the stope and found conditions satisfactory. Machineman Prentovich and helper Stupica were scraping broken ore into the cars. Prentovich operated the slusher hoist and Stupica stood at a point where there was a turn in the stope and gave signals to Prentovich. The slusher scraper and cable passed under a 14-inch stull. There was about four feet of clearance to allow the scraper to pass. The two men loaded several cars. Prentovich could not see Stupica but could see his light. He noted that the scraper had jammed on a return trip and that he did not get a light signal from Stupica. He left his machine and went to investigate. It was assumed that the scraper had hit a stull and Stupica left his signalling position to examine the situation. Prentovich found that a large slab of rock had fallen from the hanging wall and crushed his partner. The dislodged stull was probably supporting the rock and, when Stupica walked over to examine it, the rock let loose and fell on him. Dr. A. W. Souter arrived to pronounce Stupica dead. He had sustained multiple crushing injuries involving the skull, neck, thorax and pelvis.

Frank Stupica was survived by his father, Frank Sr., also employed at Teck-Hughes. The victim's mother resided in Yugoslavia.

*Headstone photograph of Frank Stupica.*
*Photographer: Author*

Interment took place in the Kirkland Lake cemetery. The inscription on the stone reads: HERE LIES FRANC STUPICA – BORN 16 JULY 1923 AT PRESKA P. SADRAZICA – KILLED AT TECK-HUGHES MINE ON 21 JUNE 1949 – LEAVES BEHIND PARENTS BROTHER AND SISTER – NOT FORGOTTEN REST IN PEACE.

## July 5, 1949
## Kerr Addison Mine

### James Willard Larmond
### (1920 - 1949)

At the beginning of the dayshift on July 5, James Larmond and his tram crew, including brother John G. Larmond and A. Sidaravicius, were in the process of drawing ore from a plugged chute. They decided to dynamite the ore to clear it.

There was nothing abnormal about their decision as this was a duty that miners perform routinely. What was abnormal was their methodology.

The normal procedure is to use a blasting pole as an extension into the plugged chute to fix the shot to the ore. But to save time, James Larmond climbed into the chute from a manway drift located halfway up, with the intention of manually fixing the shot to the ore. While in the chute, the rock let loose and trapped him inside.

Dr. D. H. Pollock was called to the mine where he went underground to the 1,300 level, finding Larmond buried in muck in the chute with only his legs visible. The doctor felt his legs but found no evidence of life and pronounced the miner dead. It took rescue workers seven hours to extricate the body. On observation, the doctor observed no major injuries and stated that Larmond died of asphyxia due to pressure on his chest and abdomen.

At the inquest held before Coroner H. F. Richardson, it was stated that Larmond, though an experienced miner, contravened established mine rules by climbing into a plugged chute to fix a charge to the ore.

James Larmond, Canadian, 28 and married with two children, had worked at Kerr Addison from December, 1945. He was buried in the Kirkland Lake cemetery. The inscription on his headstone reads: JAMES WILLARD LARMOND – 1920-1949 – IN LOVING REMEMBRANCE – FROM WIFE AND FAMILY.

## December 1, 1949
## Teck-Hughes Mine

### Stefan Szumelda (1914 - 1949)

Stefan Szumelda was a recent Ukrainian immigrant who lived at 110 Poplar Avenue with his wife Katharina and three-year-old son Taras. The family came to Canada in 1948 and he began work at Teck-Hughes in June that year.

During the afternoon shift of December 1, Machineman Eskel Farstad and partner S. Slusarz had drilled and loaded a round at the top of the angled stope above the manway. Szumelda and partner S. Siuciak were timbering in the same stope.

Along the manway leading down from the stope to the level was a slide, equipped with a bucket, cable and tugger hoist. The slide was used to convey materials up and down between level and stope. The tugger hoist was used by both crews during the shift and, at times, the bucket was disconnected by Szumelda to hoist timber. When the bucket was in use, the bucket bail was inserted into a clevis, and a bolt was passed through and supposedly nutted. However, for several previous shifts, the nut was not placed on the clevis bolt because the threads on the bolt had become burred – the nut would not fit.

Shortly before the accident, Farstad and his partner had completed drilling and loading the holes. Into the steel bucket suspended at the top of the slide, they placed the bit racks, oil bottle and an empty powder bag. The bucket and contents now weighed about 150 pounds.

The holes were lit and all four men came down the manway. Slusarz and Siuciak came down first, followed by Farstad and then Szumelda. When Farstad got to the level, he went over to the tugger hoist to prepare to lower the bucket. Szumelda, coming down the ladder, yelled, "Okay, let her go," signalling that Farstad could now lower the bucket. Farstad released the brake; the faulty bolt came out of the clevis and the bucket fell free from its

*Headstone photograph of Stefan Szumelda.
Photographer: Author*

position at the top of the slide. Falling erratically down the slide, the bucket hit Szumelda at the base of his skull and neck just as he was about to step off the ladder at the bottom of the manway. The impact of the falling bucket broke his neck and damaged his nerve cord; he died within a few minutes.

The circumstances that contributed to this accident were twofold: first, the clevis attachment was not secured properly; second, the deceased was exposed under a moving load. Szumelda contributed to both these conditions in that he was the man who last attached the bucket improperly, and, in signaling Farstad to lower the bucket, gave the impression that he was in the clear.

At the inquest held before Coroner J. F. Edis, the jury brought in a verdict of accidental death but made one recommendation: that no clevis be used that would allow a bucket to be accidentally detached. Since this accident, all clevises were replaced with a threaded type clevis in which the bolt could not be removed.

Interment was conducted in the Kirkland Lake cemetery. The inscription on the tombstone is in the Ukrainian language and is translated thus: Here lies Stefan Szumelda. Born 15th January 1914. In the village of Booshkovichi. Suddenly killed working in the Teck-Hughes as a miner. Remembered as a Ukrainian. Arrived in Canada from the fatherland. Rest in peace. Ever remembered by wife Katharina and son Taras.

### December 12, 1949

## Upper Canada Mine

## Joseph Sancartier (1884 - 1949)

Sixty-five-year-old Joseph Sancartier was injured in a small saw-mill unit on surface at Upper Canada on December 12. He died the next day in hospital.

He was employed as a sawyer's helper. He had worked almost three years on the job and was thoroughly familiar with the working conditions.

Sancartier was born in Quebec in 1884 to French-Canadian parents. In time, he married Aldea Poulin and the marriage produced ten children: Palma, Laurence, Rita, Faldora, Edna, Edgar, and four other children who died in a diphtheria epidemic.

Sancartier worked at numerous jobs throughout his life: on a log drive on the Wanipitei River, a jobber in a bush camp and an employee at the INCO smelter in Sudbury where he learned to be a blacksmith. He left INCO and started his own blacksmith business but it did not go well. He became a foreman on the railroad where his territory of responsibility on the Temiskaming and Northern Ontario was between Boston Creek and Dane. He resided in Boston Creek during the mining boom and, along with his two sons, built a hotel named The Miners Rest Hotel. When the mines closed in Boston Creek, he moved his family and his team of horses to Larder Lake where he made contracts with diamond drill companies to deliver their gear by horse to the drilling sites. Joseph enjoyed being around horses. When this business slowed down in the late '30s, he moved his family to Kirkland Lake where he became employed at Macassa, working there until the strike of '41.

He then became a forest fire fighter until he secured employment at the Perron Saw Mill in Kirkland Lake. His last place of employment was Upper Canada where he worked for three years. His last address in Kirkland Lake was 103 Second Street.

**A la douce mémoire de**

**JOSEPH SANCARTIER**

**époux de**

**ALDEA POULIN**

**décédé à Kirkland Lake, Ont.**

**le 13 décembre 1949**

**à l'âge de 67 ans.**

*Mass card for Joseph Sancartier*

At the time of the accident, a slab of lumber was being ripped when it passed on the wrong side of the splitter. Instead of reversing the carriage to release the slab, Sancartier reached out to free it. In doing so, his left hand touched the rotating blade and the three middle fingers were partially severed.

Following the accident, he was given first-aid treatment at the mine then taken to hospital in Kirkland Lake where he appeared in good spirits and showed no signs of serious shock. Dr. R. J. Neelands, assisted by Dr. M. J. Thompson, operated to amputate the damaged fingers. During the operation, Sancartier showed signs of weakness and was revived. After the operation, his condition deteriorated and he failed to respond to stimulants. He died on December 13 without having gained consciousness.

An autopsy performed by Dr. A. W. Souter revealed a deteriorated condition of the arterial system, in particular, the aorta. The conditions weren't unusual for a man of 65 years. The primary cause of death was attributed to failure of the circulatory system, aggravated by the injuries sustained and subsequent necessary surgical treatment.

Sancartier was survived by his wife and six children. Funeral services were held at Assumption Church and interment occurred in the Kirkland Lake cemetery.

An inquest was held before Coroner Dr. J. F. Edis. The jury consisted of Foreman Freeman Bigelow and jurors Bert Baker, William J. Brick, Frank Burnside and Donald Cameron. The verdict was accidental death.

Sancartier's wife died in 1968 and presently (1999) all the children but one, Edgar, are deceased. He lives in Elliot Lake and has given details and anecdotes to help create this memorial to his father. The late Leo Desjardins of Larder Lake found a photograph of Joseph Sancartier on a mass card.

Edgar remembers as he grew up being told by family of certain events occurring in the early days: the loss of siblings to diphtheria, failing businesses, loss of employment and changing times. He remembers his father as being a very generous person, a devoted husband and family man. "He would do anything to please my mother," he states.

He further reminisces, "I was told that when my dad was younger, he bet $5 that he could carry three 100-pound bags of flour a certain distance using a harness on his back. When he accomplished the feat, he wanted to bet another $5 that he could carry them back to where he started."

Edgar concludes, "The only souvenir I have of my father is his Waltham railroad watch."

*Slusher operation in a stope. Photographer: Unknown*

# Chapter V
# THE FIFTIES

### 1950

- Kalle Maki is killed at Chesterville
- Danny Starr is killed at Lake Shore
- Charles Dunn dies in hospital after an accident at Kerr Addison
- Albert Chorzepa is killed at Kerr Addison

### January 28, 1950
## Chesterville Mine
### Kalle Maki (1900 - 1950)

Forty-nine-year-old Kalle Maki, born in Finland on September 2, 1900, was married and lived in Larder Lake with his wife and son. Their adult daughter lived in Hamilton, Ontario. A drill runner at Chesterville, he was an experienced miner, employed at this mine since May 1949.

On the morning of the accident, he and partner O. Olsen were barring down in a drift on the ninth level. Broken ore was being passed from the drift into a hole in the floor at the top end of a 250-foot raise. The hole was protected by two sprags across the drift that acted as a barricade. Two safety belts were at site for use by workmen when it was necessary to go beyond the barricade to drill, blast or muck.

The ore had become hung-up near the hole and they decided to place a blasting pole of several sticks of powder into this section. Without putting on his safety belt, Maki climbed over the barricade. Olsen handed him the made-up blasting pole; Maki set it in place and lit the fuse. Before the blast, some small pieces dislodged from the hung-up section and the sound of this startled Maki. He stepped back suddenly, missed his footing, and fell into the hole. Olsen made a desperate grab for him but was unable to get a good grasp on his oilers. Olsen promptly pulled the blasting charge from the ore, yanked out the fuse and went for help. A crew under the charge of the acting Mine Superintendent Borden Bowes started recovery operations. It soon became evident that Maki had fallen to the muck 250 feet below and could not have survived. In time, his body was recovered and Dr. H. F. Richardson pronounced him dead. He had suffered a fractured skull, severe brain damage, a fractured neck, crushed chest, various internal injuries and fractured arms and legs.

The body rested at Symington's Funeral Home until the service conducted by Reverend George Soutar at the Virginiatown United Church. Interment took place in the Kirkland Lake cemetery.

An inquest was held in Virginiatown before Coroner J. F. Edis. The jury was of the opinion that Maki, though an experienced miner, was not wearing a safety belt when he fell to his death.

### May 10, 1950
## Lake Shore Mine
### Thomas Daniel Starr (1904 - 1950)

Down in a stope located more than a mile below the surface in the Lake Shore mine, 45-year-old Thomas 'Danny' Starr of 1 Brookbank died on May 10.

His partner, George Baczynski, an immigrant from Poland, was pinned waist deep in the muck and was rescued three hours after the accident. Baczynski was taken to hospital suffering from fractured ribs, a broken back, a fractured pelvis and a ruptured kidney.

Starr was employed at Lake Shore since 1930, except for five years of service in the Canadian Army during World War II.

The following statement was released by Mine Spokesman Alex Harris a few hours after the accident: "While timbering just below the 5,800 foot level of Lake Shore Gold Mines Limited, a fall of rock fatally injured Thomas Daniel Starr and

*Danny and Margaret Starr on their wedding day. Photographer: Unknown*

injured his partner, George Baczynski." The two were scaling a breast when a five-ton slab fell unexpectedly and pinned both beneath it. When they failed to report at the end of the shift, Shift Boss Joe Gibson went to look and found them under the rock.

Starr suffered a crushed chest, crushed abdomen and a compound fracture at the base of the skull. He was killed instantly.

During the Second World War, he had served with the Royal Canadian Engineers and saw service at Gibraltar where he was a member of the tunnel crews. For his service there, he was mentioned in Dispatches and was later congratulated by His Royal Highness, the Duke of Gloucester.

He was survived by his wife, the former Margaret E. Brignall, and by daughter Joan who was attending Kirkland Lake Collegiate. His parents were living in Yorkshire, England.

The body rested at Symington's Funeral Home until the funeral service at St. Peter's Anglican Church with Canon J. A. Watton officiating. Interment took place in the local cemetery. The inscription on the gravestone reads: CORPORAL DANNY STARR B28522 – 1904 - 1950 – BELOVED HUSBAND OF MARGARET E. BRIGNALL 1902 - 1980.

## August 24, 1950
## Kerr Addison Mine

### Charles David Dunn (1913 - 1950)

Charles Dunn, 37 and married with four children, died in hospital from injuries sustained in a stope on the 300 level at Kerr Addison.

He was employed at this mine as a driller since July, 1950. Before coming to Kerr Addison, he had seven years experience in mining and was considered an experienced miner.

Born in Orillia, Ontario, Dunn moved with his family to Kirkland Lake in 1943 and resided at 43 Main Street. His family included his wife Adeline and children David (13), Ann (10), Joan (9), and Charles (6).

Dunn and partner Albert Kerr started work at 8:30 a.m. on August 24, picking and shovelling ore into the path of a scraper. The stope was visited at 9:00 by Mine Captain Joe Teal and Shift Boss Duncan Johnston. The condition of the stope was considered safe by the supervisors and the workmen. At about 10:00, a large slab of rock fell from the north wall of the stope, striking the south wall and breaking into pieces. Dunn, standing near the slab, was struck by one or more of the pieces. Kerr ran to Dunn's aid and found a large chunk of rock lying across his shoulder. Unable to free him, Kerr went to level 300 to get help. Dunn was freed within a few minutes after help arrived. The first-aid officer administered treatment until Dr. G. E. Hagerman came underground.

Dunn was alive and taken to Kirkland District Hospital. He died at 2:00 a.m. on August 25, 16 hours after the accident. He had suffered a fractured and dislocated neck, a fractured shoulder blade and severe shock.

Surviving him were his immediate family, his parents, Mr. and Mrs. David Dunn of South River, six sisters: Mrs. Anthony Morano of Orillia, Mrs. Ben Welch of Kirkland Lake, Mrs. Ray Oviatt of Charlton, Mrs. Robert Neeley of Otto Twp., Mrs. James McInnis of Kirkland Lake and Mrs. Albert Goldsmith of Orillia, and brother Jack of South River.

Interment occurred in the Kirkland Lake cemetery; the gravesite is presently unmarked.

### October 11, 1950

## Kerr Addison Mine

### Albert Chorzepa (1902 - 1950)

Albert Chorzepa, 48, born in Poland on March 3, 1902, was married and had a son, Emil. He and wife Lucia lived on Lakeshore Road in Larder Lake. Their son was no longer a dependent and worked at Kerr Addison as a sampler. Chorzepa became a naturalized Canadian in 1948.

An experienced miner, Chorzepa was hired at Kerr Addison on January 17, 1939. His regular occupation was mucking machine operator.

He and partner Nicholas Rouillard were sent by Shift Boss George Rathwell to draw muck from a stope on October 10. They started passing muck at the beginning of the shift at 8:00 p.m. A tramming crew on the level below drew the muck away. In the process of passing muck, Chorzepa and Rouillard were also doing some scaling in the work area.

At 12:40 a.m. October 11, the two had filled the box-hole with muck and were waiting for the tramming crew to pull it empty from below. Chorzepa leaned his shovel against the wall and sat down on the heel of it. Rouillard looked up and noticed a small piece of loose. He called Chorzepa's attention to it and was told to scale it down. He had merely touched this small piece with his scaling bar when a large slab, weighing about three tons, fell from the back. The airlines and water lines, enclosed in a four-inch diameter pipe, were directly under the slab and fell with it. One end of the pipe struck Chorzepa, breaking his neck and killing him instantly. Rescue workers had to use a jack to lift the pipe from the body.

The funeral mass was held at Our Lady of the Lake Roman Catholic Church in Larder Lake and interment occurred in the Kirkland Lake cemetery. The inscription on the gravestone reads: WOJCIECH CHORZEPA URODZONY 3 MAR 1902 – ZMARL 11 PAZD 1950 – TU SPOCZ MAZ LUCJI A OJIEC EMILIA. Translation: Albert Chorzepa – Born March 3 1902 – Died 11 October 1950 – Here rests the husband of Lucia and the father of Emilia.

### 1951

- Leslie Loney, 41, is killed at Wright-Hargreaves
- Evan Rennie, 35, dies underground at Lake Shore

### May 17, 1951

## Wright-Hargreaves Mine

### Leslie Loney (1910 - 1951)

Forty-one-year-old Leslie Loney was born in Packenham, Ontario in 1910. He, his wife and two sons, Richard (10) and George (3), lived at 7 Brookbank Avenue.

An experienced miner, Loney was employed as a timberman at Wright-Hargreaves and had been employed there since his arrival to Kirkland Lake in 1946.

On May 17 during the night shift, he and partner P. Maru were cleaning down muck and making preparations for timbering another row of stulls in a stope. Loney noticed a large slab on the hanging wall and, taking his pick, proceeded to chip a flat spot on the slab to rest the head block of the stull. Maru was about 30 feet away at the time, standing on a muck pile. As Loney tapped the wall, the two-foot-thick slab fell off, struck the muck, toppled over and pinned Loney's head and chest against the wall, killing him instantly.

Maru went for help. Loney was freed by removing the slab with a chain block and tackle. Dr. T. F. Kelly attended the victim underground and pronounced him dead.

The body rested at Symington Funeral Home until the funeral mass at Holy Name Church. Burial took place in the cemetery at Almonte, Ontario.

### June 13, 1951
## Lake Shore Mine

# Evan Dougall Rennie

### (1916 - 1951)

Evan Rennie was born to Milne and Mabel Rennie on March 5, 1916 at Hensall, Ontario. Raised in Seaforth with three brothers, Gordon, Ross and Ronald, he was later involved in a grocery business there but left to get a job in the gold mines. On July 11, 1943, he married the former Florence Woods at Holy Name Church with Father Cavenaugh officiating. This marriage produced three children and the family lived at 111 Duncan Avenue.

As Private #B65717 throughout the war, he served overseas with the Algonquin Regiment. While there, the musical Rennie formed a band that he conducted while playing the trumpet and trombone.

After the war, he returned and got a mining job at Lake Shore. With an avid interest in sports and music, he was the catcher on the Joe Dash Baseball Team and an active band member in various local groups such as the Fay Charles Orchestra, the Kirkland Lake Citizens' Band and the Jerry Furber Band. He had many friends in the local music fraternity. His special friends were Oliver and Marie Blais, owners of an automobile dealership in town.

On June 13, 1951, Rennie, a sampler at the mine, failed to come back to surface at the usual time. Art Livingston and George Birnie searched an area on the 4,900 level that had not been worked. George Markovitch, a motorman working on this level, was the last person to see Rennie alive. He was found dead, 34 feet above the level, in a manway.

Surviving 35-year-old Rennie were his wife Florence and children: Milne Arthur (4), Betty Ann (3) and Donald Ross (five months). He was also survived by his brothers.

Funeral rites were conducted at Trinity United Church by Reverend T. V. Hart and burial occurred in the local cemetery. The epitaph on the headstone over the grave reads: 'EVER REMEMBERED, EVER LOVED'.

Details of the story of her father were given by daughter Betty Ann of Kirkland Lake. Rennie's wife and son Milne also live here presently (1999). Son Donald and brothers Ross and Ronald live in the Toronto area. Brother Gordon lives in Sundridge.

*Private Evan Rennie played trumpet for the Queen's Own Rifles during World War II. Photographer: Unknown*

*Short rest after lunch. Underground at Sylvanite 1952. Photographer Unknown*

## 1952

- Buzz Leverre is killed in a blasting accident at Lake Shore
- Robert Pitcaithley is killed in a rockburst at Lake Shore
- Jan Piekos is caught in a run of muck at Kerr Addison and dies
- Franz Ludwig dies at Matachewan Consolidated
- Antonio Poloni, Leo Kensy and Aleksei Soopalu are killed in a rockburst at Lake Shore
- Giovanni Berlingeri becomes a fatal victim at Sylvanite
- Toburn Mine closes

### January 21, 1952
## Lake Shore Mine

## Norman E. 'Buzz' Leverre

### (1927 - 1952)

Buzz Leverre, 24, was killed instantly on January 21 at Lake Shore. On the day shift, he and partner William Lacharity were doing routine blasting in a stope on the 6,200 level. After lighting the fuses of the round, Lacharity went down from the stope to guard the level and Leverre went up to the 6,025 to guard there. On reaching the lower level, Lacharity asked M. Shubat, a motorman, to guard the west side of the manway while he guarded the east. After about four minutes, three shots went off and, about four minutes later, another shot exploded. Lacharity waited for about 15 minutes more, expecting to hear his partner call him back into the stope. As it is not considered a safe practice to return to a blasting area from below before it has been checked from above, Lacharity went up to the 6,025 by means of a manway and entered the stope from there. He returned to the scene of the blast without seeing Leverre. He again went down to the 6,200 to see if his partner was there. Shubat was asked if he had seen Leverre and, since he had not, the two men returned to the scene of the blast. They were puzzled by the disappearance of Leverre but were not unduly alarmed. They decided to pull some muck from the stope.

The original blast had been set at 8:30 a.m. and it was now 10:30. Shubat informed Shift Boss Joe Gibson and Mine Captain Lou Foran of the disappearance of Leverre. Both supervisors went into the stope to investigate. Foran organized a search of the area and Gibson brought in six men to dig down into the muck. At approximately 11:45, Leverre's body was found buried under the muck in the blasted area. He apparently made an error in judgment and returned to the scene of the blast too early. He was at the scene when the last shot went off, violating Mining Act regulations by returning too soon.

Leverre lived in Kirkland Lake the greater part of his life. He was married to the former Zetta Bowers and they lived at 5 Prince Street. He had been predeceased by his father Gordon in 1948 and a sister in 1945. He was survived by his wife, his mother, Hazel Marshall Leverre (1902 – 1992), and three brothers, Harold of Larder Lake, Tom of Sesekinika and Stanley at home in Kirkland Lake.

Leverre rested at his late residence until the funeral service officiated by Reverend T. V. Hart at Trinity United Church. Interment occurred in the local cemetery. The inscription on the headstone reads: NORMAN E. 'BUZZ' LEVERRE – NOVEMBER 18, 1927 – JANUARY 21, 1952 – BELOVED HUSBAND OF ZETTA A. BOWERS – JUNE 12, 1932 – OCTOBER 3, 1952 – 'PEACE, PERFECT PEACE'.

## April 10, 1952
## Lake Shore Mine

### Robert Pitcaithley (1907 - 1952)

Showered with timber and rock when a rockburst occurred at Lake Shore, Robert Pitcaithley, a veteran miner and (ironically) a member of the Lake Shore Mine Rescue Team, was killed instantly the night of April 10.

Forty-five-year-old Pitcaithley who worked 18 years at Lake Shore had just reached his working place with four others when the burst occurred. He and partner J. Saunoris were making preparations to drill on the 5,700 level under the direction of Shift Boss Harvey Simms. A motor crew including Frank Simms and Rudy Senbergs happened to be at the area at the time the burst occurred. A severe bump was felt and a small amount of rock fell. The five workmen immediately left and headed towards the shaft station. The first bump was followed in less than 30 seconds by a violent burst that brought down 50 tons of rock and timber. Pitcaithley was not quite clear of the fall of ground as he retreated towards the shaft and was buried near the edge of the fallen debris. Frank Simms had his foot caught by a falling piece of timber but he was immediately released by his father Harvey; he was not seriously injured.

*Robert Pitcaithley. Photographer: Unknown*

Pitcaithley's absence was not noted until the four survivors reached the shaft station. After a brief interval of time, Harvey Simms returned to the scene. It was then realized that Pitcaithley was buried in the muck and immediate steps were taken to recover him. The body was recovered by a crew working under the direction of Underground Superintendent Walter Selnes. Dr. G. M. Cameron, who was present, pronounced Pitcaithley dead.

Pitcaithley was born in 1907 at Whitehead, Northern Ireland. He came to Canada in 1927, worked in Toronto and Timmins before settling in Flin Flon where he was employed by the Hudson Bay Mining and Smelting Company. He married Evelyn Robertson in 1930 and moved in 1934 to Kirkland Lake where he became a miner at Lake Shore.

He was predeceased by his parents and his only sister. Also, his son Bobby had recently died on December 23, 1950 at the age of 13. He was survived by his wife and two daughters, Glenda (16) and Marlene (13) and a brother Samuel in Ireland.

The body rested at Symington Funeral Home. With a host of friends and neighbours in attendance, funeral services were held at St. Andrew's Presbyterian Church where Reverend J. N. Hepburn officiated, assisted by Reverend John Lucas. The pall bearers at the funeral were: W. A. Livingstone, Paddy Nelson, Blake McTiernan, Frank Butorac, Orval McTiernan and Herb Allen. Mrs. Pitcaithley's sister, Mrs. Irene Stickney of Flin Flon, her brother, Joe Robertson of The Pas and Mr. and Mrs. Paddy Nelson of Toronto and their daughter Doreen, were in town for the funeral. Interment was in the local cemetery.

Presently (1999), Marlene Pitcaithley, daughter of Robert, lives at Round Lake, near Kirkland Lake. She gave details and the photographs for this memorial to her father. Her sister, Glenda Tappenden, and family live in Sudbury. Mrs. Evelyn Pitcaithley also resided in Sudbury until her passing on January 18, 2000 at Sudbury Regional Hospital at the age of 92.

*Lake Shore Mine Rescue Team getting instructions from Percy Smith (1948). Seated left to right: (two men unidentified) ,Wallace St. Cyr, Neil Culhane, Robert Pitcaithley, Pat Cain. Photographer: Unknown*

*Headstone photograph of Jan Piekos.*
*Photographer: Author*

## April 16, 1952
## Kerr Addison Mine
### Jan (John) Piekos (1902 - 1952)

Forty-nine-year-old Jan Piekos of Polish descent lived with his wife Katarzyne on Godfrey Street in Larder Lake. Their only child, Adolph, was an adult living in Hamilton, Ontario. Piekos, a timberman, was an experienced miner and employed by Kerr Addison from May 13, 1941.

On the morning of April 16, he was instructed by Shift Boss Bill Roulston to take his oil clothes and tools and go down a raise to remove some pieces of wood that had caused an obstruction in an ore chute. Since the job was of a routine nature and required only one man, Piekos told his partner Frank Grabowski to continue with the job they had been doing the previous day.

He descended the raise at about 9:00 a.m. At 9:30, John Larmond and John Wilcox, who were working nearby, heard a run of muck in the raise but did not consider it unusual. What they did not realize was that a timbered partition had collapsed in the raise and a wet fill of waste rock and pit sand had buried Piekos. At 10:00, Motorman A. Vincent observed the level in the fill compartment had dropped, which was unusual. He descended the raise and noticed that fill had spilled across the raise and into the manway. He reported this to Grabowski who immediately went to the station looking for Piekos. Not finding him, he returned to the raise where he found Roulston, who subsequently went down the raise and found that the timber had collapsed. Men working in the area reported to their shift boss that they had not seen Piekos.

At 11:15, Roulston, suspecting that Piekos was trapped in the fill, summoned aid. At noon, Dr. G. E. Hagerman of Virginiatown arrived underground. At 2:30, Piekos was recovered from the landing at the control chute and pronounced dead. Death was caused by suffocation.

The body rested at the home of the deceased in Larder Lake until funeral services were held at Notre Dame du Lac Catholic Church in Larder Lake. Interment took place in the Kirkland Lake cemetery. The inscription on the headstone reads: KU WIECZNEJ PAMIECI – JAN PIEKOS – ORODZONY W POLSCE W 1902 – ZGINOL TRAGICZNIE W 1952 – W RZALU ZOSTAWIA JAC RZONE – KATARZYNE I SYNA APOLFA – NIECH TA KANADYJSKA ZIEMIA LEKKA MU BEDZIE. Translation: In everlasting memory Jan Piekos – Born in Poland in 1902 – Lost his life tragically in 1952 – In sadness he leaves his wife – Katarzyne and his son Adolph – Let this Canadian earth be light for him.

## May 15, 1952
## Matachewan Consolidated Mine
### Franz Ludwig (1932 - 1952)

Twenty-nine and German, Franz Ludwig died on May 15 at Matachewan Consolidated. He was a recent immigrant and had been employed at this mine from February 25, 1952.

On the day of the accident, he was blasting blocked holes in a grizzly. Victor Sventoraitis was doing the same, 40 feet away. After lighting their fuses simultaneously, they went down to the level and waited about 15 minutes after the blast for the smoke to clear. They returned; Ludwig remarked that there was a missed hole in his section. After working about 20 minutes to clear away broken ore, Sventoraitis prepared another blast in his area while Ludwig prepared a blast in the missed hole. Sventoraitis shouted to Ludwig to light and then he lit his own fuse and left the area. As he was leaving, he saw Ludwig still bent over, apparently in the act of lighting his fuse. At this instant, an explosion occurred. Sventoraitis could not see because of the smoke and dust caused by the blast; he called to Ludwig but received no reply. He returned to his blasting area and cut the fuse on his potential blast. He proceeded to Ludwig's area again and found him lying on the muck, badly hurt.

Sventoraitis went down to the level and got the help of Shift Boss R. Mills and Xavier Letellier. Ludwig was removed to surface where he was examined by Dr. F. Fitch of Matachewan. The accident occurred at 2:25 p.m. and the victim died at 4:16 p.m. in the mine hospital.

From the evidence revealed at the inquest, it is likely that one of the two short fuses lit inadvertently while Ludwig was attempting to light the five-foot fuse.

## May 31, 1952
## Lake Shore Mine

### Rockburst at Lake Shore Mine Claims Three Miners

On April 10, 1952, a sharp rockburst at Lake Shore on the 5,700 claimed the life of Robert Pitcaithley. After the burst and consequent death, the mine took immediate steps to reinforce the timbering in the crosscuts from the 5,575 to the 5,825. Because of the concrete work previously done on the 5,450, it was thought that this area was secure. Another burst seven weeks later on May 31 at the 5,450 claimed three more lives.

On the 7:00 p.m. to 3:00 a.m. shift, a crew including Motorman Rene Houle, Loader Antonio Poloni and a second crew of Drill Runner Roderique Charron and Aleksei Soopalu proceeded to the 5,450 to carry out loading and tramming operations. When they arrived, they started by taking two motors from the charging station and soon discovered that the track was obstructed by several pieces of loose that had fallen from the back and walls of the cross-cut. They noticed that the ground was 'working a little' and a few pieces were still falling. A message was sent to the level above to Shift Boss Leo Kensy for instructions on what to do in this situation. Kensy was not near the phone and did not get the message but had in the meantime climbed down to this level and entered the crosscut. He instructed Charron and Houle to scale and Poloni and Soopalu to clear the track of fallen rock. Kensy also picked up a scaling bar and proceeded to help them with the scaling. About 10 minutes later, a rockburst occurred along the walls and back of this crosscut. About 200 tons of rock fell. Charron and Houle were standing by a timber ventilation doorframe that afforded them some protection. Kensy was a few feet away from the frame on the south side, and Poloni and Soopalu were north of the frame. The fall of rock resulting from the burst buried Kensy, Poloni and Soopalu. Houle and Charron escaped as soon as they were able and ran to the shaft, passing over the pile of muck that had covered the others. Help was summoned and recovery operations began immediately under the direction of Mine Captain Harvey Simms. Underground Superintendent Walter Selnes arrived shortly to assume charge of the work.

About 11 hours later, Poloni's body was recovered; an hour later, Soopalu was found and Kensy, two hours after that. Dr. C. S. Harris was in attendance and pronounced the men dead. Kensy and Soopalu sustained fractured skulls and probably died instantly. Poloni's death was caused by traumatic asphyxiation and probably occurred within a few minutes of the accident.

Rene Houle and Roderique Charron received minor injuries and required hospitalization.

*Antonio Poloni. Photographer: Unknown*

## Antonio Poloni (1916 - 1952)

Antonio Poloni was born in Italy on July 27, 1916. He was married to wife Emma who with nine-year-old son Guido came to Canada in January 1952. Poloni had arrived earlier, in 1948. He was an employee at Lake Shore from November 1949 and resided on Kirkpatrick Street.

Antonio Poloni was the cousin of Kirkland Lake's Tony Poloni who was the lessor of Culver Park and the owner of a local laundry. In the '60s and '70s, Tony Poloni was also the owner of the Queens Hotel and presently (1999) is the owner of the Don-Lou Motel, the one-time Lake Shore bunkhouse.

Many fellow miners attended the funeral held at Holy Name Church. A photo in the 1952 Northern Daily News shows the pallbearers, including Tony Poloni, carrying the casket out of the church. Interment occurred in the Kirkland Lake cemetery. The inscription on the stone marking the grave reads: ALLA CARA MEMORIA DI ANTONIO – NATO IL 27 JUGLIO 1916 – MORTO IL 31 MAGIO 1952 – MARITO DI EMMA – REPOSA IN PACE.

Antonio's son, Guido Martignago, presently lives in Kitchener (1999).

*Funeral of Antonio Poloni. Photographer: Unknown*

*The Kensy funeral was a double one as mother Agnes died of a heart attack as she listened to the CBC News announcing the presumed death of her son Leo. Photographer: Author*

## Leo Kensy (1911 - 1952)

On the Saturday night of May 31 at 8:30, a rockburst at Lake Shore killed three miners and subsequently caused the death of a fourth person who was not in the mine.

Leo Kensy was born in Poland on April 19, 1911 to Stanley and Agnes Kensy. He married the former Helen Gaynor of Winnipeg and they had a daughter Bernice, a 14-year-old student at Kirkland

Lake CVI. The family resided at 12 Poplar Avenue. Kensy worked in his father's blacksmith shop before his employment at Lake Shore in 1930.

The Kensy funeral was a double one as his mother died of a heart attack on June 1 when she heard the CBC Radio News announcing the death of her son. The funeral services for son and mother were held at Holy Name Church and interments at the Kirkland Lake cemetery occurred side by side. Surviving were his wife and daughter, his father, a 32-year-old brother Ben – also a miner at Lake Shore – and two sisters, Helen (Mrs. Robert Brough) and Cecilia (Mrs. Stanley Ratza).

Helen and Bernice Kensy later moved to Milwaukee to live near relatives there. Presently (1999), Mr. and Mrs. Brough reside in Kirkland Lake. Kensy's niece Jacqueline lives in Sesekinika and nephew Robbie Brough lives in Kirkland Lake.

## Aleksei Soopalu (1908 - 1952)

Aleksei Soopalu, born in Estonia on September 21, 1908, came to Canada in 1950 and became employed at Lake Shore in September of that year. He was married to Elena.

*Alex Soopalu and wife Elena. Photographer: Unknown*

The funeral service was held at St. Peter's Anglican Church. Interment took place in the Kirkland Lake cemetery.

An epitaph on his headstone reads: 'PUHKA RAHUS VOORAMAA MULLAS'.

*Giovanni Berlingeri. Photographer: Unknown*

### June 7, 1952
## Sylvanite Mine

## Giovanni Berlingeri (1906 - 1952)

Giovanni Berlingeri was born in Villa Santa Lucia, Abruzzi, Italy on September 4, 1906. In 1948, he left his homeland with his 18-year-old son Dominic and came to Kirkland Lake where he began mining at Sylvanite on October 15 that year. His move to Canada was sponsored by his brother Angelo who had earlier come to Kirkland Lake in 1927. Giovanni's wife Nicoletta and the two other children came to Canada months later. By 1952, the re-united family was living at 52 Woods Street.

Berlingeri was employed in various underground jobs during his four years at Sylvanite and, at the time of his death, was working as a slusherman.

On June 7, 46-year-old Berlingeri was working in a stope at the 1,250 level with partner Emidio Celli. Berlingeri noticed a piece of loose over the slusher travelway near the end of the stope and proceeded to scale it with a five-and-a-half foot scaling bar. He tried to scale from one side and, unable to get good leverage, attempted to walk under or around it to get to the other side. Suddenly, a two-ton mass of rock fell, crushing him. He called for help to Celli, who was a few feet away. Celli wasn't able to move the rock from his partner and immediately went for help to the level below. Dr. C. S. Harris was summoned and immediately went to the scene of the accident and pronounced Berlingeri dead. He had a fractured clavicle and a fractured leg but his death came from a crushed chest and internal injuries.

He was survived by his wife, two sons: Dominic (21) and Tony (14), daughter Lucy (17), brother Tony of Niagara Falls and brother Angelo.

The funeral mass was held at Holy Name Church and interment occurred in the local cemetery. A second interment years later took place in Port Colborne, Ontario.

Some of the information and the photograph were given by Don Berlingeri, a nephew of Giovanni. Berlingeri, his wife Sue and their two children presently (1999) live in Kirkland Lake; his brother Ralph and his wife Helen also reside in town.

### 1953

- Eino Sandell is killed at Kirkland Lake Gold
- John Stinkowji is killed at Lake Shore
- Leo Murray and Leon Wojcechowski are killed by a rockburst in Wright-Hargreaves
- Gerald Boudreau dies after an accident at Young-Davidson
- Joseph Jendral is killed at Kerr Addison
- Julian Antuchiewicz is killed at Kerr Addison
- Kirkland Lake Gold pays its last dividend

### January 20, 1953

## Kirkland Lake Gold Mine

## Eino Sandell (1897 - 1953)

Eino Sandell, born in Finland in 1897, emigrated from his homeland to Canada. By 1923, he was working in the silver mines of Cobalt. He remained there for a few years then moved to Kirkland Lake where he gained employment at Kirkland Lake Gold in 1929. At the age of 55, Eino and his wife Aino (née Halonen, 1894 – 1964) lived at 89 Third Street. He had one stepson, Dick.

Sandell was an experienced miner whose recent duties were maintenance and repair on levels, raises, manways, and drifts throughout the underground workings. He and partner John Zalec had worked together for several years at the same type of work.

Several days before the fatality, the two were engaged in scaling and timbering the 41-6 west drift to make it safe for future diamond drilling operations. This drift was driven six years previously and had been idle since.

On the morning of January 20, scaling and timbering was almost completed to the end of the drift. Mine Captain Tom Elliott, Underground Superintendent Len Dixon and Safety Engineer Ronald Goodman visited the area and were satisfied that work was progressing safely and satisfactorily.

Sandell and Zalec were attempting to scale a large piece of loose near the end of the drift. The piece failed to come down and Sandell decided to place a post and a head block underneath it as a precaution. He was standing under the piece and was in the process of measuring the length of post required, when a slab suddenly snapped from the back and buried him. Zalec was standing nearby but was not hit.

Zalec, unable to dislodge the fallen rock from his partner, went to the shaft station for help. Under the direction of Dixon, several men uncovered the body. Dr. A. B. Ostrander examined Sandell underground and pronounced him dead. A detailed examination of the deceased later showed multiple lacerations of the scalp, a fractured lower jaw, all ribs fractured and the fracture of both legs. Death was probably instantaneous.

An inquest was held and it was determined that Sandell was killed accidentally. It was always his habit to do excessive scaling to make the mine safe for others. It was stated at the inquest that scaling accidents are one of the chief causes of injuries and deaths in mines.

Sandell was survived by his immediate family in town and his father, two brothers and four sisters, all living in Finland at the time.

Interment took place in the local cemetery. The epitaph on the headstone reads as follows: RAKUS AITI – LEPPA RAUHASSA.

### March 11, 1953

## Lake Shore Mine

## John Stinkowji (1921 - 1953)

Thirty-one-year-old John Stinkowji of Ukrainian descent immigrated to Canada with his family in December 1950 and was hired immediately at Lake Shore on December 6. At the time of his death, he was employed as a drill runner. He lived at 53 McCamus Avenue with his wife Anna and their nine-year-old son Emel.

Before starting the night shift of March 11, he was instructed by Shift Boss William Hastie to pick up a set of drill steel on the 4,450 level and to take it up to the 4,325 to drill in a stope there.

The shift was lowered in the shaft by cagetender Art Royle. Stinkowji got off at the 4,450 and instructed Royle to pick him up on the return trip. Meanwhile, Stinkowji picked up three drill steels and waited at the station for the cage to return.

On its return, he entered with the steel in his hands. Royle instructed him to lay them onto the floor of the cage, then proceeded to ring the cage up to the 4,325. Stinkowji placed the two shorter steels on the floor and was in the process of doing the same with the longer one when the cage started to move upwards. The bit end of the long steel slipped underneath the rear door of the cage. As the cage ascended, the protruding end hit the wall of the shaft. The floor of the cage acted as a fulcrum, causing the other end of the steel to strike Stinkowji in the face and knock him across to the other side of the cage. The cage continued upwards with the offending steel whipping wildly up and down in it.

*Headstone photograph of John Stinkowji. Photographer: Author*

It was amazing that Royle and another passenger, J. Baczynski, were not injured. The force of the impact to Stinkowji's head caused fatal injuries.

The funeral mass was held at Sacred Heart Ukrainian Catholic Church on Second Street. Interment occurred in the local cemetery. The inscription on the headstone is partially in Ukrainian and translates as follows: IN MEMORY BY HIS WIFE ANNA AND SON EMIL – DEAR HUSBAND JOHN STINKOWJI – BORN IN BORSHEWVIHAPEOONA 1921 - 1953 – HUSBAND AND LOVING FATHER

Presently (1999), Anna lives in Kirkland Lake, and Emel and his family live in New Liskeard.

### April 27, 1953

## Wright-Hargreaves Mine

## Rockburst Traps 2

On April 27, The Northern Daily News headline read in large bold letters: ROCKBURST TRAPS 2.

The shift was just 55 minutes old when the burst shook the whole town. Previous bursts had occurred in old abandoned workings at Lake Shore but this one happened in the Wright-Hargreaves where men were working.

Within minutes, the miners came to surface but two were unaccounted for: Leo Murray and Leon Wojcechowski. On checking the 4,900 level, rescuers noticed that 50 feet of rock plugged the area to the stope where these men had been working. Rescue took not hours but days. With each day, hope of finding the men alive faded.

Ironically, on the day of the burst, Wojcechowski's regular partner, John Silvakovs, had an ailing back and was replaced by Murray. A few days later, Silvakovs was one of the working miners participating in the rescue operation.

When the two were found, it was revealed that Murray had suffered multiple fractures of the skull with brain lacerations; Wojcechowski had multiple fractures at the base of the skull, a fractured neck and other crushing injuries to the body. Dr. A. B. Ostrander, who examined the bodies, stated these injuries would have caused instant death to both.

*Leo Murray. Photographer: Unknown*

## Leo Germain Murray

(1928 - 1953)

Leo Murray, 25, a native of Fisherville, Quebec, lived with his 18-year-old wife, Alma (née Paquette), and a child at 33 A Government Road West. He was employed as a machine helper at Wright-Hargreaves and had worked there since November 17, 1949.

On April 27, he started his first shift in the No. 4902-A-58 stope along with Leon Wojcechowski who had been working in this stope for some weeks. The two were instructed by Shift Boss H. Gauld to set up, drill and blast a breast. Their shift commenced at 7:00 p.m. and it was estimated that they arrived at their place of work at about 7:30. At 7:55, a severe burst was felt throughout the mine and at surface; it was centred in the raise of the stope. A rescue crew was put to work immediately under the direction of Underground Superintendent W. T. Sampson. Recovery operations were long, difficult and hazardous. Murray's body was recovered on May 5.

When the pregnant Alma Murray learned of her husband's death, she tearfully said that if she had a son, she would name him Leo.

The funeral mass was held at Assumption Church. Interment took place in the local cemetery.

Later in 1953, Alma Murray gave birth to a baby boy; he lived for only one day. She named him Joseph.

The inscription on the stone over the grave of Leo Murray reads: 1928 LEO G. MURRAY 1953 – EPOUX DE ALMA PAQUETTE – JOSEPH MURRAY 1953 – A 1 JOUR.

## Leon Wojcechowski (1914 - 1953)

Thirty-nine-year-old Leon Wojcechowski lived with his wife Lidia and their 16-year-old daughter Tania at 26 Wishman Street.

After coming to Canada with his family from Europe, Leon gained employment at Wright-Hargreaves in 1950. He had been working at the mine for three years before the burst. He and his replacement partner for that evening, Leo Murray, died instantly when they were buried at their work site by tons of rock.

Wojcechowski was recovered 11 days after the burst in a manway raise, about 20 feet above the point where Murray's body had been found two days earlier. The body was removed through a special emergency entry made by blasting through 50 feet of solid rock.

*Leon Wojcechowski and family were recent Polish immigrants and survivors of war and wartime imprisonment. (Headstone Photograph) Photographer: Author*

Wright-Hargreaves General Manager R. L. Healy praised the employees responsible for the recovery. "Great care had to be taken during the process as the rescue crews were working in areas of crushed and broken timbers and loose rock." On behalf of the mine, Mr. Healy expressed his deepest sympathy to the relatives of the two dead miners.

Wojcechowski was buried in the Kirkland Lake cemetery. Lidia died in 1962 and is also buried there.

### May 9, 1953
## Young-Davidson Mine
## Gerard Boudreau (1923 - 1953)

Thirty-year-old Gerard Boudreau, Canadian and married with six children, worked as a chute-loader at Young-Davidson near Matachewan. He had been employed there from June 13, 1952.

On the afternoon shift of May 9, ore tramming proceeded on the fourth level. Boudreau and helper G. Bouchard were part of a loading crew on this level. Motorman Cleo Harvey was nearby.

At the time of the accident, four cars of a five-car train had been dumped into the ore pass. Harvey was on the motor, Boudreau was operating the air cylinder controls and Bouchard was attaching the chains to the door. A large piece of rock became lodged in the box of the last car and would not roll out. Boudreau wrapped a chain around the rock and attached the lifting cable to the chain to dislodge the piece by lifting it out. During the delay, M. Cormier, a loader from another crew, gave him a hand. Boudreau stood beside the car to steady the lifting cable while Cormier operated the airlift. The attempt to remove the rock from the box of the car was not successful; it remained in the box and the force of the lifting cable raised the front wheels of the car from the track. The car swung towards the rock wall and squeezed Boudreau against it before he could step clear. He was able to push the car away after being struck. He walked a short distance, then collapsed.

He was taken to surface and was examined there by Dr. Reed of Matachewan. The doctor diagnosed the injuries as five fractured ribs on the left side, a fractured scapula, and a crushing and bruising of the left lung. Arrangements were made immediately to have Boudreau taken to hospital in Haileybury.

There, Boudreau was under the care of Dr. J. A. Joyal. On May 14, five days after the accident, Boudreau's condition deteriorated and he died after contracting pneumonia.

### July 29, 1953
## Kerr Addison Mine
## Joseph Jendral (1900 - 1953)

Joseph Jendral was born in Czechoslovakia on March 10, 1900. His wife remained in Europe while he and his son Steve came to Canada. The mine accident report listed Joseph as a widower.

Hired in 1944, he was an employee at Kerr Addison for nine years and was considered a careful workman. His son was also employed at the mine.

On July 29, Jendral and partner Eugene Boissinot were instructed to build a water launder (wooden conduit) under a chute. The launder, to be built of one-inch boards and roofing paper, was to be inclined 34 degrees, the same as the chute bottom.

At one time, the chute bottom was supported by six posts from the floor of the drift but now only

*Headstone photograph of Joseph Jendral.*
*Photograph: Author*

three of the six remained in place, two at the lower end and one at the upper.

Gaston Gauthier inspected the area around the chute where the workmen were cleaning at this time. He instructed Jendral, the leader of the crew, to place a horizontal sprag across the centre of the chute for support to take the place of the missing posts.

Jendral did not put in the sprag as instructed and, shortly after Gauthier left, he began to chop out the two posts at the lower end. Boissinot told him the chute bottom would fall out if he continued. Jendral heard what Boissinot said but continued to chop. When he cut through the first post, the bottom of the chute began to move. As he began to cut the other post, Boissinot again told him the bottom would fall. Jendral answered him by saying the posts were not supporting the bottom of the chute. The reason why he wanted these posts removed was that they were in the way of the planned construction of the water launder.

As he cut through the second post, the lower plank section of the chute fell, knocking him on his back, causing his head to strike some timber. Also, two or three tons of fine wet muck fell with the planks and built up to a depth of one to two feet over his legs, abdomen and chest. He appeared to be knocked unconscious by the mishap.

Boissinot removed muck from Jendral's face but saw no evidence of life. He realized that he needed more help to release him so he ran to an area where a slushing crew was working. These men in turn called others and Jendral was released. When Dr. Manachi examined him, there was no sign of life and he was pronounced dead. The cause of his death was given as asphyxia and probable concussion.

The body was interred in the Kirkland Lake cemetery. The epitaph over his grave reads: IN MEMORY OF BELOVED FATHER OF STEVE JENDRAL.

### October 5, 1953

## Kerr Addison Mine

## Julian M. Antuchiewicz

### (1920 - 1953)

Kerr Addison Mine backfilled its abandoned stopes by pumping hydraulic fill from the mill into them. Hydraulic fill consisted of a mixture of screened sand, classified mill tailings and water. As the water slowly seeped out of the fill, a solid residue remained that maintained the strength of the mine workings.

On October 5, a stope on the 1,750 level was to be backfilled. Henry O'Reilly, fillman on duty, telephoned the mill, leaving instructions to start pumping the fill that ran directly from the mill to the stope by means of a pipeline.

Once the fill was running satisfactorily, O'Reilly went down to check the bulkheads at the 1,750. He inspected two of them and they were fine. As he descended towards the third, he heard liquid running in excess of the usual amount. He did not enter this area but returned immediately to the telephone and notified Mine Captain Doug Campbell of the leaking bulkhead. Campbell telephoned the mill and had the fill turned off. Meanwhile, O'Reilly descended to the failing bulkhead with tools and a plank to repair what he expected to be a minor leak. When he arrived, he saw a considerable amount of fill escaping through the bulkhead. All he could do was grab a ladder and escape to safety as the bulkhead suddenly failed

*Headstone photograph of Julian Antuchiewicz.*
*Photographer: Author*

completely. On the 1,750, he met Captain Otto Bjaranson who turned in the alarm to surface.

About 150 men were working in the mine at the time. The ethyl-mercaptan stench warning was immediately introduced into the mine on instructions from Underground Superintendent Cam Yule. Many of the miners clambered up ladders in their escape as cages were put out of action by the torrents of liquid and debris that now percolated down through the mine workings.

Water and fill ran down the 1,900, and past the 2,050, stripping out timber as it passed through. The bulk of the escaped fill and water ran towards the shaft on the 2,200 level. Most of the solids were deposited along drifts and crosscuts but the water ran into the shaft through ore and waste passes. The heaviest flow passed through the crusher station located below the 2,500 level. Gates were knocked out, causing muck to be carried into the shaft with the water; linings were torn out of shaft compartment partitions. A cage sent down to lower levels to evacuate the men was immobilized at the 2,500 station by the rush of water. Workmen below the 2,650 were obliged to climb manways because of the limited hoisting facilities; some climbed all the way to surface. The shaft was flooded to the depth of 242 feet.

Some miners waited underground for they did not know what was happening in the mine. Perhaps, they thought, the mine workings had collapsed beneath the waters of Larder Lake and the lake was now filling up the mine. The rushing waters subsided after about an hour and most miners got to surface by climbing ladders in manways and raises.

Four men, Julian Antuchiewicz, Bruno Milani, Peter Wirzman and Russel Hunter, were working in a drift at the 2,050 level about 300 feet below the point where the avalanche of slime and debris originated from the broken bulkhead. When they heard the noise of the cascading liquid, they feared for their lives. Wirzman ran through the drift and found safety at the mine station. Hunter was caught in a current of slime, swept down 25 feet to a lower level but miraculously escaped.

Preparing to drill, Antuchiewicz and Milani heard the water running and ran to a raise about 30 feet away to investigate. Seeing the rush of timber, debris, tailings and water coming towards them, they turned and ran the other way along the drift until they came to a partially filled stope, 40 feet deep. There was a remote chance of escape. This stope was attached to another drift at right angles to the drift that they were in. With heavy boots and wet clothing, Milani leapt about seven feet across the open stope to safety. He turned around and saw Antuchiewicz still in the other drift, standing upon a plank barricade. The water and fill had risen to a height of three feet and Antuchiewicz was left stranded above it. He told Milani to go to the station and phone for help.

A short time afterwards, Milani returned but there was no sign of the planks or Antuchiewicz; it appeared obvious he had been washed away.

Recovery operations for Antuchiewicz began. Doug Campbell, Gaston Gauthier and Joe Teal found him dead the next day in a stope at the 2,200 level. He had fallen approximately 90 feet and sustained a fractured skull, multiple bone fractures and internal injuries.

In total, four other men reported minor injuries sustained in their rush to safety but fortunately and amazingly no other deaths occurred.

Julian Antuchiewicz, 35, came to Canada from Germany in 1949. His wife and two sons came a year later. Their third son was born in Virginiatown.

Antuchiewicz, born in Poland in 1920, served in the Russian Army during World War II. He was captured and placed into slave labour where he worked in a bakery. He had spent eight years in Germany before coming to Canada. He was happy in Virginiatown, and spent much of his free time with his children. His fellow workers said 'Gypsy' was a fine man and very popular at the mine.

Surviving were his wife, the former Maria Sas, and sons Marion (7), Teddy (6), and Victor (2).

After the funeral service at Holy Name Catholic Church, the interment occurred in the Kirkland Lake cemetery. The inscription on his headstone reads: SW PAMIECI JULIAN ANTUCHIEWICZ – UR 6 CZERWA 1920 – ZAKONCZY SWEZYCIE DN SX 1953 – MAZ MARIA ANTUCHIEWICZ – ORAZ OJCIEC – MARION TADEUSZA I VICTOR. Translation: In holy memory of Julian Antuchiewicz – Born 6 June 1920 – His life ending on October 1953 – husband of Maria Antuchiewicz – father of Marion, Ted and Victor.

An inquest was held before Coroner J. F. Edis in the McGarry Council Chambers in Virginiatown on October 14. The jury brought in a verdict of accidental death but made a recommendation: "We the jury, recommend, as was presented in the evidence that after having two such failures in present construction of bulkheads that future construction be designed using a safer method of operation for the safety of all concerned."

### 1954

- William Bendnsond is killed at Teck-Hughes
- Clarence Whalley is killed at Lake Shore
- Fred McNaughton, a long-time miner in this area, is killed at the Roy Silver Mine at Elk Lake
- Aurel Nadeau is killed at Young-Davidson
- Harold Linney is killed at Kerr Addison
- Karl Lenk is killed at Kerr Addison
- Robert Paget, injured on surface at Teck-Hughes in December 1953, dies on January 23, 1954

### January 23, 1954
## Teck-Hughes Mine
## Robert Paget (1885 - 1954)

Robert Paget died in Kirkland District Hospital while undergoing treatment for a fractured right leg sustained in a fall while he worked as a surface labourer at Teck-Hughes on December 7, 1953.

An epitaph on the headstone over his grave in the Kirkland Lake cemetery reads: "WHATEVER THE TASK WHATEVER THE JOB, IF YOU WANT IT WELL DONE JUST CALL ON BOB."

Sixty-eight-year-old Paget, married and with one adult son, was employed at Teck-Hughes from January 1946. He worked under the supervision of the Mechanical Department, doing odd jobs on surface.

On December 7, 1953, while placing a winter covering box on a surface fire hydrant, he slipped on the ice and fell, fracturing his right femur. First-aid was administered at site; he was later attended by Dr. W. T. Boyd and sent to hospital.

His general condition improved, and the fracture seemed to be mending satisfactorily until December 28 when he began suffering from abdominal discomfort. At this time, the doctor noted signs of internal hemorrhaging. His blood count had deteriorated and several transfusions were administered. Doctors Boyd, Souter and Messenger consulted on his case and considered him a poor surgical risk because of his age, his deteriorated circulatory system, a perforated duodenal ulcer and the existence of the unhealed fracture.

Paget failed to respond to treatment and died on January 23, 1954. Surviving were his wife Alice (1885 – 1956) and his son.

### February 25,1954
## Teck-Hughes Mine
## Wasil 'William' Bendnsond (1924 - 1954)

Twenty-nine-year-old William Bendnsond was of Ukrainian birth. Before immigrating to Canada, he was a coal miner in Belgium. In Canada, he became

*William Bendnsond. Photographer: Unknown*

employed at Teck-Hughes on June 27, 1952 and was killed at this mine on February 25, 1954. He was survived by his wife Hilda and three children: Billy (2), Elaine (3) and Nicholas (5).

On the day of the accident, he and partner Enno Merila were mucking, scaling and tramming on the 4,100 level. They stopped to have lunch; during this time, they idly talked about death. After lunch, the two continued mucking and scaling in the main drift. Bendnsond suggested to Merila that they fill another two ore cars to add to the nine they already mucked. While they were mucking, he asked Merila to move the ore car closer to the muck. While doing this, Merila noticed a slab of rock on the drift wall slowly moving out from the wall and he shouted, "Watch out." Bendnsond looked up and jumped back but was a second too late. The rock pinned his head against the opposite wall. Merila came to his aid and tried to move the rock with his scaling bar but it was too heavy. He picked up a sledgehammer and started swinging at the rock but it was too thick and would not break. It looked hopeless. He checked Bendnsond's pulse and realized that he was dead. He left to get help.

Merila later stated that his partner was a good worker and a friendly man to work with.

Interment occurred in the Kirkland Lake cemetery.

### March 29, 1954
## Lake Shore Mine
## Clarence W. Whalley
### (1904 - 1954)

The Whalley family was among the first to settle in the Magnetawan, Ontario area. In fact, the nearby Whalley Lake was named for the family.

Clarence Whalley was born in Magnetewan in 1904 and came to Kirkland Lake to earn a living in the gold camp. He began his employment with Lake Shore on April 29, 1927. He and his wife, Lily, the former Lillian F. Anderson, had three children: Robert T. Whalley (1942 – 1943), Jean F. Whalley (1932 – 1987), and Donald R. Whalley who is a retired teacher presently (1999) living in Thunder Bay.

Whalley was an avid fisherman and hunter. He also spent quiet, enjoyable times in his garden at his residence on Prospect Avenue.

Lily predeceased him in 1949. In 1954, young son Donnie was left without parents when his father was killed.

At about 9:00 a.m. on March 29, Mine Captain Clarence Whalley, Shift Boss Angus MacMillan and Contract Engineer Ray Milton entered a stope on the 4,200 level where Albert Bergeron and S. Fowler were working. Whalley and McMillan checked the condition of the stope and decided a slab of loose on the breast should be taken down before drilling operations commenced. Fowler and MacMillan wanted to drill a hole in this slab and blast it apart at lunch time but Whalley decided that this was not necessary. Taking an ax, he proceeded to chop out a short piece of blocking that was holding the slab in place. Fowler and MacMillan cautioned him against this but Whalley insisted that it would be safe to do so.

After chopping three-quarters of the way through the blocking, he handed the ax to Fowler and told him to finish the job. Fowler succeeded in chopping the piece out after a few blows and then stepped back. When the block came out, the slab dropped and an estimated 25 tons of loose broke away with it. This knocked out the centre post of the set causing a whole row of sets to collapse. Whalley fell to a set below and was buried in loose rock and timber. Fowler saved himself by wrapping his arm around a post. MacMillan, who was standing in the safety of the manway, helped Fowler reach safe ground. Seconds earlier, Milton had walked up the manway to the next level. Bergeron was not in the stope at the time of the accident.

The accident occurred at 9:30 a.m. Recovery operations for Whalley's body began immediately under the supervision of Underground Superintendent Walter Selnes. During the operation, it was necessary to replace the collapsed square sets with temporary timber to protect the men engaged in the rescue work. The body was recovered at 2:20 p.m. Dr. G. M. Cameron was on the scene and pronounced Whalley dead. He had sustained a crushed chest, broken ribs, a fractured clavicle, fractured right knee, fractured elbow, fractured femur and numerous face and body lacerations.

Interment occurred at the family plot in the Kirkland Lake cemetery.

The photograph was provided by son Donald. The personal history of Clarence Whalley was contributed by nephew, Floyd 'Fuzz' Whalley, a retired man who at the age of 65 lived in Swastika. He died in 1998 during the writing of this book.

*Clarence Whalley (on the left), underground at Lake Shore with two other miners. Photographer: Unknown*

*Fred McNaughton. Photogrpher: Unknown*

May 4, 1954

## Roy Silver Mine

### Fred G. McNaughton (1914 - 1954)

Fred McNaughton was born to Charles and Minnie (née McMillan) McNaughton in New Liskeard on March 13, 1914. The family soon moved to Tantallon, Saskatchewan but returned to New Liskeard when Fred was seven. He had two brothers, Harvey and Allan, and six sisters: Jean, Lorene, Lillian (Fred's fraternal twin), Dorothy, Hilda and Rete. He was raised in New Liskeard and, as a young man, employed at a butcher shop owned by Mr. Holmes. In 1934, he gained employment at the Four Nations Mine in Kenogami, where he earned his 'mining papers'.

In 1936, he married Nora Deighton of Swastika. This marriage produced four children: Gail, Judy, Thomas (Buddy) and Sally.

He worked at Lake Shore and Upper Canada before World War II, then served in the Canadian Army during the war. In 1946, he returned to Upper Canada. Later, he worked at the Levack Mine in Sudbury and then returned to Kirkland Lake to be employed at Kirkland Lake Gold. He left the area again on April 13, 1954, to work in the Roy Silver Mine near Elk Lake. He was killed at this mine three weeks later.

On May 4, he was working in a box hole as a driller. Caught in an ore run down a chute, he was carried with the ore, then buried and trapped beneath the rubble. A doctor arrived, went underground and administered sedatives to McNaughton by means of his exposed leg. However, he died about an hour after the accident.

McNaughton was an active person who enjoyed life and his family. At 40 years of age, he was survived by his wife and children, aged nine to 14.

The funeral service was held at Holy Name Catholic Church in Kirkland Lake. The church was full as McNaughton was well known and popular in town. He was buried in the local cemetery.

The details of this story and the photograph were given by Allan McNaughton, Fred's brother, who presently (1999) resides in Kirkland Lake.

June 18, 1954

## Young-Davidson Mine

### Aurel Nadeau (1918 - 1954)

Aurel Nadeau, 36, Canadian, married with two children, was killed at Young-Davidson on June 18.

On the day of the accident, he was working with his partner, Donat Morin. They arrived at the bottom of a 58-degree service raise on the sixth level and decided that Morin should go up the manway to the top of the raise and use the tugger hoist to bring Nadeau up in the bucket; this was done. When the bucket reached the top, and Nadeau was just ready to step off, the cable slipped through the cable clamps, causing him to fall down the raise, a distance of 67 feet.

Hector Girard, R. Ferguson, V. Sventoritis and Morin all came to Nadeau's aid. They found him at the bottom of the raise, alive but unconscious.

He died within an hour without regaining consciousness. Dr. C. E. Reed arrived and pronounced him dead from brain injuries and a fractured skull.

### October 16, 1954
## Kerr Addison Mine

## Harold Linney (1919 - 1954)

The body of 35-year-old Harold Linney was found in a stope 17 hours after his accident. Rescuers had to remove about 20 tons of muck in order to get to the body.

The accident occurred on October 16 while he was standing on ore that filled a millhole at the 2,350 level. As he stood there, the doors were opened from below to allow a load of ore to pass to a lower level. Fellow miners said that he tried to jump clear but was too late. As he was swallowed into the hole, he was rapidly covered by falling rock.

Linney frequently told his friends that he feared he would be killed down in the mine. He had already given notice to leave Kerr Addison and was planning to open a bake shop in New Liskeard. It was believed that the shift on which he lost his life would have been his last.

The accident was reported immediately. Dr. J. F. Schubert of Virginiatown reached the scene within about ten minutes accompanied by Mine General Superintendent Bob Beggs. Beggs set rescue operations underway immediately but the work in the millhole was done cautiously due to the fear of another movement of ore. The rocks had to be removed slowly, by hand.

The body was recovered 16 feet below the place where he had been standing. Dr. G. E. Hagerman, on the scene at the time of the recovery, made an observation of the body and pronounced Linney dead. It appeared that Linney had fallen feet first. The only bone injury was a broken leg. It is believed that he died of suffocation from the weight of the rock pressing against his chest.

An inquest conducted by Coroner J. F. Edis was held in Virginiatown. OPP Constable C. A. Cousins made detailed inquiries into the cause of the accident; also involved in the investigation was Mine Inspector Robert H. Galway. The jury returned the following verdict: "We find that Harry Linney came to his death by being caught in a rockfall on Saturday the 16th day of October, 1954. This jury has agreed that Harry Linney came to his death while attempting to release a hang-up of rock with the use of drill steel while standing in the immediate vicinity of millhole No. 1, strictly against safety regulations."

Linney was married in 1944 to the former Dorothy Elizabeth Everred of Manchester, England. An Englishman, he came to Canada with his wife in 1952. He started working at Kerr Addison almost immediately after coming to this country. In England, he had been a manager of a chain of bake shops and it was his ambition to return to this business in Canada. Canadian-born were his two children, Judith Elizabeth (2) and David Harry (five weeks).

During World War II, Linney served with distinction as a chief steward in the British Merchant Navy. He was aboard one of the 70 ships that set out on the first convoy from Britain to the U.S. He also took part in the invasion of Italy. Once, while docked in a merchant ship at Halifax, he risked his own life by diving into icy waters and rescuing a fellow seaman.

In Virginiatown, Harry was well known for his excellent baking. Many people went to him with orders for wedding cakes and pastries for important

*Harry Linney and his wife and daughter.*
*Photographer: Unknown*

occasions. In Manchester, England, he was chosen to make all the cakes and pastries for a large-scale reception in honour of a visit by Queen Elizabeth II and the Queen Mother.

Surviving were his wife and two children, his mother Annie, and two brothers, Jim and John. Burial occurred in the Kirkland Lake cemetery. The inscription on the gravestone is the following: IN LOVING MEMORY OF HARRY LINNEY – BORN: SEPTEMBER 20, 1919 – DIED: OCTOBER 16, 1954 and IN LOVING MEMORY OF ANNIE – BORN: JUNE 18, 1893 – DIED: AUGUST 2, 1967.

Dorothy Linney re-married Jim Linney, the brother of the deceased. Jim and Dorothy presently (1998) live in Hamner, Ontario.

### December 24, 1954

## Kerr Addison Mine

### Karl Lenk (1923 - 1954)

Karl Lenk was born in Germany on the 23rd of October, 1923, and died at Kerr Addison on December 24, 1954. He was a recent immigrant, unmarried, and had been employed at the mine since August 29, 1951.

After five hours of testimony and continued cross-questioning of two doctors, the tense inquest session in the Virginiatown council chambers concluded with the verdict of five jurors: "On the 24th of December, 1954, Karl Lenk died while at work in the 1,621 – 64 raise at Kerr Addison Gold Mines Limited at approximately 2:30 a.m. It is the opinion, from the evidence submitted, that death was caused by a deficiency of oxygen."

The verdict rested on the testimony of Dr. G. E. Hagerman of Larder Lake who was the first to examine the body and Dr. A. W. Souter of Kirkland Lake who performed a thorough autopsy two days later. Both doctors admitted to uncertainty as to the cause of death. It was this uncertainty that was responsible for what Coroner J. F. Edis described as the "justifiable timidity" of the jury's verdict.

Testimony was given first by Dr. Hagerman who said that when he viewed the body, brought to surface on December 24, he could find no external evidence of the cause of death. In a detailed summary of his findings, Dr. Souter said that he had examined all the vital organs to discover the cause of death. He could only say that his findings were in accordance with death from oxygen deficiency. He emphasized, however, that there was no definite evidence of this.

Lawyer Fred Parry, who represented Karl Lenk's family, told the Northern News after the verdict that the family would have to apply to the Compensation Board for a decision on what compensation was payable. "The jury's decision imputes no negligence to anyone," said Mr. Parry. "It is assumed that mining is fraught with certain hazards. On very rare occasions, a jury may add a rider holding some party responsible for the accident. There is no such rider in this case."

Other evidence was given by Gerry Wilson of the Mines Rescue Station in Kirkland Lake. He testified that he was called to the mine soon after the discovery of Lenk's body and found that there was less than 16 per cent oxygen at the site where the miner met his death (normal air contains 19 per cent oxygen). Sixteen per cent is the minimum basic requirement. A safety lamp used to test the area went out, showing there was less than the basic requirement.

Bill Hargrave, an engineer, testified that shortly after the discovery of the body, he went to the site and sat down for a quarter of an hour. At first he felt dizzy but later found himself breathing normally.

In answers to questions put by Edis, Mine Inspector Galway and Crown Attorney Howard Walter, medical testimony was given that the same degrees of oxygen deficiency could affect different men in different ways. Wilson testified his tests showed a poisonous gas was present at the time of his visit to the site of the accident.

Karl Lenk's job at the mine was to do the rounds, checking valves on the hydraulic backfill system. He went alone but as a safety measure, called back regularly to the mill to report the results. He made his final report at 1:30 a.m. At 3:00 he failed to turn up at his lunch break. A search was started and, after a period of time, his body was found.

Witnesses at the hearing included: William Dewolfe Smith, surveyor; Gerry Smith; Duncan McDougall, ventilation engineer; Vernon Massicotte, mill man; Bill Roulston, shift boss;

Randal Larocque, shift boss; Henry Wagner, room mate of Karl Lenk; G. Pichette, working partner; Bill Hargrave; and Drs. Souter and Hagerman. Members of the jury were David Mann, Wallace Wilke, Earl Brooks, Steve Felesko and James Taylor, foreman.

Surviving Lenk was a sister, Herta, who had been in Canada for seven months and was working in Toronto, as well as a father and three brothers, all in Germany.

The funeral service was held in the chapel of Symington's Funeral Home. Reverend Robert Bartlett, United Church minister from Virginiatown, officiated. Interment of the body was conducted in the Kirkland Lake cemetery. The inscription on the headstone reads: HIER RUHT IN FRIEDEN UNSER LIEBER SOHN U BRUDER KARL LENK – GEB 23 OCTOBER 1923 – GEST 24 DECEMBER 1954 VIRGINIATOWN – IN UNVERGESSLICHEM GEDENKEN FAMILIE JOHANN LENK – RUDENAU A/M GERMAN. Translation: Here rests in peace, Our beloved son and brother Karl Lenk – Born 23 October 1923 – Died 24 December 1954 Virginiatown – Always remembered never forgotten by the family of Johann Lenk of Rudenau Germany.

### 1955

- Alexander Bowman, 27, is killed at Kerr Addison
- Romain Beaulieu, 29, is killed at Kirkland Lake Gold
- Louis Keller, 23, is killed at Kerr Addison
- St. Joseph's School is built in Kirkland Lake

### August 9, 1955

## Kerr Addison Mine

## Alexander Henry Bowman

### (1928 – 1955)

Alex Bowman was born to Alexander Bowman and Beatrice P. Hamilton in Haileybury on June 13, 1928. There were three children in the family, including older sister Margaret and younger brother George. When Alex was about six months old, the family moved to the Kirkland Lake area where his father worked as a bookkeeper at the Teck-Hughes office. Alexander Bowman Sr., a veteran of the British Army and the Canadian Army, died on February 16, 1935.

Alexander Bowman Jr. attended elementary school in Bourkes and high school in Kirkland Lake. Later, he got a job working on the railroad and took up residence in Larder Lake. He became employed at Kerr Addison on December 2, 1947 and worked there on three different occasions until August 9, 1955, the date of his death. He was 27 and single.

At the start of the shift on August 9, he and partner Raymond Kellington were instructed to move a 3,600-pound slusher hoist to another millhole in a stope. Bowman, being the senior man, was directing operations and standing on the operator's platform. The hoist moved forward but the front end of the skids caught an obstruction, causing the platform to tilt upwards at a steep angle. Bowman jumped and fell to the floor of the stope. The hoist remained in the tilted position momentarily then dropped back to its original position, crushing Bowman beneath.

Kellington descended to the next level and notified the first aid attendant who in turn summoned a doctor. Shift Boss Gerry Ryan heard the call; he went immediately to the scene and directed recovery operations. Jacks were used to

*Alexander Bowman. Photographer: Unknown*

remove the body from beneath the hoist. Dr. J. F. Schubert arrived during recovery operations and pronounced Bowman dead. The body was brought to surface and examined by Dr. H. F. Richardson of Larder Lake, finding that the cause of death was a broken neck.

An inquest was held that night before Coroner J. F. Edis in the Virginiatown Municipal Hall. The jury returned with a verdict of accidental death.

Bowman was survived by his mother, sister (Mrs. Carl Johnson), a brother, nieces and nephews.

The headstone marking his grave in the Kirkland Lake cemetery has the following inscriptions: IN MEMORY OF ALEXANDER BOWMAN – BORN JUNE 13, 1928 – DIED AUGUST 9, 1955. BEATRICE P. BOWMAN – BORN MAY 16, 1904 – DIED OCTOBER 18, 1996 – WE WILL MEET AGAIN.

Sister Margaret remembers her brother as "a great guy, a comedian, a fun-loving man who loved to play his guitar." Niece Judy remembers her uncle as a man who was fun to be with.

The personal information and the photograph were given by Margaret Johnson of New Liskeard and Judy Johnson of Sesekinika.

## September 9, 1955
## Kirkland Lake Gold Mine

### Romain Beaulieu (1926 - 1955)

Twenty-nine-year-old Romain Beaulieu lived with his wife and family on Comfort Street. He and Shirley (née Cooper) had two daughters, Suzanne (2) and Carole (nine months).

He was struck and killed by the descending cage of a winze in Kirkland Lake Gold on September 9. He was hired at this mine on October 6, 1952. At the time of his death, he had at least eight years of mining experience and was working as a cagetender's helper.

Beaulieu and partner Ivan Samojlenko were working the graveyard shift, 11 p.m. to 7 a.m. At 1:00 in the morning, they were called to the 5,010 by a motor crew that requested the hoisting of a trainload of muck. They took the cage down, got off and proceeded to load the muck. Samojlenko turned around and noticed Beaulieu leaning over with his head in the shaft. Samojlenko, knowing that the other cage would soon be coming down, yelled a warning to his partner. Beaulieu started to straighten up but before he could get clear, the cage caught him on the top of his head and crushed his skull.

At the station, E. Hotts and L. Trottier witnessed the accident; neither could explain Beaulieu's actions. Samojlenko stated that Beaulieu had lost a package of cigarettes earlier in the shift and may have been looking for them. Alternately, he may have thought the cage was below the level rather than above and was watching for the skip to come to the loading position from below.

He was predeceased by his mother, Alma Poudrier, in 1953 and survived by his wife and daughters and his father, Hercule, of Kirkland Lake. He was also survived by brother Harvey of Kirkland Lake and three sisters: Mrs. Allen Beauregard and Mrs. Lucien Lacasse, both of Kirkland Lake and Mrs. Kenneth O'Grady of Sarnia.

The body rested at Symington's Funeral Home until the funeral mass at Assumption Church. Interment was in the local cemetery.

An inquest was held before Coroner J. F. Edis. The jury returned the following verdict: "Romain Beaulieu came to his death through negligence on his own part, struck by a cage while looking down the shaft at the 5,000-foot level, September 9, 1955."

*Headstone photograph of Romain Beaulieu.*
*Photographer: Author*

November 24, 1955

## Kerr Addison Mine

### Alois 'Louis' Keller (1932 - 1955)

Twenty-three and single, Louis Keller of Virginiatown was killed instantly at Kerr Addison on Thursday, November 24. His marriage was to have taken place at Holy Name Church on Saturday and a reception was planned for Virginiatown later in the day. But the wedding date became the miner's funeral date and the service was conducted by Father Kinlough in Assumption Church. A friend had been asked to sing at the wedding but instead she was a member of the choir singing the Requiem Funeral Mass.

Keller was born in Germany in 1932 and his next of kin was listed as his father, Alois Keller Sr., of Germany. He had been an employee at Kerr Addison since 1951 and was considered an experienced miner.

On the day-shift of November 23, Keller and partner Paul Yergeau* were instructed by their shift boss to clean up some excess fill at a stope and dump it through an opening into a raise. They were also told to remove the planks covering the opening. The ore car, when it was in the dumping position, would cover this opening. When no fill was being dumped, a barricade was placed across so that no one would inadvertently walk into the opening. On this shift, five loads were dumped.

The night shift carried on with this work and, on the next day shift, Keller and Yergeau were instructed to continue. The first car was dumped into the raise. As there was some fill lingering in the car, Keller used a shovel to clean it out. He was standing with one foot on a lower sprag and, according to Yergeau, the planks suddenly collapsed and he slipped between the end of the car and the opening of the raise. He fell 150 feet down the raise into a stope. His body was recovered an hour later.

An inquest was held before Coroner J. F. Edis in Virginiatown. The jury returned the following verdict: "Alois Keller died instantly of a fractured skull by falling 150 feet through a stope raise while engaged in dumping a two-ton muck car. We assume the cause of this accident was due to an insecure barricade. The barricade, presumably there for the protection of the workmen, fell away when the workman braced his foot against the same, while cleaning the remaining fill from the car. We the jury feel that safety precautions were not executed prior to the re-opening of this working place."

Interment took place in the Kirkland Lake cemetery. The epitaph on the headstone is: ZUM-LIEBEN ANDENKEN AN.

** Paul's brother, Roger, would die in a mining accident at Kerr Addison in 1959.*

## 1956

- Janis Gulbis dies instantly at Kerr Addison
- Kalman Szucs dies in hospital after an accident at Kerr Addison
- Jan Kubalak dies underground at Kerr Addison
- William Fortey is killed while at work in Lake Shore
- Kerr Addison Mine 1,220 men employed, Kirkland Lake Gold 284, Lake Shore Mine 464, Macassa 242, Sylvanite 281, Teck-Hughes Mine 211, Upper Canada Mine 226, Wright-Hargreaves 445
- Young-Davidson Mine closes

January 21, 1956

## Kerr Addison Mine

### Janis Gulbis (1924 - 1956)

Born in Latvia on May 19, 1924, Janis Gulbis worked in the coal mines of Belgium for two-and-a-half years before immigrating to Canada. He was hired at Kerr Addison on June 10, 1953. He was married with two children, Benita (2) and Arthur (1).

On the day shift of January 21, Gulbis and partner Marchand were instructed to go into a stope to scale, support with sprags and install rock bolts.

During their shift, they scaled and supported the workings in the stope. A large piece of rock that had been scaled down was now resting across a millhole. Gulbis went over to observe this rock. While standing in this position, another piece of rock fell, striking and killing him instantly. Marchand stated later that this area had been sounded with a scaling bar and was thought to be safe.

An inquest was held before Coroner J. F. Edis in Virginiatown. The jury returned the following verdict: "That Janis Gulbis came to his death caused by traumatic asphyxia caused by a large piece of loose rock falling on him. From the evidence, we are of the opinion that death was accidental and can find no blame attached to anyone."

The body rested at Symington's Funeral Home and interment occurred in the local cemetery.

### February 9, 1956
## Kerr Addison Mine

### Kalman Szucs (1923 - 1956)

While a distraught wife sat at home, watching the slow progress of the clock, miners at Kerr Addison worked feverishly to rescue her husband, trapped by sliding muck in a millhole just below the 2,200 level on February 9.

Thirty-three-year-old Kalman Szucs, Hungarian and a two-year employee at Kerr Addison, was working in a stope, drilling holes for the placement of rock bolts. He and partner Franz Sachs had built a staging on top of a millhole through which muck was to be pulled to a lower level. Both men believed that the millhole was full of muck at the time.

Suddenly, however, the ore settled, drawing Szucs into the opening along with more broken ore; he fell to a depth of about 25 feet. Only part of the ore slipped down into the millhole; the rest jammed at the opening, leaving a space of about 15 feet between Szucs and the hung-up ore above him.

Buried in muck to his chest, he was completely immobilized and suffering a fractured leg and hand. Partner Sachs called for help and rescue operations began immediately. Through the overlying rock, the miners could hear Szuc's calls for help. The rescuers had to proceed with extreme caution to not disturb the rocks hung up above him.

One man who had helped with the rescue work stated to a Northern Daily News reporter that three fairly big rocks had jammed in the millhole, holding all the muck above the miner. "It was like a miracle," the man said. "If those rocks shift, the falling rock will kill Szucs."

The millhole was circular, five feet in diameter and surrounded by back fill. The walls of the millhole consisted of 12-inch-thick concrete.

The rescue workers dug a tunnel down parallel to the millhole, then made a horizontal tunnel towards it. These tunnels were in sand and had to be progressively timbered to protect the workers.

Dr. J. F. Schubert, who had been called shortly after the accident, spoke to Szucs in Hungarian and German, trying to calm him and reassure him that the rescue operation was progressing well. Szucs was concerned about his partner and, when he realized a rescue was being attempted, he cried out, "Work fast." During the digging of the tunnels in the sand, a quiet operation, the trapped miner became anxious, thinking that the rescue operation was being discontinued but Dr. Schubert reassured him again.

When the miners got to the wall of the millhole, they used diamond drills to poke holes into the concrete. They inserted steel rods to prevent the rock from falling further into the space above the trapped man. A two-foot square opening was cut, hole by hole, and using steel bars they broke through the millhole wall. Through the opening, they saw the trapped miner wedged in and half buried in the muck. A large rock pressed against his left leg and his left arm was pinned down by the ore.

While the workers at the mine were proceeding with their delicate and complicated work, Mrs. Szucs sat at home and waited in quiet desperation. The mine had informed her of the accident but as the hours passed, her agony increased. "We must wait and hope," she repeated again and again. "Why haven't they reached him yet? If I could only help him."

The rescue workers were concerned about the suspended rocks penetrating through the steel rods so they built a timbered bulkhead beneath the rods. Now that the danger of the falling rock was eliminated, they slowly removed all the hung-up rock, cut the rods with a torch, dismantled the timbered bulkhead and, after 15 hours of unremitting toil, were able to reach the trapped miner.

The rocks around Szucs were removed by hand and then hauled up in buckets. Finally, he was free. He was placed on a stretcher and taken to surface where Dr. Hagerman observed that he was conscious and had a fractured leg and injured hands. The rescue was completed at 4:00 in the morning, about 20 hours after the accident. Officials at Kerr Addison had the highest praise for the rescuers who diligently toiled throughout the day and night to free the trapped miner.

Szucs was rushed to Kirkland and District Hospital where he was given a blood transfusion and x-rays. Unfortunately he died at 5:00 a.m. on February 10. The news of his death came as a great shock to the men at Kerr Addison. They had made a heroic effort to save their fellow worker but now he was dead.

Szucs was born in Hungary on April 2, 1923. He was married in Germany to the former Helene Friege. They had three children: Kalman (9), Imre (8) and Elizabeth (3). The family had lived in Virginiatown for three years and just bought a home in 1956.

Interment took place in the Kirkland Lake cemetery. The epitaph on the headstone reads: RAM UBERFTANB'NEN FMMEREN LIEBEN BIN ICH VERFEBT IN HOHEN FREUBEN.

An inquest was held before Coroner J. F. Edis in Virginiatown. The jury returned the following verdict: That Kalman Szucs died in hospital from a pulmonary embolism from thrombosis of the left femoral vein, due to a crushing injury of the left leg with fracture of the tibia. Injuries resulted from falling into a hung-up millhole as the muck dropped from a stope at the Kerr Addison Mine.

After the accident and inquest a mine rule was added, requiring that in any stope, any millhole that is covered by muck during a blast is to be considered hung up and no man under any circumstances may work on top of the muck.

*Headstone photograph of Jan Kubalak.*
*Photographer: Author*

### September 27, 1956

## Kerr Addison Mine

## Jan Kubalak (1930 - 1956)

Born in Czechoslovakia on May 15, 1930, Jan Kubalak immigrated to Canada in 1951 and became a resident of Virginiatown and an employee at Kerr Addison in 1954. The 26-year-old married man resided with his wife on Connell Avenue.

Kubalak was the leader of a three-man stope crew. His partners were miner Nick Cilluza and Mine Labourer Falece Gallina.

On September 26, the three came on shift at 7:00 p.m. and were given instructions by Shift Boss Metro Schaetzler to finish drilling holes on a breast of a stope, to slush muck under the breast and to blast the breast in two sections, top and bottom. At 8:30, the shift boss visited them, noting that work was proceeding satisfactorily in a stope in good condition. At 9:30, they had finished drilling and began to slush. They obtained powder and fuse at 10:30 to load 30 holes in the lower half of the breast. The lower holes were blasted at 11:30 and they left for lunch at midnight. At 12:40, they returned to the stope, washed the breast and started to scale the loose.

The breast started to 'work and crack' so they left it alone for a while. In the meantime, Kubalak and Gallina went to get the explosives needed to blast the top half of the breast; Cilluza was left to watch the breast. At 12:55, the shift boss visited the stope again. Kubalak and Gallina were still fetching the explosives but Schaetzler called Cilluza away from the active breast and went to observe it himself. Kubalak and Gallina returned with powder and fuse, and the shift boss called them over to point out a crack that had developed across the stope, starting at the breast. He advised the men to stay away from the breast until the cracking stopped; he instructed them to work in another area of the stope until then.

At 2:00, the breast stopped 'working'; the three men picked up powder and fuse and proceeded carefully towards the breast. Cilluza and Gallina carried the materials while Kubalak approached from a different direction. He had a scaling bar with him and was sounding the rock in the stope as he advanced. He tapped a section when suddenly large chunks crashed down on him.

Cilluza and Gallina summoned help and the body was soon retrieved. Kubalak suffered severe crushing injuries to his chest, skull and back and probably died instantly.

The two remaining men now realized that the cracking, heard previously, had probably created the loose that killed their leader. Kubalak did not scale any of the loose but attempted to judge its condition by sounding with a scaling bar.

He was survived by his wife, the former Anna David of Virginiatown.

The body rested at Symington Funeral Home until time for the funeral mass at Assumption Church with Reverend Father Kinlough officiating. Interment was in the Kirkland Lake cemetery.

### October 15, 1956
## Lake Shore Mine

## William A. Fortey (1894 - 1956)

Sixty-two-year-old William Fortey, employed at Lake Shore for 26 years, was killed almost instantly on October 15 by an overflow of sand.

Fortey of 36 Dixon Avenue was a native of England who came to Canada in 1930, starting work at Lake Shore that year. He was an Imperial Veteran of World War I and a member of the Masonic Order.

*William Fortey. Photographer: Unknown*

It was the long-time practice at Lake Shore to backfill used stopes and raises with sand, transported by rail, from pits near Crystal Lake. Fortey and partner Nick Pochopsky were employed as fill-pass tenders whose job it was to ensure that backfill would flow at the proper rate and direction. This was full-time work; Fortey had been doing it for the past 10 to 15 years.

The sandfill passageways at Lake Shore extended from surface to the 7,700 level and consisted of a series of raises, each extending from a dump at one level to a chute at the next level. On each level, the chute at the bottom of one raise was directly opposite the dump of the raise below. A hinged extension from each chute could be connected to the dump opposite it. In this way, sand from the raise above could be passed across the level to the raise below, all the way down to the bottom of the mine. Equipped with cramp chain gates, the chutes on each level controlled the flow of sand.

At Lake Shore, this dump chute system of transporting fill was interrupted at the 3,325 level. A rockburst years before had wrecked the 'chute and dump' on this level. As a result, the dump had become very enlarged. Therefore, the opening was guarded with rails and a grizzly was constructed over it. It was here that Fortey was working on the day he was killed.

He and Pochopsky had been engaged in passing sand through the 3,325. Pochopsky stayed on the level above to pass sand down to Fortey's station. During this process, the chute at Pochopsky's level became plugged. To get help to bar and blast the chute clear, Pochopsky went down to tell Fortey of the problem.

He walked into the crosscut on Fortey's level and found it blocked tightly with sand. Fortey was not in sight. Pochopsky called but did not get an answer. He reported to the cagetender by phone that a spill of sand had occurred and that Fortey was missing.

Assistant Underground Superintendent Morris Seymour took the cage down, examined the sand spill and immediately directed men to start digging for Fortey. They soon found the back of his head protruding from the sand. His face and head was cleared as quickly as possible but he was dead.

Apparently as he was working in this area, the grizzly became plugged with flowing sand. He may have been attempting to keep the grizzly open as the sand ran (he had a shovel in his hand when his body was recovered.) A quantity of wet sand probably came from the chute, blocked the grizzly, then overflowed. Of course, this was all speculation as there were no witnesses to the accident. The body was examined by Dr. C. S. Harris, who said death was due to asphyxiation.

Fortey was survived by his wife, son William of Brantford, daughters Rena (Mrs. Edward Orton of Galt) and Edith (Mrs. Frank Mantia of Joliet, Illinois). Five grandchildren also survived him, as did two brothers and three sisters, all siblings were of London, England.

The body rested at Symington Funeral Home with the funeral service being held at the chapel with Reverend B. S. Morwood officiating. Interment was at Galt in the family plot.

### 1957

- A fall of ground kills Paul Melnyk at Kerr Addison
- A strange accident takes the life of Rolland Marcouiller at Kerr Addison
- Jean-Paul Morissette and Max Parker die after an accident at Teck-Hughes
- Antonio Castagna is instantly killed at Kerr Addison
- Sylvester O'Reilly falls to his death at Teck-Hughes
- Russian Orthodox Church of the Holy Configuration is built on Taylor Avenue

### April 9, 1957

## Kerr Addison Mine

## Pawlo 'Paul' Melnyk (1927 - 1957)

Thirty-two-year-old Pawlo Melnyk, a native of Ukraine and married with no children, had been employed at Kerr Addison just a few months, since January 4, 1957. Earlier, he had been a miner for five-and-a-half years at Upper Canada Mine in Dobie.

On the night shift of April 8, he and partner B. Dreier reported for work at 7:00 p.m. Their shift boss, Metro Schaetzler, instructed Stope Leader Melnyk to complete scaling No. 2,606-20 stope, to install rock bolts and to drill off a breast. The two men arrived at the stope at about 7:30, scaled for two hours, covered the millhole with planks, erected a drilling staging and installed rock bolts. Melnyk started drilling horizontal holes in the breast while his partner installed rock bolts into the back of the stope. They stopped at 11:20 for lunch.

At 11:50, Melnyk resumed drilling. Dreier installed a rock bolt then went to get some drill steel. Melnyk shut off his machine. Suddenly, a large piece of loose fell from the back and struck him. Dreier rushed to his aid and asked, "Are you okay?" Melnyk replied, "No, I'm not okay." He then collapsed and appeared to be dead. Dreier went to another level to summon help. The body, removed from the stope at about 12:30, was taken to surface and examined by Dr. R. F. Colapinto and pronounced dead. Melnyk received crushing injuries to his abdomen and probable internal injuries.

Interment occurred in the Kirkland Lake cemetery. The gravesite is presently not marked.

## June 21, 1957
# Kerr Addison Mine

## Rolland Marcouiller (1925 - 1957)

Rolland Marcouiller was a married man who came from Montreal to work at Kerr Addison in May 1953. He became a cyclone operator at the mine in October 1954. He left in 1955 and was rehired as a labourer in June 1956. He was promoted to cyclone operator, Class B, in January 1957 and held that job until the time of his death. He lived with his wife and two children in Virginiatown.

On the day of the accident, he reported to work on the afternoon shift of 4:00 to 12:00 p.m. He and partner Tom Denner were working at the cyclones at the south end of the mill near the No. 1 thickener tank. Denner last saw Marcouiller at about 6:30 and, when he had not returned by 7:45, he started to look for him. There was no reason for Marcouiller to go to the thickener tank, for he had nothing to do with it. Filter Operator J. Dame, who looked after the No. 1 thickener, stated he saw Marcouiller go past the thickener tank on the way to the washroom sometime between 6:30 and 7:00 and did not see him again. Denner, not finding him, notified Shift Foreman Max Michiel at 8:30. The whole mill area and gatehouse was checked but there was no sign of Marcouiller. Two sumps were drained with negative findings. It was then decided to check the thickener tanks to see if he had fallen into one of them.

The thickener tanks were full of a grey-coloured cyanide solution that concealed the rotating rakes. The walkways around the tanks were protected by adequate railings and wooden fences. A long rake was used to search for a body in the opaque liquid of the No. 1 thickener. At 3:00 a.m., a body was located in the middle of the tank using a rake. It was pulled to the edge and identified as that of Marcouiller.

Dr. R. F. Colapinto observed the body after recovery and stated that rigor mortis had set in, indicating that death had occurred several hours previously.

*Empty thickener tanks at the Kerr Addison Mill.*
*Photographer: Eddie Duke*

No one had seen Marcouiller enter the tank. Although the tanks contained cyanide, a person "would have to drink more than a quart to get a fatal dose." The tanks have a depth of about five feet on the outside to six feet in the centre; a person falling in would be able to get out by himself. Mill Superintendent George Hawkes was informed by Marcouiller's mother-in-law that Marcouiller could not swim.

An autopsy performed by Dr. A. W. Souter on June 22 disclosed the following: There were no injuries other than small scrapes on the elbows. There was no brain damage. The heart was normal. The lungs were congested but not diseased. There was no evidence of death by cyanide poisoning. Marcouiller was alive when submerged and the cause of death was given as drowning.

An inquest was held before Coroner J. F. Edis in Virginiatown. The jury returned the following verdict: "We find that the said Rolland Marcouiller came to his death on June 21, 1957, sometime about the hour of 7:00 p.m., at the mill of the Kerr Addison Gold Mines Limited, Virginiatown, by drowning when he accidentally fell into a thickener tank. Safe working conditions appear to be adequate in the vicinity of this tank."

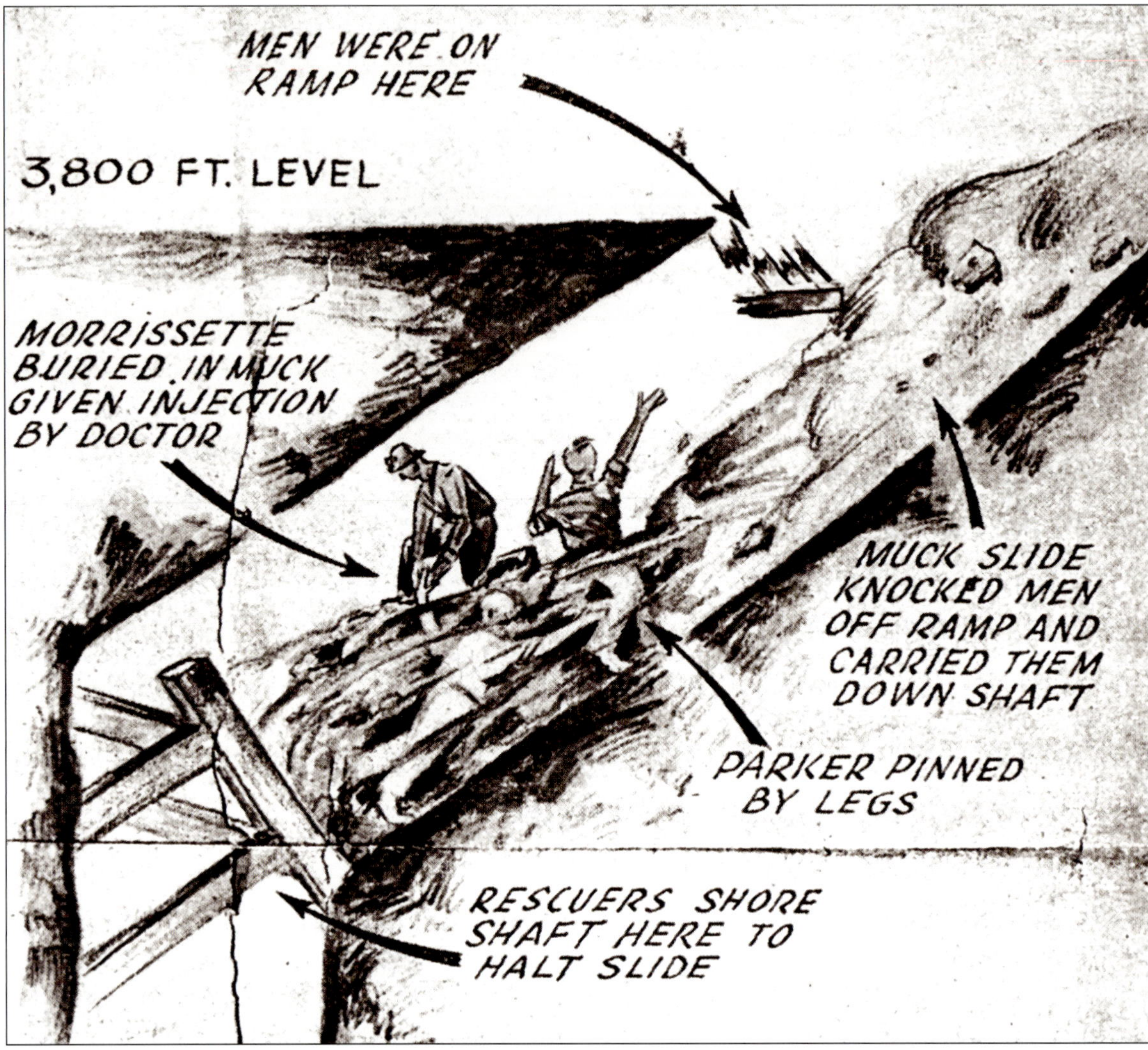


*Two miners trapped by slide of muck. Artist: Ed Parker, Toronto Star 1957*

### August 28, 1957

## Teck-Hughes Mine

## Two Miners Trapped By Slide of Muck

At the beginning of the shift on the day of the accident, Jean-Paul Morissette was sent into the middle raise at the 3,800 level to slush ore. At about 8:30 a.m., Shift Boss Maxwell Parker went up into Morissette's working place. W. Osborne, a trammer, was loading cars below the raise. He loaded three or four cars and took them out to the ore-pass. At about 9:00, he returned to the middle chute and heard yelling from above. He went up the access raise and discovered that the staging had been knocked down and that two men were caught by muck about eight feet below. He summoned help and Acting Superintendent A. McDonald came down from an upper level to direct rescue operations.

A four-ton rock had pinned Parker's legs against the back of the raise and trapped Morissette against the wall. It was also discovered that the raise below them was empty. The men were in a confined space and care was taken to fill the empty raise below them with old timber so that the large rock could be braced and the men released. During rescue operations, Parker was conscious all the time and Morissette part of the time.

Dr. Boris Nickoloff went underground at 10:30 a.m. and remained there during the day. He administered morphia to both men.

Parker had his head and arms visible. Morissette was completely hidden except for one hand; however, he could talk to his fellow miners through the coarse rubble. Morissette realized that the situation was serious and asked for a priest. Reverend Father Maurice Magnen was summoned to go underground to attend to him. Throughout the ordeal, Morissette showed great courage in the face of much suffering and the priest accorded him the comforts of his faith in his final hours.

Family and friends waited at the pithead for the two miners. Mrs. Morissette, with eight young children to care for, had to wait patiently mostly at home.

Parker was finally freed by drilling and chipping the rock pinning his legs. At about 9:00 p.m., he was taken to surface and rushed to Kirkland Lake hospital by ambulance. Morissette was released at about 9:30 p.m., after being buried for 12 hours. On the way to surface, he died.

There were no witnesses to the accident but it would appear that the two were standing on the staging in the raise when an unusually large rock came down, knocking out the staging and trapping them.

An autopsy on Morissette was performed by Dr. Souter and he gave the cause of death as traumatic shock due to multiple compression injuries, particularly to the chest and right thigh.

On August 30, Parker was recuperating in hospital with cuts and bruises and damaged legs. Amputation was at one time considered but he was fit and his condition improved daily. On the same day, the Morissette family prepared for a funeral.

Parker's condition took a turn for the worse and he was rushed to Toronto on September 4. He died in Toronto General Hospital on September 11, two weeks after the accident.

A report signed by Dr. W. G. Bigelow of Toronto General Hospital stated that an autopsy at the Department of Pathology had determined the primary cause of death as kidney failure with contributing factors being bruising, infection and coronary thrombosis.

## Jean-Paul Morissette

### (1919 - 1957)

Jean-Paul Morissette was born in 1919 to Palma Morissette and Amanda Boivenue in Gatineau, Quebec. He had four brothers: Albert, Lucien, Gaston and Yvon; and three sisters: Thérèse, Pierrette and Rolande.

As a young man, he came to Ontario and worked several years for the Morrissette Drilling Company of Timmins. He came to Kirkland Lake in 1942, marrying Lucienne Lessard in 1945. In 1947, he began work at Teck-Hughes and, at the time of the accident, was an experienced underground miner. The family of eight children Madeline (10), Roger (9), Pauline (7), Michelle (6), Jean-Guy (5), Louise (4), Nicole (2), and Richard (13 months) lived at 70 Second Street.

Jean-Paul Morissette, 38-year-old father and husband, was a member of the League of the Sacred Heart and of the Loyal Order of the Moose. Surviving him was his immediate family, his parents, sisters and brothers.

The funeral mass was held at the Church of the Assumption in Kirkland Lake. Interment was in the local cemetery. The inscription on the gravestone reads: IN MEMORY OF JEAN-PAUL MORISSETTE 1919 - 1957 – BELOVED HUSBAND OF LUCIENNE LESSARD 1922 - 1987.

*Mr. and Mrs. Jean-Paul Morissette and their eight children. Photographer: Unknown*

The photographs and some personal information were provided by Jean-Paul's daughter, Pauline Ring, who presently (1999) lives with her husband and family in Cambridge, Ontario.

## James Maxwell Parker

### (1892 - 1957)

Born in 1892 at Nine Mile River, Nova Scotia, Max Parker was from a large family of seven children. His brothers and sisters were Sidney, Orin, Catherine, Lena, Alma, and Arthur who was living in Kirkland Lake at the time of the accident.

Sixty-five-year-old Parker lived at 606 Government Road with his wife, the former Crawford Tobit (1898 – 1977). They had a daughter, Helen (Mrs. Herbert Collins).

He left Nova Scotia for Cobalt in the early 1900s to work in the silver mines. During World War I, he joined the Canadian Army and subsequently became stationed near Stanley, Perthshire, Scotland. There he met Crawford and was married on July 31, 1918. Their daughter was born on June 16, 1919 and, within six weeks of the birth, Parker returned with his new family to Canada where they took up residence at the family farm at Nine Mile River.

After a period of time, the family moved to Halifax and later to New Liskeard where Max worked on the Shepherdson farm for a few years.

In 1925, he began his employment at Teck-Hughes where the family took up residence on the mine property for a few years. Parker, however, never lost his desire to farm; he had a small holding

*Soldier Max Parker (World War I).*
*Photographer: Unknown*

in nearby Dane where he grew vegetables. Being very handy with tools and enjoying carpentry, he built a few houses in Chaput Hughes and ultimately one of these became the family home.

A kind and helpful person, Parker was a shift boss who enjoyed his work and was well respected by his co-workers. At home, his family was the centre of attention. Sunday family drives were frequent as he enjoyed driving his car on the back roads that he knew well. If the weather was fine, the drive would often end at a lake for a picnic.

During the summer, Max and his family enjoyed camping, swimming, picking berries and travelling. On holidays, they went to many parts of Ontario and visited their former home in Nova Scotia many times; they also went to Scotland to visit friends and relatives. Parker especially liked horses and would go with the family on July 1 to New Liskeard to watch the horse races and to partake in the holiday festivities.

In the fall, Parker would go partridge hunting and usually came home with his limit. In winter, he and his daughter enjoyed skiing together. He enjoyed going to the movies, especially the Saturday Westerns, and would come home after to have kipper for supper. Later that evening, he would relax by listening to a Leafs' game on the radio. He also enjoyed playing cards, particularly 500 and Crazy Eights.

Shift Boss Parker and miner Jean-Paul Morissette were caught in a fall of loose in a narrow raise on August 28, 1957; both were buried. A Northern Daily News photograph on August 30 showed Parker lying in a hospital bed; it seemed his recuperation was progressing well. A few days later, his condition changed and he died. He had been an employee at Teck-Hughes for 32 years.

Funeral services were held and interment took place in the Kirkland Lake cemetery.

*Max and Crawford Parker (1956).*
*Photographer: Unknown*

Parker loved his grandchildren and spent many enjoyable hours with them. Today, they remember a smiling man with a cap on his head and a pipe in his mouth – a kind gentle man who enjoyed life.

Much of the information and the photographs for this account were given by his grandchildren: Heather McDougall of Chelmsford, Ernie Collins of Cheltenham, Stan Collins of King Kirkland, and Lynne Stabryla of Kirkland Lake. Their mother, Helen Collins (Parker), passed away in 1994.

*Headstone photograph of Antonio Castagna. Photographer: Author*

## September 25, 1957
## Kerr Addison Mine

### Antonio Castagna (1925 - 1957)

Thirty-two, single and of Italian descent, Antonio Castagna was hired in January at Kerr Addison as a mine labourer with no previous experience. Nine months later he was dead, killed in an accident at the mine.

The day shift crew of September 24 had just blasted and, when coming off shift, reported that the millhole was hung up. William Zapotochny and partner Castagna came on shift that night and were instructed to blast down the millhole and then to slush. They blasted the millhole clear and slushed ore into it. When this was completed, they covered it with planks, and checked the back and walls of the stope with long bars. Shortly after 1:00 a.m., they hooked up a jack-leg machine to drill a large piece of ore partially buried in the fill and not moveable with the slusher. Zapotochny had just begun to drill when suddenly a large piece of loose dropped. His partner was hit and knocked down, and Zapotochny injured his ankle. Help was summoned and Castagna was moved to a safe location. Dr. G. E. Hagerman examined him underground and pronounced him dead. The cause of death was multiple fractures of the chest; death was probably instantaneous.

Interment was conducted in the Kirkland Lake cemetery.

## October 26, 1957
## Teck-Hughes Mine

### James Sylvester O'Reilly
### (1911 - 1957)

Sylvester O'Reilly was born to Edward James O'Reilly and Bridget Helferty in Eganville, Ontario, on December 16, 1911. He was one of four children.

He married Sarah Kirwin in 1937 and the marriage produced a family of seven children: Nancy, Edward, Alphonse, William, Sharon, Sylvia and Penny.

Employment opportunities in the mines brought him to Kirkland Lake. He initially worked at Lake Shore and then Toburn until it closed in 1953. He worked at Teck-Hughes from December 1953 to the time of the accident. The O'Reilly family lived at 43 Dixon Avenue. He is remembered as a loving father and a writer of short stories.

At the time of the accident, he was a welder in the mechanical department. On the morning of October 26, preparations were made for maintenance on the sheave wheel 101 feet high in the south shaft headframe. O'Reilly and partner A. Hodgins were instructed to take welding equipment up to the catwalk. They placed an acetylene cylinder on the hood of the cage and had the deckman ring them up slowly.

*Sylvester O' Reilly (circa 1947). Photographer: Unknown*

When the cage arrived, Hodgins told O'Reilly they would take the cylinder onto a special staging platform that had been constructed specifically for this job. O'Reilly stepped onto the platform and Hodgins turned to take the cylinder off the cage. He heard a noise, looked around and then below, and saw O'Reilly lying face down on top of another skip 32 feet below. The deckman, also hearing the noise, went with Hodgins to where O'Reilly had fallen. When they got there, he was either unconscious or dead.

Dr. A. W. Souter was summoned to the scene and pronounced him dead. The cause of death was traumatic shock due to a compound fracture of the skull with probable concussion; death was probably instantaneous.

An investigation showed that the planks on which O'Reilly had stood were in good condition and not slippery. There was no indication as to whether he fell over or under the guard rail, which was still intact.

An inquest was held before Coroner J. F. Edis. The jury returned the following verdict: "Sylvester O'Reilly came to his death accidentally. We find that blame is attached to no one. We the jury recommend that staging in mines be built with more railings for security."

The funeral mass was held at Holy Name Church and interment took place in the local cemetery.

Sylvester O'Reilly's son Edward presently (1999) resides with his wife Barbara in Kirkland Lake. Another son, Alf, lives in New Liskeard. One daughter, Sharon, is in Sudbury. The other children of Sylvester and Sarah reside in southern Ontario and Calgary, Alberta. The personal information and photograph were submitted by Ed and Barbara.

*Sylvester O' Reilly was killed while doing maintenance on the sheave wheel, 101 feet-high in the Teck-Hughes south shaft headframe. Photographer: Eddie Duke*

## 1958

- Niklos Rendes, a recent Hungarian immigrant, is killed by a fall of loose at Sylvanite
- A fall of rock kills Germain Beland at Kerr Addison
- Maria Goretti School at Swastika opens
- Church of the Assumption in Kirkland Lake burns to the ground but is rebuilt within a year

## September 11, 1958
# Sylvanite Mine

## Niklos Rendes (1930 - 1958)

Twenty-eight and single, Niklos Rendes had worked at Sylvanite as a mucker for only six months when he was killed on the morning of September 11. He had no previous mining experience.

A recent Hungarian immigrant, he came to Canada in 1957 and before coming to Kirkland Lake had been in Winnipeg. He was employed at Sylvanite on April 17, 1958.

Mine Captain Max Leavens visited the stope in which four men were working on the day of the accident. He instructed the men (Joseph Harhaj, Garhard Most, Vilmos Vecsey and Rendes) to scale, load and tram. Noticing a lot of loose including a very large piece, he told Rendes and Harhaj to build a staging from which they could scale properly. The staging was built and Harhaj went into the drift to get some timber; he left Rendes scaling the loose. Most and Vecsey watched him scale down a small piece; suddenly a large one fell onto the muck pile, cartwheeled down, struck the staging (knocking him off) and then rolled onto him. The three men pried up the loose and released him but he showed no signs of life and died instantly.

An inquest was held before Coroner Peter J. Neelands Jr. in the Township of Teck municipal building. The jury returned a verdict of accidental death.

The body rested at Symington Funeral Home and interment took place in the local cemetery. The inscription on the headstone reads: ITT NYUGSZIK RENDES MIKLOS – SZULETETT 1930 BA – MEGHALT 1958 BA – NYUGODJON BEKEBE.

## October 8, 1958
# Kerr Addison Mine

## Germain Arthur Beland (1922 - 1958)

With six years of mining experience, thirty-six-year-old Germain Arthur Beland was hired on August 15, 1958 as a mucking machine operator. He was an employee at Kerr Addison for less than two months when he was killed October 8 by a fall of rock.

On the morning of the accident, Beland and partner L. Hogan, accompanied by Mechanics A. Bond and Bob Lapierre, went to a drift to dismantle and remove a scraper. Before lunchtime, the hoist was rigged and the scraper dismantled. Immediately after lunch, Beland and Hogan were in the process of moving the scraper to another level when suddenly, loose fell in the drift. Hogan was thrown clear as the rock fell but Beland was knocked to the floor. Unable to move a large piece of loose from Beland, Hogan obtained help from men from the level above. Dr. G. E. Hagerman was notified and proceeded underground where he examined the body and pronounced the miner dead. The cause of death was given as asphyxiation due to a fractured windpipe. Death was probably instantaneous.

Beland lived with his family at 744 Laliberte in Rouyn. He was survived by his wife and seven children. The Darby Funeral Home in Rouyn handled the funeral arrangements.

## 1959

- Roger Yergeau is killed at Kerr Addison
- Two miners, Gilles Huard and Peter Moras, are killed by a rockburst at Wright-Hargreaves
- Guiseppe Gallina dies at Kerr Addison
- Stoel Bentley is killed at Lake Shore

## March 19, 1959
# Kerr Addison Mine

## Roger Yergeau (1923 - 1959)

Roger Yergeau was born in 1923 to Ernest Yergeau and Hectoryne Pepin of Trois Rivieres, Quebec. He was raised there along with his five brothers: Jean-Louis, Paul-Emil, Paul-Ernest, Gaston and Louis-Philippe. The family moved to Rollet, Quebec in 1932 to make a living at farming during the Depression. It was here where Roger's sixth brother, Maurice, was born.

Two of his brothers were already working at Kerr Addison when Roger left his job in the lumbering industry and moved to Kearns on June 4, 1954 to also become an employee at this mine.

On the night shift, he and A. Starasts were instructed by Shift Boss Leo Gramoli to check the No. 1916-55 stope, and scale and slush enough muck so two square sets could be installed. At about 2:30 a.m., March 19, Starasts was operating a slusher and Yergeau was standing a short distance away when suddenly 450 tons of falling rock hit him. He was crushed and received a skull fracture, broken ribs, a broken leg and internal injuries. Starasts, who was not injured, escaped through a chute and went for help. He later stated that the timbers suddenly collapsed without warning.

Rescue operations began immediately and Yergeau's body was recovered within three hours. Dr. G. E. Hagerman, who had proceeded underground, stated that death was probably instantaneous. After examination at the Kirkland and District Hospital, the cause of death was given as traumatic asphyxiation.

Yergeau was survived by his wife, the former Thérèse Maitais, and three children, Marie, Lise and André.

The funeral mass was held at the Catholic church in Rollet; interment also occurred there.

One of Roger's brothers, Jean-Louis, presently (1999) lives in Kearns. His wife, Florange, gave some details for this story.

### May 28, 1959

## Wright-Hargreaves Mine

## Severe Rockburst Kills Two Miners

Two miners, 18-year-old Gilles Huard and 54-year-old Peter Moras, were killed when buried under rock dislodged by a rockburst in No. 2705 – 12 stope at Wright-Hargreaves on May 28. The stope had been idle for a month. A three-man timber crew consisting of Huard, Moras and Frank Uramowski was sent to the manway to reinforce the centre stull supporting the bulkhead.

Shortly before 10:00 a.m., Shift Boss G. Mondoux visited the men at the top of the manway. He left and had just started walking along the drift when a severe burst occurred, causing the waste pillar in the stope to collapse, trapping the three men. Mondoux tried to get up the manway to the stope but couldn't because of debris and rocks. He went to the station for help and Underground Superintendent W. Sampson took charge of rescue operations. The 2,500 level was checked to see if entry into the stope could be made through another manway but the drift there was also blocked due to the burst.

It appeared that the only possible way to get to the trapped men was by the first manway. At considerable risk, crews of miners and supervisors eventually worked to a point near the top of the manway where Uramowski could be heard yelling. Gordon Cameron, safety director, managed to chop an opening through the lining of the manway and Uramowski was removed, taken to level, treated by Dr. F. Walk and sent to hospital.

Cameron, Sampson, and members of the rescue crew heard no sound from Huard or Moras and presumed that both were buried in the muck of the stope.

Recovery operations continued daily on a three-shift basis under hazardous conditions. On June 1, four days after the burst, the body of Huard was recovered at 4:00 a.m. and another two days later, the body of Moras was found. Dr. C. S. Harris examined the bodies and stated that death to both was probably instant. Huard had a fractured neck and Moras, a fractured skull.

## Gilles Huard (1940 - 1959)

Gilles Huard had just turned 18 on September 23, 1958. He had been attending Kirkland Lake CVI, enrolled in a commercial option. In October he began work at Wright-Hargreaves as a timber helper. He was living with his parents and older brother Jean-Louis, two years his elder, at 39 McKelvie Avenue.

He was born in Matachewan to Ernest Huard and Yvonne Anglehart, originally of Gaspe, Quebec, and was raised and received his Roman Catholic elementary education in Matachewan

*Headstone photograph of Gilles Huard.*
*Photographer: Author*

where his father was employed at the Ventures Mine (Matachewan Consolidated) as a hoistman.

Later, the family of four left Matachewan to live in Holtyre and within a year moved again, this time in 1956 to Kirkland Lake.

Huard was gifted with a musical talent inherited from his father who was a 'stepper' and a mouth organ player. He loved to dance and play the guitar. A teenager in the late '50s, his rock n' roll talent made him very popular, especially on weekends at local dance halls when he would jive the night away. All this came to an abrupt end when a rockburst at the mine killed him on May 28.

The funeral mass at Assumption Church was attended by a huge number of people including many of Gilles' young friends. He was buried in the Kirkland Lake cemetery.

The personal information for this account was given by Huard's brother Jean-Louis who now lives in Gloucester, a suburb of Ottawa. Huard's mother lives near Ottawa, in Orleans. His father passed away in July 1999.

## Peter (Pietro) Paulo Moras

### (1905 - 1959)

Peter Moras, of Italian descent, came to Kirkland Lake to work in the gold mines. While here, he met Gabriell Vaillant from Notre Dame du Nord, Quebec. They were married and had a son, Emilio. At the time of the accident, the family was living on Anderson's Road in Goodfish; Emilio was eight and his dad was 54.

Surviving Moras were his wife and son, as well as brothers and sisters in Italy. Interment occurred in the Kirkland Lake cemetery; the gravesite is unmarked.

Relatives surviving today (1999) are his wife in southern Ontario, his son in Burnaby, B.C., his brother-in-law Leo Hicknell of Kirkland Lake and his niece Theresa Hicknell of Swastika.

### August 26, 1959

# Kerr Addison Mine

## Guiseppe Gallina (1928 - 1959)

Born in Italy in 1928, Guiseppe Gallina came to Canada in 1951 and gained five years of

*Headstone photograph of Guiseppe Gallina.*
*Photographer: Author*

mining experience before becoming employed at Kerr Addison in 1957. He was considered a qualified miner.

Thirty-one-year-old Gallina lived with his wife, the former Lidia Maria Pellizer, in the village of Kearns. They had one child, Stefano, who was eight months old at the time of his father's death.

On the day shift of August 26, Gallina and partner Robert Cousineau were instructed by Shift Boss Robert Wilson to scale and rockbolt No. 2515-57 stope where necessary. They washed the face and back of the stope and then began scaling. Gallina was using an eight-and-a half-foot scaling bar about 15 feet from Cousineau when a large piece of rock suddenly fell from the back and struck him. Cousineau heard the noise but did not hear any breathing when he approached his partner. He went up to the 2,350 level for help. Dr. G. E. Hagerman was summoned underground and pronounced Gallina officially dead.

Funeral arrangements were provided by Symington Funeral Home and services were held at St. Peter the Apostle Church in Virginiatown. Interment took place in the Kirkland Lake cemetery.

## November 3, 1959
## Lake Shore Mine

### Stoel Smith Bentley (1910 - 1959)

"As a husband and father, he was one in a million," said Stoel's wife Hilja in 1999. Hilja is a long-time Kirkland Lake resident who came to Kirkland Lake in 1915. Her father, a Finlander, was a base metal miner in Michigan who left the state with his family to try a new life in Cobalt in 1911. The family left Cobalt in 1915 and came to Kirkland Lake where her dad became employed at Teck-Hughes; they lived in one of the homes on the property until 1918 when they moved to a farm on Round Lake.

Stoel Bentley was born in Burnley, Lancashire, England on June 29, 1910, to Thomas Bentley and Betsy Smith. The family consisted of eight children: Sybil, Eliza, Nellie, Elsie, Thurston, Stoel, Mercer and Lester. The family members gradually left their original home; Eliza came to Canada and became Mrs. Frank Ellis, Nellie came to this country and became Mrs. Busby and Sybil never left. Thurston, Mercer and Stoel left England and came to Kirkland Lake.

In 1925, Stoel was 15 and living in Kirkland Lake. One week after his arrival, his 17-year-old brother Mercer drowned in Kirkland Lake at a beach near Wright-Hargreaves.

Later, Stoel's parents and younger brother Lester moved from England to Kirkland Lake where his father found work at Wright-Hargreaves. By 1930, 20-year-old Stoel was working at Lake Shore.

He met Hilja Mommo in a candy store: she was cleaning and he was playing cards with his buddies. Their marriage produced two daughters: Irene, born in 1929, and Ruth in 1932. Both attended Kirkland Lake Collegiate and Vocational Institute and later married, Irene becoming Mrs. Ivan Teeple and Ruth, Mrs. Harry Brookmyre.

Stoel was a dedicated and faithful employee at Lake Shore for almost 30 years. He was also president of Local 240 of the United Steelworkers of America for some years. Hilja remembers that if Stoel wasn't at home or at work, he was at the union hall. He was very supportive of his fellow workers.

On November 3, 1959, the body of 49-year-old Shift Boss Stoel Bentley was recovered ten hours after an accident at the mine. Wolfgang Polenz (22) and helper Donald Farrow (18) had bulkheaded a chute in an open section. When Shift Boss Bentley visited them, they were preparing to pass sand fill

*Stoel and Hilja Bentley. Photographer: Unknown*

into the open section. They blasted a hang-up and were preparing to blast it again when it let go, knocking out a fence and trapping Bentley and Polenz. Rescue operations began immediately and Polenz, who remained conscious throughout, was removed after about nine hours of confinement. Bentley was removed shortly after but he had been killed instantly. The primary cause of death was given as severe crushing injuries to the chest.

Dr. F. Walk was in attendance throughout the rescue operation and administered injections to Polenz twice during the night but he could do nothing to help Bentley.

Interment occurred in the Kirkland Lake cemetery.

The personal history and photograph were submitted by 89-year-old Hilja Bentley who still lives in Kirkland Lake (1999). Presently, the two daughters live in Sudbury and Ottawa.

*Former Teck-Hughes assay office, Kirkland Lake. Photographer: Lawrence Chrismas*

Chapter VI

# THE SIXTIES

## 1960

- Dean McDonald dies at Kerr Addison
- Armand Robidas is killed by a fall of ground at Kerr Addison
- Edwin Glover, a long-time employee at Lake Shore, dies
- Robbie Sheme is electrocuted at Kerr Addison
- Lake Shore is milling only 480 tons a day
- Wright-Hargreaves is North America's deepest mine at 8,222 feet
- Kirkland Lake Gold closes after producing $39 million in gold

### January 14, 1960

## Kerr Addison Mine

## Dean Ellery McDonald

### (1939 - 1960)

Twenty-year-old Dean McDonald had been an employee at Kerr Addison since May 18, 1959. He was an apprentice miner who was interested in his work.

He was born on December 5, 1939 at Kirkland District Hospital to Scottish and Finnish parents whose families were early pioneers in the Cobalt and Larder Lake areas. He attended Larder Lake Public School and his recreation time was spent hunting, fishing and downhill skiing. He grew up in Larder Lake with his buddy, Rod Blackburn. Today Rod states, "His dad was a part-time trapper. Dean and I spent many hours in the bush together. Thinking back, it seems that we were always in the bush, hunting or fishing somewhere."

McDonald didn't go to high school but got odd jobs around town. When he turned 18, he left to live with his brother Rollo in Elliot Lake and got a job there, working in a mine. In the spring of 1959, he returned to Larder and became employed at Kerr Addison.

On January 14, 1960, while working underground, he told his partner Tom Pascoe he was going to look for a shovel in another part of the mine. It was about 10:00 p.m. and he was later seen on the 2,350 level by Shift Boss Hans Kneirman. When he didn't return, Pascoe, Kneirman and Tram Leader Gerald Yandeau went to look for him.

At about 11:45, Yandeau approached a bulkheaded manway that led to a stope abandoned since January 1959. He noticed a reflection of his light on an object beyond the bulkhead. He called to Knierman who was nearby. Knierman gave Yandeau a boost up over the bulkhead to see if it was McDonald. Yandeau peered into the darkness, noticed a body, took a breath, yelled "Gas!" and momentarily passed out into Knierman's arms.

Knierman immediately notified mine first aid and called for a doctor. Dr. A. S. McPhail arrived on scene at approximately 1:30 a.m. and, at 3:00, when the site was deemed safe to enter, pronounced McDonald dead.

*Dean McDonald. Photographer: Rod Blackburn*

At the inquest, a five-man jury learned that McDonald left his partner to look for a shovel. Pascoe stated, "It is a normal practice for men to hide their shovels so they know where to get them the next night." In his search, McDonald apparently squeezed through the timber bulkhead that closed off the abandoned manway. He crawled into the area and immediately was overcome by gases in this unventilated portion of the mine. Dr. A. W. Souter performed the post-mortem; he said, "The body had all the symptoms of being in a place where there wasn't enough oxygen. The whole body was blue – although usually just the lips are blue. The lungs were inflated and not deflated as in a normal dead person."

Air samples taken and analyzed by the Department of Mines after the accident confirmed Dr. Souter's statements. In air suitable for healthy breathing, the oxygen content is about 20 per cent. Between 16 and 19 per cent, air is unhealthy for breathing. Between 10 and 15 per cent, "a life would be snuffed out very easily." The oxygen reading in the manway where McDonald was found was only 7.6 per cent. Dr. Souter added, "In that atmosphere, his death probably took place in less than a minute. Death was due to asphyxiation."

It was assumed that the chamber in which McDonald was found had an abundance of carbon dioxide created by the decomposition and oxidation of the old timbers and a lack of ventilation (dead air).

The jury reviewing the case included R. Burnett, G. Stevens, N. Gravel, E. Morton and A. Baxter, all of Virginiatown. At the end of the inquest, they recommended "that all abandoned parts of a mine should be completely sealed off and signs placed about the mine to let the men know that there is danger ahead."

McDonald was single and was survived by his parents, Mr. and Mrs. William McDonald of Second Avenue in Larder Lake. His brother was living at Elliot Lake. On the day of the accident, William McDonald was in hospital at North Bay.

The funeral service was held at Symington's Funeral Home and interment occurred in the Kirkland Lake cemetery.

The personal details and the photograph for this memorial account were donated by Rod Blackburn of Larder Lake.

### January 30, 1960
## Kerr Addison Mine

## Armand Gerard Robidas
### (1937 - 1960)

Twenty-three and single, Armand Robidas of Kearns was crushed to death on January 30 by a fall of rock at Kerr Addison. He had worked at the mine from August 1955 to March 1956 and was rehired in December 1959 as an apprentice miner.

On the day of the accident, he and another apprentice miner, Jacques Millette, were working with Stope Leader H. Pleau, drilling, slushing, rockbolting and scaling. Pleau and Robidas had just finished scaling when Pleau heard a "snapping and cracking" sound; suddenly, a large piece of rock fell from the back of the stope, striking Robidas and pinning him to the floor. Pleau and Millette were able to pry the flat piece of rock from him but no sign of life was evident. Dr. A. S. McPhail came underground to the scene and pronounced Robidas dead. He stated that death was probably instantaneous from a skull fracture and cerebral hemorrhage.

Armand Robidas was survived by his parents living in Field, Ontario.

### February 1, 1960
## Lake Shore Mine

## Edwin Glover (1899 - 1960)

Edwin Glover, listed on the Miners' Memorial, died on February 1 in Kirkland Lake at the age of 60.

He was born in Churchtown, near Penzance, Cornwall, England in 1899. He came to Cobalt in 1922 and to Kirkland Lake in 1926 when he began work at Lake Shore.

Glover was survived by his wife, the former Gwendolin Courts, two sons, John of Toronto and Jeffery of Peterborough, a daughter, Mrs. Lindsay Tuck of Brampton, and two grandsons.

The body rested at the Symington Funeral Home. Interment was in the Kirkland Lake cemetery. The epitaph on the headstone reads simply, 'PEACE'.

## April 12, 1960
# Kerr Addison Mine

## Robbie Sheme (1938 - 1960)

Twenty-two, single and a resident of the bunkhouse in Virginiatown, Robbie Sheme died in an electrocution accident at Kerr Addison on April 12.

He was hired for the second time on September 16, 1958. He had previously worked at Kerr Addison from September 1957 to July 1958. At the time of his death, he was an apprentice miner and was considered to be a competent workman who knew the mining regulations.

Miners were advised at safety meetings never to poke anything into electrical connector receptacles. The safety rules called for the disconnect switch to be open whenever the cable was connected or disconnected.

On the night shift of April 11, Sheme and partner Alfred Stembridge did some drilling in a stope and were planning to slush. The slusher had not been used for the past two weeks and, as a result, the electrical cable to the slusher had to be reconnected. The reconnection was made and Stembridge closed the switch but found that no power was delivered to the slusher. He subsequently noticed that a connector was damaged. It was midnight and the two men broke for lunch.

After lunch, Stembridge went to another level, leaving Sheme at the work place. Before leaving, however, he mentioned to Sheme that they would have to replace the cable on the damaged connector to get the slusher to work. At that time, Sheme was standing in the stope. When Stembridge came back 15 minutes later, he found Sheme lying on the muck pile. He checked his pulse but could find none. He left the stope to get help and first aid was notified. Sheme was removed to the first-aid room on surface and artificial respiration began. Later, a resuscitator was used but he could not be revived. Dr. A. S. McPhail was in attendance and pronounced him dead.

Dr. A. W. Souter performed an autopsy and found minor lacerations about the face, the left elbow and wrist. There was also a slight scalp laceration at the back of the head. According to the doctor, these injuries were not sufficient to cause death. Burns on both hands, however, indicated an electrical shock and the doctor gave the cause of death as electrocution.

Electrical Underground Foreman Ed Massicotte discovered that at the time of the accident, the disconnect switch was closed, allowing the current to flow. It would have been possible for a person to get a full 550-volt shock by poking a small conducting instrument into a receptacle if he were holding the grounded connector at the same time.

A file that Stembridge used before lunch was found near the body. It is probable that Sheme, for some unknown reason, inserted the tang of the file, which had no insulating handle, into the female end of the connector and received an electrical shock.

According to the shift boss, both Stembridge and Sheme were aware of company regulations stating that the power must be shut off before working on a connector.

The inquest into the accident was headed by Dr. J. F. Edis. The Crown Attorney in attendance was George Black. The members of the jury (Foreman Dave Mann, Jack Crozier, William Lalonde, Rolly Acton and Pearly Woodrow, all of Virginiatown), returned the following verdict: "That Robbie Sheme met his death by electrocution, through the insertion of a metal object, presumably the small bare end of a steel file, into the female end of a live electric slusher cable socket."

## 1961

- Arthur Comptois is killed in July at Sylvanite and the mine closes at the end of the year
- Stan Korytko is killed at Kerr Addison

## July 11, 1961
# Sylvanite Mine

## Arthur Comptois (1930 - 1961)

Thirty-one-year-old Arthur Comptois and his wife Thérèse lived at 51 Third Street with their children, Girard, Francine and Suzanne. Another daughter, Marie, had died as an infant.

In 1930, Arthur was born to Mr. and Mrs. Henri Comptois. He had six brothers and three sisters: Maurice, Alfred, Laurent, Ernest, Real, Jean, Cecile, Jeanette and Helene.

He became employed at Sylvanite in 1957 and was considered an experienced miner. He was killed on July 11 when he fell from a staging. Death was probably instantaneous.

On July 11, the dayshift crew was instructed to erect a new staging in stope 807. The safety engineer instructed them to wear lifelines at all times during installation. Similar instructions were left in writing for the night-shift crew. The day crew had almost completed the job before the end of the shift but an eight-foot plank was left loose on the staging. During the construction, the miners used lifelines at all times, as instructed.

When the night crew came on shift, Shift Boss John Chornie read aloud the written instructions left by the safety engineer. The crew consisted of Ferdinando Longo, Steve Stypulkowski and Comptois. Chornie told them to proceed to stope 807 and to wait there until he arrived with working instructions. The crew went underground to its working place and waited. Longo later stated that Comptois didn't wait long and went down to where the new staging was being built. He had told Comptois to wait until the shift boss arrived but shortly afterwards, he heard a noise and caught a glimpse of a plank tilting and Comptois falling. Chornie was on his way into the stope when he met Stypulkowski, who informed him of the accident.

Chornie had Dr. C. S. Harris notified and then directed rescue operations. Comptois' body was located on the muck, 80 feet below the staging. An eight-foot plank was nearby. The doctor went immediately to the scene and pronounced Comptois dead.

An inquest was held before Coroner J. F. Edis. The jury returned the following verdict: "Arthur Comptois came to his death by falling down an open stope. Death was caused by a fractured skull and a broken neck. The verdict we reach is accidental death."

Symington Funeral Home handled the funeral arrangements and the funeral mass was held at Assumption Church. The burial took place at Laverlochere, Quebec.

** Authors note: In 1997, I interviewed Shift Boss John Chornie about this accident. He was very disappointed that the dayshift left a loose plank on the newly built staging. He stated that if a few spikes had been driven into the plank at the end of the shift, the accident would never have happened.*

### October 18, 1961

## Kerr Addison Mine

## Stanislaw Korytko (1913 - 1961)

Stanislaw Korytko, 48, a miner at Kerr Addison was killed almost instantly on October 18. He was one of a crew of three, mucking out a drift round on the 5,600 level. At the time of his death, he was switching when he was crushed between the wall of the drift and the train. The other two who were in the same working area were William Skawinski, the loader operator, and J. Avens, the locomotive driver. Korytko was an experienced miner, employed at Kerr Addison since July of 1949, a period of 12 years.

Born in Poland in 1913, he came to Canada in 1949 where his family made their home in Larder Lake. Korytko was survived by his wife, the former Alexandra Capenko, and three sons: Edward (13), Richard (9), and Victor (2), and two daughters: Helen (10) and Jane (7), as well as a brother and two sisters in Poland.

The body rested at Symington Funeral Home until time for the funeral mass in St. Joseph's Polish Roman Catholic Church in Kirkland Lake. Burial took place in the Kirkland Lake cemetery. The epitaph on the headstone reads: IN MEMORY OF OUR FATHER.

At the inquest held before Coroner T. W. Armstrong in Virginiatown, the jury brought in the following verdict: "That it is agreed, that Mr. Stanislaw Korytko was accidentally killed from injuries sustained by being crushed between the muck car and the face of rock in a drift. Recommendation: It is recommended that safety stations be spaced to conform to mining regulations and that the switchman be in position in a safety station and from that position signal the motorman to proceed."

## 1962

- Anton Corak is killed at Upper Canada
- Donald Perrier dies in a Noranda hospital after an accident at Kerr Addison
- Two are trapped by a rockburst at Macassa. Richard Witczak is killed instantly and Russel Baskin is buried alive and rescued
- Northern Ontario Institute of Technology opens in Kirkland Lake
- Jones and Laughlin Steel Corporation of Pittsburgh announce plans to develop a huge iron deposit near Dane

### March 27, 1962

## Upper Canada Mine

### Anton Corak (1934 - 1962)

At Lika, Croatia, in 1934, Anton Corak was born to Mile and Marija Corak. He was raised with one brother, Stevan, and two sisters, Kata and Manda.

In 1957, 23-year-old Corak left Croatia and came to Canada. While in Kirkland Lake, he met Mary Viskovich and, on April 14, 1958, they were married. He became the proud father of two sons, Miles, born in 1958, and David in 1960. He was a devoted family man who loved his wife and children very much. In 1962, Mary and Anton joyfully anticipated a third child. This time, they hoped for a daughter.

Corak was employed as a miner at Upper Canada since February 13, 1961 and was considered an experienced workman. His previous underground experience included mining at Kirkland Lake Gold and, before that, working in mines at Elliot Lake and Steep Rock.

Life was good in March of 1962; Anton and Mary were living happily with their two young boys at 25 Queen Street. A proud moment also occurred on Thursday, March 22, when Magistrate E. W. Kenrick officially presented Anton with his Canadian Citizenship papers. However, four days later, he was killed instantly by an explosion while drilling underground at Upper Canada.

He and partner Stefan Pietrasik were doing routine mining work at their workplace at the time. While drilling, they alternated the use of the machine because they only had one. Leaving his partner drilling, Pietrasik went to the station for lunch. Having finished his lunch, and noting that Corak was late for his, he returned to the stope. On reaching the work area, he heard air blowing from an open air line. He shut off the air and the header and just beyond this point he discovered Corak lying in a crumpled position and showing no signs of life. He immediately went to the station for help. Shift Boss William Smith arranged for a stretcher, a doctor and ambulance. Dr. M. J. Thomson from Kirkland Lake arrived at the scene and pronounced Corak dead.

He was survived by his wife and two sons, two sisters, one brother and both parents. Requiem Mass was held at Holy Name Church and interment occurred at the Kirkland Lake cemetery.

An inquest was held before Coroner J. F. Edis in the Ontario Provincial Police Headquarters, Swastika. Five witnesses were called: E. W. Ritchie, D. R. Thompson, S. Pietrasik, J. T. Coughlan and Dr. M. J. Thomson.

*Headstone photograph of Anton Corak.*
*Photographer: Author*

Upon questioning the jurors, the Coroner realized that all five were miners employed at Upper Canada. "I can't have those at the mine on the jury," said the doctor emphatically. "I am going to have the inquest and render the verdict myself," he said. After the proceedings, the coroner's findings were as follows: "That Anton Corak met his death, about 11:05 a.m., March 27th, 1962 on the 3,350 level, L2-8.5 stope at Upper Canada Mines at Dobie, Ontario, from multiple head and chest injuries sustained by drilling into a missed hole."

The hoped-for daughter, Marianna, was born to Mary Corak and the late Anton Corak six months later.

Mary Corak, now (1999) Mary Grgich, lives in Victoria, British Columbia; she donated letters and newspaper clippings to aid in the research. As for the other children, Miles lives in Ottawa and Marianna in Richmond Hill. David is deceased and buried with his father. Julio Viskovich and Nevio Viskovich, brothers-in-law to Anton Corak, live in Kirkland Lake.

### April 3, 1962

## Kerr Addison Mine

## Donald Joseph Perrier

### (1933 - 1962)

Donald Perrier was an underground miner at Kerr Addison; he lived with his wife and child in Rouyn, Quebec.

On April 3, Perrier, 29, was working in a stope with Claude Favron and Ronald Cloutier. They noticed a two-and-a-half ton piece of loose on the wall. Unable to scale it down and after checking the cribbing supporting it, they left it, thinking the situation was safe.

At 11:00 a.m., while they were doing their regular work, the piece of loose fell and pinned Perrier to the timber. Shift Boss Jack Barbour and the Mine Captain Murray MacAulay heard cries for help and rushed into the stope, finding the miner pinned at the hip by a corner of the rock. Using planks as levers, the men lifted the rock and freed him.

The first-aid man, Eric Lowe, was already calling doctors but the Virginiatown and Larder Lake physicians were attending medical cases in Kirkland Lake and were unavailable. Although the Kirkland Lake hospital was four miles closer to the mine than the hospital in Noranda, the quick and clear-thinking Lowe contacted the Youville Hospital in Noranda because there was heavy construction on Highway 66 between Virginiatown and Kirkland Lake at the time.

Perrier was taken by stretcher to surface where the waiting McGarry Township ambulance rushed him to Noranda.

On the morning of April 5, Perrier died in hospital because of internal injuries received in the accident. The cause of death was severe shock and anuria.

An inquest into the death was held in Virginiatown and was conducted by Coroner T. W. Armstrong. The jury foreman was Charles Locke and the jurymen were Howard Barnard, John Campbell, Clifford Lamoureux and William Zapotochny. After listening to the testimony of the witnesses and the doctors, the jury returned with the verdict of 'accidental death.'

### April 18, 1962

## Macassa Mine

## Richard Witczak (1924 - 1962)

A rockburst near the 4,200 level trapped two men in a narrow sub-drift at Macassa on April 18. The trapped miners were Richard Witczak, 37, of 33 Main Street and Russel Baskin, 34, of 52 Taylor Avenue. The rescue crews were hampered in their operations by the narrowness of the tunnels. Two-and-a-half days elapsed before they found Baskin* buried alive. He was immediately taken to surface and to hospital where he was treated for shock and multiple pressure sores. Officials at the mine stated that he was kept alive by natural air seepage through the rubble.

Eight hours after Baskin was taken to safety, the rescue crews found Witczak face down in a rock pile located in the opposite side of the drift from where

*Macassa Mine Resue Team was involved in retrieving the body of Richard Witczak and in saving the life of Russel Baskin. (Left to right: Supervisor Robert Walker, Vice-Captain William Haskins, John Jones, Ronald Quevillon, Bryon Young, John Klawitter, Captain Guy St. George and Supervisor Oscar Mumford. Photographer: Eddie Duke*

*Headstone photograph of Richard Witczak.*
*Photographer: Author*

Baskin was found. Dr. C. N. Brebner pronounced him dead and determined he had died instantly.

Witczak was born on May 12, 1924 at Skierniewice, Poland. His parents were Jan Witczak and the former Jozefa Dybiec. There were seven children in the family including three other brothers and three sisters. After World War II, at 21 years of age, he worked as a shoemaker and a bricklayer, trades he learned from his father.

On January 16, 1946, Witczak married; in time, two daughters were born. During the late summer of 1949, the young family left their home in Germany and arrived in Canada on August 4. Soon, Witczak found work at Macassa. While in Kirkland Lake a third daughter was born.

Witczak loved his family and enjoyed the north, where he became an avid fisherman and hunter.

Left to mourn his passing were his wife Luba and their daughters, Anne (15), Linda (14) and Jean (8). The Requiem Funeral Mass was held at St. Joseph's Polish Roman Catholic Church with Reverend Father Anthony Hylla officiating. The pall bearers were Stanley Niedzielski, Louis Zapora, Leo Strojny, John Gurniela, Ed Dolata and Max Kitaj. The funeral arrangements were handled by Symington Funeral Home and burial took place in the local cemetery.

Particular information for this story was contributed by Richard Witczak's daughter, Jean, who now lives in the Toronto area (1999).

** Authors note: A description of Russel Baskin's ordeal appears on page 145 of Kirkland Lake – On the Mile of Gold by Michael Barnes.*

### 1963

- Rockburst at Kerr Addison kills Roy Fletcher and Aksel Meister
- Roy Davis, employed for 18 years at Kerr Addison, dies

### February 1, 1963

## Kerr Addison Mine

## Rockburst at Kerr Addison Kills Two Miners

Two miners, Roy Fletcher and Aksel Meister, were working in a stope located 120 feet above the 1,750 level when a heavy rockburst in that area caused a fall of ground to hit and bury them at about 12:25 a.m., February 1.

Shift Boss Jack Barbour had visited the stope an hour-and-a-half before the burst. He found it in excellent condition as there was no indication of ground movement. The two miners were removing steel forms and planning to move the slusher and scraper to a new position during their shift.

Revisiting the stope again at about 1:55 a.m., Barbour noticed evidence of recent ground movement in the raise that led to it. When he was unable to get a reply from the two men, he suspected an accident and went for help. Re-entering the stope by means of another raise, he found both men lying under broken rock that had fallen from the back; both appeared to be dead. Dr. G. E. Hagerman arrived underground shortly after but was unable to examine the men until they were removed to surface at about 6:00 a.m. Death to both was caused by multiple fractures and was probably instantaneous.

## Roy Fletcher (1928 - 1963)

Born on August 26, 1928, 34-year-old Roy Fletcher, employed at Kerr Addison since 1959, was considered an experienced miner. He, his wife – the former Isabella Graham – and their two young daughters, Annette (10) and Elga (2) lived on Dorfman Avenue in Virginiatown.

Interment occurred in the Kirkland Lake cemetery. The epitaph on the headstone reads: BELOVED HUSBAND AND LOVING FATHER.

## Aksel Meister (1928 - 1963)

Thirty-five-year-old Aksel Meister, of Estonian ancestry, was a well-known sportsman from Larder Lake. He, his wife Winnifred and their three young daughters, Elly (12), Hemli (9), and Patricia (3), lived on Sixth Avenue.

Meister was an avid skier and spent many hours teaching young skiers at the Larder Lake Ski Club. After his death, a ski run at the club was named the 'Aksel Run' as a memorial to him.

### October 29, 1963

# Kerr Addison Mine

## Roy Davis (1909 - 1963)

*Roy Davis and wife Violet with their children Gerry and Susan (circa 1949). Photographer: Unknown*

Roy Davis was born in Quyon, Quebec in 1909 and came to Kirkland Lake in the early 1930s. In Sudbury in 1938, he married the former Violet Hills. At the time, he was working in the gold mines at Shining Tree. When the mines there closed down, the young couple moved to Toronto but soon returned to the north when Davis got a job at the Young-Davidson in Matachewan. After the outbreak of the war, they moved to Ottawa where Roy was employed at Ottawa Car. He wanted to enlist in the armed forces but due to his age this never happened. Instead he aided in the war effort by working at Malton, Ontario for Victory Aircraft, a company involved in the construction of bomb carriages for Lancaster bombers.

At the end of the war in 1945, Davis, Violet and their newborn son Gerry left Toronto and moved to Virginiatown where Roy became employed at Kerr Addison. A daughter, Susan, was born in 1947. Davis worked at the mine for 18 years until his death in 1963.

He died suddenly at home at 32 Waite Avenue in Virginiatown on October 29. He'd had a heart attack while at work and came home feeling ill. His 16-year-old son was present with him in a room when he had another massive attack and died. Davis was 54 at the time.

He was survived by his immediate family and by three sisters: Annie, Ruby and Millie, all of the Ottawa area.

Davis had been a member of the Kerr Addison Pioneer Club and the Virginiatown United Church.

The body rested at Symington Funeral Home until funeral service at Virginiatown United Church. After, the body was taken by train to Shawville, Quebec for burial in the family plot at Norway Bay cemetery.

The information and photograph for this account were given by Violet Linney, widow of Roy Davis. She presently (1999) lives in the Kirkland Lake area. Her son Gerry lives in Calgary and daughter Susan is deceased.

## 1964

- Huge rockbursts shake Kirkland Lake; two Wright-Hargreaves miners, Edouard Duchesne and Donald Yaworski, are killed
- Paul St. Gelais is killed at Kerr Addison
- Mines in the Kirkland Lake area close but the Adams Mine, on an iron ore deposit, opens and gives hope to experienced miners
- Some unemployed miners get jobs at the newly opened Jaymar Plywood plant at the end of Main Street
- Tough-Oakes Mine vault is preserved as a tourist attraction

## The Wright-Hargreaves Rockburst of August 14, 1964

The day-shifters at Wright-Hargreaves on August 14 noticed nothing unusual about the 4001 stope. On shift, Machineman A. Levstik noted this stope was in good shape. Safety Inspector Gordon Cameron also visited during the day and found everything in order, observing no cracking or signs of undue pressure.

The night-shift crew in this stope, Edouard Duchesne and Donald Yaworski, was instructed to drill off the rest of the holes. Harry Gauld, the night shift captain on his beat, came through, noticed the men were set up and had started drilling. Everything seemed to be in order so he left to inspect other areas.

At 9:27 p.m., Kirkland Lake was rocked by the effects of a rockburst that rumbled through town. Alarms about town were set off, sprinkler systems turned on, electrical failures occurred, windows and dishes rattled and homes shook to their foundations. Anxious wives phoned police and contacted mine officials to get more details. The mining town became quite nervous.

A burst had occurred in Wright-Hargreaves at the 4,000 level. Gauld had left stope 4001 about 25 minutes previously and was on another level when he felt an extremely heavy bump. The shift bosses quickly searched each work area and told the miners to evacuate and get to surface immediately.

Working in a stope on the 4,000 at the time of the burst, A. Pervik was suddenly blown back about 10 feet and his light went out. He was partially trapped by broken rock but was freed by his partner and they proceeded by another level to escape. Pervik later confirmed there was no prior warning of a burst.

A winze cage with a load of men was stranded in a shaft because of a power failure and knocked out signals. Another cage in No. 4 shaft could not be used because of damage.

On returning to the 4,000, Gauld learned that two men were unaccounted for. He went back to stope 4001 but found all entries blocked by fallen ground. There was no sign of them; he yelled but got no reply. Edouard Duchesne and Donald Yaworski were missing.

Mine Superintendent W. Sampson was called and proceeded underground to direct rescue operations but smaller bumps were occurring, causing heavy ground movements and making operations too dangerous. All able men were evacuated from the mine. Rockbursts continued for hours. Due to the severity of the continued unrest in the ground, it was decided to postpone any recovery operations until August 17.

Starting on this date and working around the clock, recovery operations continued until August 24, at which time the bodies of Duchesne and Yaworski were found. Dr. C. S. Harris examined them, finding severe crushing injuries; he said death to both was probably instantaneous.

An inspection of the mine later, revealed large scale damage over an area extending from the 4,500 to the 2,000-foot levels. All access to the 4001 stope had been blocked. New access and clearing of debris had to proceed to recover the bodies.

Mine General Manager Frank Buckle reported that he had received information from both Montreal and Ottawa concerning seismographs that had been made in these cities on August 14. These indicated heavy rockbursts in Kirkland Lake at 9:27 and 11:51 p.m. The report from the Dominion Observatory indicated that the reading at 11:51 was equivalent to an earthquake of magnitude four on the Richter scale.

All this signalled the end of Wright-Hargreaves. Plans were previously made to close but now it would be sooner rather than later.

## Edouard Duchesne (1909 - 1964)

Edouard Duchesne was born in Sturgeon Falls, Ontario in 1909 to a family that would later comprise three brothers and five sisters. He was the first born of the boys and, as such, learned responsibility at a young age.

The hard work on the farm made 'Eddy' healthy and strong. In fact, he became the annual 'Hay Harvest-Time Champion' three years in a row. The goal of this contest was to find the individual who could load the most bales of hay onto a haywagon in one day. From sunrise, through the heat of the day, until sundown, he and other competitors pitched bales onto the wagons. He outdid all others and was declared champion.

In the fall of the year after the hay and crops were harvested, he was responsible for the winter supply of firewood. Then when there were snowflakes in the air and the ground was frozen, he would leave the farm, go into the bush and work as a lumberjack for the winter months. This was a job for the hardiest of men, as long hard hours of work were expected of them.

At Christmastime, he would load up his packsack with his belongings and head for home. Only sleigh and horse tracks were present in the deep snow of the bush road, making it very difficult for one to walk such a long distance. In the darkness of the morning, Duchesne would leave the bush camp with his heavy laden sack on his back and walk the whole distance to home by nightfall.

In the summer of 1934, he married Evelyn Lapierre of Cache Bay, Ontario, who also came from a large family and knew about hard work.

The Depression presented difficult times for everyone. Duchesne wanted to work but could only get jobs here and there. Brother Rolly had a car and went up north to Matheson and Timmins looking for work but no one was hiring. On his way home, he stopped at Kirkland Lake and was hired at Kirkland Lake Gold. Rolly worked there for six months and during this time was recognized as an excellent worker. He was asked by management if he had any brothers looking for work. As a result, Duchesne was hired at the mine. This was very fortunate, as long lines of unemployed men were waiting at the gates of all mines in town.

*Edouard Duchesne. Photographer: Unknown*

Duchesne and his wife came to Kirkland Lake in 1938. Here they would raise five children: Fernande, André, Yvon, Maurice and Normand. He remained in mining for the rest of his life. The working conditions were very demanding but he was glad to have a job to support his large family.

Involved in the mine strike of 1941/42, he returned to Kirkland Lake Gold when it ended and worked there until the mine closed. He got a job at Wright-Hargreaves in 1960 and worked there until the rockburst took his life in 1964. He was buried in Sturgeon Falls.

This personal history was given by Edouard's son André who lives with his wife Elsie in North Bay. Fernande and Maurice are now deceased. Yvon lives in Oakville and Normand lives in Mississauga. Grandson, Leo Duchesne lives in Kirkland Lake (1999).

In telling the story about his father, André included a couple of interesting anecdotes: "When I was a young boy, I lived near the Toburn Mine. During the quiet of the night while lying in bed, I could hear the men working underground below me."

"In my early youth, I attended a Catholic elementary school. Many mines were in production in Kirkland Lake at the time and it was a relatively common event for an airblast (rockburst) to shake the town. It didn't matter what the students were doing in class, when they heard an airblast, they would stop and say a prayer for the miners working underground."

## Donald Yaworski (1919 - 1964)

Donald Yaworski was born on February 18, 1919 in Cobourg, Ontario to White Russian parents; he was raised in rural Southern Ontario. He enlisted in the Canadian Army on June 5, 1941 at the age of 22. He became part of the Fort Garry Horse 10th Armoured Division and was shipped overseas. While in England, he met Olive Cullingworth of Blackpool. They were married during the war and when the war ended, the couple and their young daughter Christine came to Canada. Like other soldiers returning after the war, Yaworski was looking for work and came to the gold mines of Kirkland Lake with his family.

*Donald Yaworski. Photographer: Unknown*

A son, David, was born on October 3, 1947 and son Arthur was born on October 4, 1948. Daughter Christine later married Jim Webster.

The family was living at 17 McKelvie in 1963 when Olive passed away, leaving Donald a widower. Forty-five-year-old Yaworski was instantly killed a year later when crushed in a rockburst at Wright-Hargreaves. He was a machine runner and had worked at the mine since 1946. David (17) and Arthur (16) were now orphan teenagers.

The funeral service was held at St. Peter's Anglican Church and interment took place beside his wife at the Kirkland Lake cemetery. The inscription on the gravestone reads: YAWORSKI – IN LOVING MEMORY OF OLIVE CULLINGWORTH – 1921 - 1963 – BELOVED WIFE OF DONALD YAWORSKI.

Much of the information on the Yaworski family and the photograph was provided by Yaworski's son David, who lives in Kirkland Lake (1999).

### November 24, 1964
# Kerr Addison Mine

## Paul St. Gelais (1926 - 1964)

Paul St. Gelais, a 38-year-old Kearns man and father of nine, was killed instantly when he fell down an ore pass at Kerr Addison. He was hired at the mine in December 1959.

On the night-shift of November 24, he and helper Laurent Rene were pulling muck from a millhole into a train consisting of two dump cars and a locomotive. Locomotive Operator St. Gelais stopped the train at the ore pass. Rene was walking behind and proceeded to the platform to operate the air valve that would open the ore pass door and place the ramp in a vertical position. The door opened but the ramp did not completely move into position. St. Gelais got off the locomotive, walked along the track and stooped over to remove a plank that appeared to be jammed in the ramp. Rene told him to wait and had just started to turn the air valve into the closed position when the ramp suddenly sprung upright. Rene saw his partner fall backwards into the ore pass. He notified Shift Boss Russell Graham and rescue operations began immediately.

*Paul St. Gelais. Photographer: Unknown*

The body was recovered from the top of a muck pile after falling down the ore pass, a distance of 118 feet.

Dr. G. E. Hagerman examined the body and stated that death was probably instantaneous. Dr. D. G. Howatson, pathologist, performed an autopsy the following day and gave the cause of death as massive internal injuries and a fractured skull.

After the accident, it was found that the plank did cause the ramp to jam but it wasn't necessary for St. Gelais to place himself in such a dangerous situation. The ore pass door could have been closed and the plank removed safely. Safety belts were located on both sides of the track at this site but were not used. Company regulations required that a safety belt be worn when working around any dangerous opening.

St. Gelais was survived by his wife, Jeannine, and their children: Réal (11), Leon (10), Alain (8), Lina (7), Roger (6), Dianne (4), Nicole (3), Carole (1) and Marcel (five months old).

The body rested at the Darby Funeral Home in Rouyn. The funeral mass was conducted in the Roman Catholic Church at McWatters, Quebec; burial took place in the cemetery there.

An inquest was held in Virginiatown before Coroner J. M. Leblond. The jury brought in the following verdict: "We the jury, find that Paul St. Gelais came to his death at the Kerr Addison Mine due to his failure to abide by the mine's safety rules. The death was accidental."

### 1965

- David Dool and Rolland Trudel are killed in separate accidents at Kerr Addison
- Wright-Hargreaves closes after producing almost 5,000,000 ounces of gold
- Lake Shore closes after producing 8,499,000 ounces of gold
- Upper Beaver Mine (the old Argonaut) reopens in Gauthier Township
- Population of Kirkland Lake is 16,445
- Kerr Addison Mine personnel team goes to Europe to search for 100 miners and tradesmen

### June 30, 1965
## Kerr Addison Mine

## David Adam Dool (1932 - 1965)

David Dool was originally from North Bay, Ontario. Around 1957, he worked and lived in Kirkland Lake, and was employed at Wright-Hargreaves and Macassa. He was hired at Kerr Addison on October 28, 1963, as a mine labourer. He advanced to apprentice miner in December and became a miner in 1964. He married the former Suzy Roux and they had two children, Suzanne and

*David Adam Dool and Suzie Roux (headstone photograph). Photographer: Author*

John. At the time of the accident, his parents were living in North Bay. Dool was fatally injured on June 30 when struck and pinned beneath collapsed square-sets in a stope at the 1,750 level.

He and partner Marcel Coté* were slushing ore at the time. Coté was operating the slusher hoist and Dool was directing the movement of the scraper using light signals. When the 1,665-pound scraper was within seven feet of Dool, three rows of square-sets collapsed without warning and fell on him. Coté went to see if Dool was injured after the sets fell. Dool was partially visible but could not be approached because timber was still falling. Coté went for help and V. Pederson came into the stope to begin rescue operations but Dool appeared to be dead.

Dr. G. E. Hagerman was called and went immediately to the scene of the accident. He stated that death was instantaneous. Dr. J. W. Clegg, pathologist, later performed an autopsy and gave the cause of death as a heavy, comprehensive injury to the chest, which caused a cracked spine and instant death.

An inquest was held before Coroner J. M. Leblond. The collapse of the timbers could have been caused by the scraper catching and knocking

*David Dool was killed during a slusher operation. Photographer: Unknown*

out a post in the lower sets. Coté stated that in slushing operations, the scraper quite often hit posts but had never before caused any damage. He said there was no indication of the scraper hitting a post this time. He also stated they checked the timber at the start of the shift and everything appeared to be in good order. Shift Boss V. Pedersen visited the stope two-and-a-half hours before the accident and he also didn't detect any problems. The jury returned the following verdict: "That David Dool came to his death from a complete fracture of the spine causing shock, resulting in instant death. The accidental death was by falling timbers from the square-set staging collapsing due to undetermined causes. We recommend that a great deal of care should be taken in the original erection of square-set staging."

Interment occurred in the Kirkland Lake cemetery. A photograph of Mr. and Mrs. David Dool is fixed to the headstone.

** Marcel Coté's brother, Lucien, would die in a mining accident at Kerr Addison in 1968.*

### October 9, 1965

## Kerr Addison Mine

## Rolland Trudel (1938 - 1965)

Twenty-seven-year-old Rolland Trudel, Canadian, married and a father of three, was killed at Kerr Addison. He had been working at the mine for just five months – hired May 22, 1965 without previous mining experience.

On the 8:00 a.m. – 4:00 p.m. shift of October 9, he and partner Donald Marceau removed blocking, did considerable scaling and installed headcovers over the sets already installed. Shift Boss Frank Kaiser visited the stope at about 10:30 to find the ground 'working' and ordered the men to retreat from the stope for a short period.

Kaiser re-visited the stope at 1:10 and stayed with the crew until 2:00. The ground was now quiet. After Kaiser left, Marceau and Trudel worked at erecting a set. Marceau had just stepped back to obtain wedges for Trudel when a portion of the stope above the set collapsed, knocking down the set, breaking the stage on which Trudel was standing and breaking the cover over the millhole. Trudel, the collapsed rock and the staging were swept into the millhole.

Marceau immediately went to the level above for help from miners Monty Sampson and Lou Gunsinger. Gunsinger went to the station to summon more assistance, and Sampson and Marceau returned to the stope.

Rescue operations began immediately. The loose rock around the millhole was shored up and the body was recovered at 5:45. Dr. A. S. McPhail of Larder Lake examined the body and pronounced Trudel dead. The cause of death was multiple skull fractures. The dead man's watch had stopped at 2:53.

An inquest was held before Coroner T. W. Armstrong. Present were District Engineer of Mines Fred Nabb, Crown Attorney J. Spence, and Jury Foreman Dan Story. The jurors were Ron Bracher, Fred Francoeur, Mrs. Rollie Acton and Mrs. F. Turcotte. The jury returned a verdict of 'accidental death due to a fall of rock.'

### 1966

- Long-time employee at Macassa, Roy Sipprell, is killed underground at the mine
- Richard Jones, 33, a father of seven, is killed at Kerr Addison, two days before Christmas
- Kirkland Lake takes over the Wright-Hargreaves property

### May 25, 1966

## Macassa Mine

## Roy Logan Sipprell (1908 - 1966)

Roy Sipprell of 602 Government Road West in Chaput Hughes was fatally injured on the 5,150 level of Macassa.

Prior to employment at Macassa, Sipprell obtained mining experience at East Malartic, Siscoe and Hollinger mines; he was a graduate mining engineer. The 57-year-old worked at Macassa for 27 years, beginning work on October 1, 1938. He left Macassa on November 5, 1942 and went into the army, returning after the war on December 15, 1945.

On May 25, General Superintendent Charles Green arrived at the 5,150 station at about 10:20 a.m. and met Sipprell. They discussed the ground conditions and Sipprell mentioned there was an ore car in the way of the working place. At noon, the supervisors including Sipprell went for lunch to an upper level. After lunch, he took the cage and returned to the 5,150. No one saw him alive after this time.

At 1:45, the supervisors stopped the cage at the 5,150 for Sipprell but he wasn't at the station; they proceeded to surface. He could have been in any one of several places at this time as he had been in the past, owing to the nature of his duties.

At 2:15, Shift Boss H. Ducharme reported that Sipprell hadn't come to surface. No action was taken at this time because occasionally Sipprell would stay underground to supervise the hoisting of men at the end of the shift.

At 3:00, Green phoned Cagetender W. Lawrence to ask if Sipprell had phoned for service. Lawrence replied that he hadn't phoned or been seen. Green instructed the cagetender to go down to the 5,150 to look for him. He did so but informed the superintendent that Sipprell was not around.

Green, Ducharme and First Aid Attendant E. Haskins proceeded underground immediately and started their search on the 5,150 level. While walking down a drift, they found Sipprell lying half out of the motor cab of a locomotive. His right arm at the armpit was jammed under a chute lip. The motor controls were in the full-open forward position with the brake released. The control switch was on, but due to the overload, the circuit breaker had cut off power to the wheels. Dr. E. Yamka was called and came underground to the scene and pronounced Sipprell dead.

Pathologist Dr. J. W. Clegg performed a post-mortem that evening. He stated that Sipprell suffered fractured ribs on the right side, a laceration of a lung, abrasions to the body and a bump on the right side of the head. He gave the cause of death as asphyxia due to external pressure on the chest.

It would appear that Sipprell took the locomotive shortly after 12:15. It would also appear that he was standing in the cab rather than sitting on the seat. If he had been sitting, there would have been ample clearance between the locomotive and the lip of the chute. The control was engaged in the full-on position but he may have opened the control the wrong way by mistake.

Sipprell was survived by his wife, the former Nancy Powell, and by six children, aged 10 to 23. They were John, Robert, Edward, Alfred, Mary and George.

The funeral services were conducted at St. Andrew's Presbyterian Church with Reverend K. J. Wilson officiating. Interment took place in the local cemetery.

An inquest was held before Coroner J. M. Leblond in the Ontario Provincial Police building at Swastika. The jury returned a verdict of accidental death with blame attached to no one but they did make three recommendations: (1) a workman does not work alone; (2) consideration of a dead man control on battery motors; (3) that supervisors should be aware of the location of workmen at all times.

Presently (1999), Mrs. Nancy Sipprell lives in Kirkland Lake. She verified the accuracy of this account.

### December 23, 1966

## Kerr Addison Mine

## Richard 'Radar' T. Jones

### (1933 - 1966)

Margaret E. Jones left North Wales and came to Kirkland Lake where she established a boarding house at 45 McCamus Avenue. Dave Jones, a Lake Shore miner and one of the boarders, met, courted and then married her daughter, Selwin. The marriage produced two sons, Joe and Richard.

Richard was born in Kirkland Lake in 1933 and was educated here as well. Aged 33, he had almost 11 years experience as a miner. He had worked as a motorman and miner at Lake Shore, Teck-Hughes, Macassa, Sylvanite and Wright-Hargreaves. He was hired at Kerr Addison on September 15, 1964 and worked there until the accident on December 23, 1966.

Jones lived at 47 Queen Street with his wife, the former Joan Montague, and their seven children.

*Richard Jones with his son. Photographer: Unknown*

His death was all the more distressing in that it occurred just two days before Christmas.

He and partner Gaston Yergeau* scaled the stope in which they were working and then began to slush but had problems with the pull back cables of their hoist. They went to the face of the stope to adjust and re-attach the cables. Suddenly, 30 tons of ore fell from the face and back while they were there. Jones was struck by the falling ground while Yergeau jumped clear of the main fall and was only partially buried. He managed to dig himself out and reported the accident to Clement Pepin of an adjacent stope. Pepin called to his partner, John Hester, and told Yergeau to go to the station for more help. Pepin, Hester, Captain Murray MacAulay and Shift Boss Jack Turner proceeded with rescue operations. Jones was brought out of the stope and pronounced dead by Dr. G. E. Hagerman. The cause of death was a very acute hemorrhage from a ruptured heart.

He was survived by his wife and children: Debbie, Linda, Christine, Ricky, Shawn, Mark and David. He was also survived by his father, David, his grandmother, Margaret and one brother, Joe, all of Kirkland Lake. The body rested at Symington Funeral Home where the funeral service was held in the chapel and conducted by Reverend J. M. Finlay. Interment occurred in the Kirkland Lake cemetery.

Richard's brother Joe presently (1999) lives in Kenogami with his wife Mary (1999). He supplied the information concerning the family. Richard's brother-in-law, Gary Montague, supplied the photograph.

** Gaston Yergeau's brother, Roger, died in a mining accident at the Kerr Addison in 1959.*

### 1967

- No mining fatalities occur in the Kirkland Lake area but a well-known local student, Patrick McCarthy, is killed at the Renabie Mine during his summer employment
- Teck Centennial Library opens locally as the Canadian Centennial is celebrated across the country
- First Museum of Northern History opens in the vacated Wright-Hargreaves assay office

### August 1, 1967

## Renabie Mine

## Patrick Leonard McCarthy

### (1949 - 1967)

Patrick McCarthy was born in Kirkland Lake to Jim and Mary McCarthy on September 20, 1949. He had two older brothers, Jack and Jim Jr.

He attended St. Jerome Separate School and graduated from grade 12 at Kirkland Lake CVI. While at high school, he was involved in many sports and excelled at football and basketball. In town, he played midget fastball and was captain of the Kirkland Lake Lions Juvenile Hockey Team. He was a quiet young man with a number of very close friends.

After graduating, he left town in search of a summer job and it was his intention to return to KLCVI in September to complete his grade 13. He was hired at Renabie Mine in Wawa, on the understanding that his birth date was September 20, 1948.

*Patrick McCarthy. Photographer: Unknown*

Single and 17, McCarthy died on the afternoon of August 1 when he fell down an ore pass from the 2,325 to the 2,475 level of Renabie. He had worked for a month in the mill and was transferred underground as a motorman-trainee on July 26. For four days, he was part of a three-man mucking crew engaged in loading ore cars from a box-hole, pulling the cars by locomotive to the 2,325 dump and discharging the ore into an ore pass.

On August 1, an ore car had its load dumped but retained a large rock. One member of the crew attempted unsuccessfully to bar the rock with a four-foot steel and force it through the car side opening. The employee backed away when McCarthy took the bar from him and tried to lever the rock over the top edge of the car. McCarthy was standing near the ore pass opening and, as he levered, he lifted himself off the ground several times. Suddenly, the bar came loose and he slipped feet first through the opening and dropped 150 feet to the level below. This occurred so quickly that no one was able to help.

A rescue was arranged, and a stretcher and a doctor were summoned. At the 2,475 level, men went into the ore pass and brought out McCarthy's body. Dr. Young from Chapleau certified the death.

The body of Patrick McCarthy was transported to Kirkland Lake where family, friends and townspeople were saddened by the sudden death of this young man. Funeral services were held in town and burial occurred in the local cemetery. The inscription on the gravestone reads: McCARTHY – IN LOVING MEMORY OF PATRICK L. McCARTHY 1949 - 1967 – BELOVED SON OF JAMES J. McCARTHY – 1901 - 1989 – BELOVED HUSBAND OF MARY T. FOLAN – 1912 -.

After the accident at Renabie, safety lines were placed at the ends of the ore pass dump and one over it so men working near the ore pass in similar circumstances would hook the safety line to their personal safety belt. A triangular wood partition was erected on each side of the dump to close the vertical opening.

An inquest was conducted at Renabie on August 30, 1967. The jury returned to Coroner J. A. Pidutti with the following verdict: "That Patrick Leonard McCarthy came to his death at 2:05 p.m. on August 1, 1967 at Renabie Mine and that the death was caused by multiple fractures of the ribs and bones and a torn aorta was sustained as a result of falling down the ore pass between the 23rd and 24th levels. The accident occurred while the deceased was barring an oversize piece of ore from an ore car. We the jury have determined that McCarthy lost his footing and slipped through the opening into the ore pass. We recommend the use of the safety equipment provided."

At KLCVI, the students and staff established a basketball tournament held in December 1967. The Patrick McCarthy Memorial Basketball Tournament was held annually at the high school for almost 20 years.

Presently (1999), McCarthy is survived by his mother Mary, a brother Jim in Haileybury and brother Jack in Kirkland Lake. Jim supplied the information and donated the photograph for this memorial to his brother.

## 1968

- Doug McFarlane, injured at the Upper Beaver, dies in hospital two days later
- Tex Clark drowns in a thickener tank at Adams Mine
- Emmett Harney is killed at Kerr Addison; two months later, another miner, Lucien Coté, is killed here in separate accident

- Herb Bowers is killed on surface at Upper Canada
- Teck-Hughes closes in its 51st year of operation
- Macassa is the only producing gold mine left in town and is down to 6,700 feet
- Adams Mine of Jones and Laughlin Mining Company process about 20,000 tons per day and during the year shipped 1,000,000 tons of pellets containing 65 per cent iron
- Crescent Gold Mine, formerly the Swastika Gold Mine, becomes a tourist attraction

*Doug McFarlane worked in his father's grocery store for 17 years before starting at Upper Canada in 1964. Photographer: Unknown*

### February 7, 1968

## Upper Beaver Mine

## Douglas H. McFarlane

### (1928 – 1968)

Doctors fought to save the life of 39-year-old Douglas McFarlane of King Kirkland who was severely injured in an explosion at the Upper Beaver Mine, six miles north of Dobie.

McFarlane, married and the father of four, was listed in critical condition at Kirkland and District Hospital on February 7. He had multiple injuries including massive external hemorrhage, multiple lacerations of the thighs and severe fracture of the right femur.

A second miner, James Kasner, also injured, was taken to hospital with thigh lacerations, a broken leg and a severe cut over his left eye but his injuries were not life threatening.

McFarlane had been instructed to timber in a stope and to help blast as Kasner's regular partner was off work that day. The two were having difficulty lighting the round on the drift face as the lit leads kept going out. They kept re-lighting them, when finally Kasner told McFarlane that it was time to go and started to walk away. When he was a distance down the drift, he noticed that McFarlane still had not started out, so he went back to get him and was about six feet from him when the first shot went. Kasner, now bleeding and suffering a broken leg, crawled a distance of 900 feet to the station where the cagetender noticed him. He was taken to surface in a stretcher and rescue workers went for McFarlane. He was conscious and still had a pulse. He was brought to surface and rushed to hospital where he never did regain consciousness. The struggle by doctors to save his life ended two days later, on February 9.

He was the son of the late William B. McFarlane and the former Janet Wattam. Born in New Liskeard in 1928 and after receiving his elementary education, he later attended high school in Kirkland Lake. After leaving school, he worked in his father's grocery and butcher shop business in King Kirkland for 17 years until his father's death in 1964. He then became employed at Upper Canada and worked there until the accident in 1968.

He was married in 1952 to the former Nancy Lindsay. He was a member of the King Kirkland United Church, the Doric Masonic Lodge No. 63 and the Dobie Curling Club.

Surviving McFarlane were his wife, two sons, Lindsay (12), and David (4), two daughters, Mary Jean (14), and Laurie (10), his mother, Mrs. William McFarlane of New Liskeard, brother Donald of Sturgeon Falls and two sisters, Evelyn (Mrs. Bernard Wignal) and Jean (Mrs. Donald White), both of New Liskeard.

An inquest was held before Coroner J. M. Leblond. The jury returned the following verdict: "That death was due to a blasting accident at the Upper Beaver Mine. The cause of the accident was a delayed leaving of the workplace after the Thermalite was lit. A recommendation was made that miners leave the working place after the Thermalite is lit. We the jury commend the heroic efforts of Mr. James Kasner in obtaining assistance in spite of his serious injuries."

Interment took place in the Kirkland Lake cemetery. The epitaph on the headstone reads: BELOVED HUSBAND AND OUR LOVING FATHER.

The personal information and the photograph provided by wife Nancy.

*Tex Clark. Photographer: Unknown*

## May 13, 1968
## Adams Mine

### Tex Malcolm Edward Clark

(1931 - 1968)

Tex Clark was born on January 3, 1931 in Penetanguishene, Ontario. His parents, Allen and Ellen Clark, moved in the 1930s to Kirkland Lake where his father set up a company, Northern Mining Supply. At 16, Clark became involved with the company and remained so until his death in 1968. Northern Mining Supply was engaged in buying and selling used machinery, with the business situated in Chaput Hughes.

The marriage of Clark to the former Irene Birnie produced three sons: William Allen (b. June 23, 1954), James Andrew (b. October 25, 1959), and Richard Boyd (b. October 5, 1961).

"Tex was a great guy," remembers Irene. "He was always helping out people, no matter what the time of day or night."

One time, he rushed to a fire in Chaput Hughes. Irene recalls, "There were deaths involved. He used his car as an extra ambulance to transport the bodies to Symington's Funeral Home."

Another time Clark received a phone call from authorities asking if he could come immediately to Aidie Creek where a man had drowned following a car accident. "Tex recovered that body also."

When Clark died, his eyes were donated to the Eye Bank and given to a young person in Toronto. "He would have been pleased. That's the kind of guy he was," she said.

Clark and his friend Jack Royal were fellow members and founders of the Northern Aquatic Group. In 1959, Clark was employed in Huntsville, Ontario, recovering logs from a lake. In 1968, he was in perfect health and had no history of illness.

At the time of the accident, he was doing contract diving to detect suspect solids that were clogging and impeding the movement of mechanical rakes in an outdoor thickening tank at Adam's Mine of Jones and Laughlin Mining Company Limited (see photograph).

The tank was 280 feet in diameter with a depth of 14 feet at the perimeter and 26 feet at the centre. The tank's contents were a slime and water mixture, containing fine tailings, grey in colour. Clark, an experienced scuba diver, had been contracted to work at this thickener several times previously, particularly during the start up of the mine in 1964/65. He had also supervised the installation of the rubber lining during the construction of the tank. At about 9:30 a.m. on May 13, Assistant Manager Bob Rouleau telephoned Clark, explained to him that the thickener had become stuck and asked him to dive into the thickener and spot the trouble. Clark packed a lunch, left his home at 828 Government Road West and arrived at the mine alone in his own automobile at 11:00 a.m. He had brought his equipment with him.

*Aerial view of Adams Mine property. Note the circular thickening tank at left centre. Photographer: Eddie Duke*

He put on his rubber suit, a pair of canvas boots, a weight belt, a rack containing an air cylinder and a safety belt attached to a rope some 150 feet long.

Clark was to go into the thickener and locate the clogged area so it could be cleared by a crane and clamshell. He walked over to the edge of the tank, put on his mouthpiece, tried the respirator, tested the valve and entered the tank with his scuba gear. He went down for a couple of seconds and came up to let some air out of his suit by means of a valve. He submerged again and walked circumferentially along the bottom, stopped, turned to his left and headed to the centre of the tank. His progress was followed by the appearance of intermittent bubbles. He would have to examine the clogged rakes by feel, as visibility was nil.

He reached the centre and started to come back. At about 15 feet from the centre he became stationary, surfaced and in the process removed his mouthpiece. He appeared to be struggling with something around his middle. A taut lifeline, tended by Rolly Duroucher, vice-president and general manager at Adams, and Doria Blais, the crane and clamshell operator, was maintained during Clark's travel in the tank.

Clark surfaced at the centre well, appeared to attempt to climb out, then submerged. He again came to the surface quite quickly, and tried to get out. There now seemed to be more anxiety in his actions.

It was now realized that he was in trouble and the lifeline was pulled in. However, it had been disconnected by Clark and just the belt came in. Men tried to reach him by lowering themselves into the tank by rope but they were unable to locate him in the murky liquid. Several minutes were required before recovery was completed with the aid of a ladder and a grappling hook. Artificial respiration was given but to no avail.

A post-mortem was performed by Dr. John Morphet, who stated that death was by cardiac arrest caused by anoxia due to drowning. An inquest was held before Dr. J. M. Leblond. The jury returned the verdict: "That Tex Malcolm Clark came to his death at 11:30 a.m. on the 13th day of May 1968 and that the death was due to accidental drowning due to a scuba diving accident in the thickener tank on the Jones and Laughlin Steel Corporation Mine site. We the jury unanimously agree that the death was caused by the loss of his compressed air tank on which the clasp of the tank brace was accidentally released. We feel that there was no negligence on the part of the deceased or the company concerned."

The following recommendations were made: 1) appropriate life saving apparatus be installed at the thickener tank; 2) in future operations of a similar nature, a boat should be in the water in the diving area with two people aboard; 3) a lifeline should be maintained and affixed to the diver from the boat.

Interment occurred at the Kirkland Lake cemetery. The epitaph reads: IN MEMORY OF BELOVED HUSBAND AND FATHER.

In 1969, Irene Clark moved with her three boys, aged 14, 10, and 8, to Timmins where she felt she could get a better job in order to provide for the children. "Jones and Laughlin never offered assistance, financially or otherwise; however, many of the employees donated money. Dick Flewelling came to my door with over $500 collected from the workers, many of whom Tex knew," recalls Irene.

A non-jury trial was held in April 1972 at the Supreme Court of Ontario in Haileybury. Jack Royal, former scuba diver and friend of Tex Clark, was called as a witness on behalf of the defendant, Jones and Laughlin Mining Company Limited. He had observed the air cylinder that had been recovered after the thickening tank was drained, and, upon seeing it, stated that the 'reserve knob on the top was not operating.' Royal went on to testify that a scuba diver should use flippers; that a rope around the waist would act as an anchor; and that it would have been difficult for Clark to ascend to the surface, as he was attached to a rope and wore no flippers.

Royal went on to explain that the non-return valve in the mouthpiece was not operating and thus there was a leak in the air system; this allowed fluid from the thickening tank to get into the mouthpiece. In response to questioning by lawyer Murray Ellies, Royal said: "The contents of the thickening tank were getting into the mouthpiece and he was choking on the liquid."

At the trial, Judge J. O'Driscoll stated in his report: "Upon reviewing all the evidence, I can come to no other conclusion other than that the deceased was an independent contractor at the time of his unfortunate death and that this death came about as the result of his own negligence in entering the tank, as he did, with faulty scuba diving equipment and without taking the precautions of having either a tender on the surface in a boat, or having a 'buddy' dive with him; it would also appear to me from the evidence I heard from Mr. Royal and Mr. Davis (scuba diving expert) that it was negligence in this case for the deceased to have worn the canvas boots and not the regular scuba diving flippers." He continued, "On the evidence before me, I cannot find that the Defendant (Jones and Laughlin) or any of its servants, agents or employees in any way worsened the position of the deceased; indeed it would appear that the only thing that the deceased asked of the Defendant was that they tender the safety line, which was done."

The Clark family was very disappointed with the trial and its outcome.

It was in Timmins where Irene met David McKay. They were married in September 1972. David and Irene presently (1999) live in Cobble Hill, B.C.; Bill lives in the Kirkland Lake area; Jimmy lives in Calgary; and Richard lives in Victoria, B.C. Much of the information for this account was given by Irene.

### June 28, 1968
## Kerr Addison Mine

# Emmett Joseph Harney

### (1929 – 1968)

Emmett Harney was fatally injured at Kerr Addison on June 28. He had been employed previously at the mine from 1962 to 1966 and was rehired on January 3, 1968. Raised in Latchford, Ontario, he was a 38-year-old father of six and the husband of the former Jennifer Woloschuk.

Beginning the dayshift, Harney and partner Steve Malciw Sr. were instructed by Shift Boss Richard Mayer to slush broken muck from a stope into a five-foot diameter concrete millhole.

Mayer made his second visit about five hours later and the men were still slushing ore but there were a number of timbers accumulating in the stope. Mayer instructed Harney to hoist this excess timber up a raise to the next level. Harney asked him if he could simply leave the timber in the stope and it would be buried in the fill but Mayer said no because much of the timber was still good. Harney then requested that the millhole be emptied of ore and he be allowed to throw the timber down and remove it from the chute. Mayer agreed to this, proceeded to the level below and instructed mine labourers, Neil Patterson and Ronald Sparling, to have the millhole pulled empty and to remove the timber after it was thrown down.

Harney and Malciw began to clear the stope of timber. Malciw piled the old broken timber in the corner while Harney selected the good timber and dropped it down the millhole. At approximately 3:20 p.m., Sparling yelled up the millhole that it was quitting time. Harney replied that he would throw down a few more timbers before leaving. A couple of timbers came down and there was a long silence.

Sparling and Patterson looked up into the chute and saw Harney lying on top of the timbers that were previously thrown down. In the stope, Malciw heard the men yelling and began to look for his partner. Patterson ran to the station where he informed Mayer that Harney had fallen down the millhole. Mayer immediately called surface for a stretcher, a first aid man, a doctor and more supervisors.

Harney was removed from the chute, placed on a stretcher and taken to the first-aid room on surface. An examination was made by Drs. MacPhail and Hagerman and he was pronounced dead. An autopsy performed by Dr. J. A. Morphet at the Kirkland and District Hospital revealed Harney died of shock, hemorrhage, visceral and vascular trauma, broken ribs, broken pelvis and broken legs.

Interment occurred in the Kirkland Lake cemetery.

An inquest was held before Coroner J. M. Leblond with the jury returning the following verdict: "We the jury say that Emmitt Harney came to his death due to a 73-foot fall down a millhole causing internal hemorrhage and multiple injuries. He neglected to carry out safety rules by not putting on his safety rope when working around an opened millhole."

*Lucien R. Coté. Photographer: Unknown*

### August 31, 1968

## Kerr Addison Mine

## Lucien R. Coté (1940 - 1968)

Lucien Coté's parents, Elise Coté and Alphonsine Ross, were descendants of early settlers of St. Ulric, Quebec, located in the part of the province known as Gaspé. Most of the ancestors of the family were farmers and bush workers. Seven children were born to Mr. and Mrs. Coté; in order of birth: Ghislain, Noella, Lucien, Fernand, Marcel, Cécile and Thérèse.

By the time Lucien was born on June 24, 1940, the family had already left Quebec and moved to Ontario. He was born at his home in Kirkland Lake. He received his early schooling, attending Assumption Separate School in town.

Coté's father was a miner in Kirkland Lake; he later worked at Kerr Addison at Virginiatown where the family moved in 1958. At the age of 18, Lucien also became a miner there. At 23, he left Ontario, seeking new opportunities in the Yukon but returned eight months later on September 15, 1966 and he was re-hired at the mine. In 1967, he married the former Gisele Bureau.

He was very active in baseball and hockey and his younger sister, Cécile, remembers him as a good sport.

On the dayshift of August 31, Coté and partner Clement Goulet were instructed by Shift Boss Ron Dagelman to wash, scale and rockbolt the breast area of stope 4,621. After scaling for about 15 minutes, the stope back started to 'snap and crack', so the two men went to a safe location to wait it out. After a period of time, the rock quieted down and they continued with the scaling. As they were doing this, there was a loud cracking noise and a large slab of rock fell from the back. Coté attempted to push his partner clear but the slab hit Goulet on the side and knocked him down to the muck pile. Coté himself was completely covered by the rock. Goulet, realizing that he alone could not help his partner, climbed a ladderway to the next level, took a motor to the station and phoned the skiptender. Soon George Rathwell, Jack Sagin, Murray MacAuley, Underground Superintendent Joe Teal and Hugh Armstrong, first-aid man, were proceeding to the accident scene. Along the way they were joined by Shift Bosses Gerry Ryan and Ron Dagelman, and Electrician Joe Lemire.

During recovery operations, the large slab was braced, jacked up and Coté was extricated. Dr. G. E. Hagerman performed an examination and pronounced him dead.

Goulet was taken to Kirkland and District Hospital where he remained for 13 days, recuperating from his injuries.

An autopsy was performed by Dr. T. W. Armstrong. His findings were that Coté sustained fractured ribs, neck, femur and pelvis, and had massive internal injuries. Death was instantaneous.

The funeral mass was held at the Catholic Church in Virginiatown and the funeral arrangements were handled by Symington Funeral Home. Interment took place in the Kirkland Lake cemetery. Much of the personal information for this account and the photograph were supplied by Lucien's sister, Mrs. Cecile Roddick of Larder Lake. Presently (1999), Noella lives in Toronto, Ghislain and Marcel in Timmins, Fern in Sudbury, and Thérèse in Cochrane. A cousin, Rolly Coté, lives in Kirkland Lake.

### December 5, 1968

## Upper Canada Mine

## Herbert James Bowers

### (1915 - 1968)

If you lived in Dobie during the 1950s and '60s and needed a helping hand, you could always count on Herb Bowers being there.

He was born on May 25, 1915, on a farm just north of Haileybury. The family moved to Rosegrove, Ontario after their farm was destroyed in the Great Fire of 1922. Their stay was very short and within a year they moved again to another farm located at Boston Creek. In 1932, 17-year-old Bowers got a job working there at the Miller mine.

His marriage to Violet E. Blakeman in 1936 took place in the Swastika United Church. Also about this time, he left the Miller and gained employment at the Argonaut, north of Dobie.

In 1940, he left the Argonaut and got a job at Lake Shore in Kirkland Lake where the family, now with three children, took up residence.

In 1942, he spent a short time in the Canadian Army but received an honourable discharge for

*Herb Bowers (wearing safety hat) on the Upper Canada property with Pa Drury and grandson Terry Bowers. Photographer: Jim Bowers*

health reasons. It was about this time he began his employment at Upper Canada. The family moved to Dobie in 1950. Bowers was very active in community affairs and a member of the Masonic Lodge.

In 1968, Bowers, a 27-year employee at Upper Canada, was 53 and a surface superintendent. On December 5, he and R. Poudrette, a heavy equipment operator, went to the surface warehouse with a front-end loader to get two filter rolls for the mill. The rolls, crated in wooden boxes, weighed 800 pounds each and were 17 feet long. A chain sling was placed around the centre of the rolls and, with the travelling crane, the load was lifted to clear the top of two stored electric motors. The load also had to pass by a ball mill scoop-feeder weighing approximately a ton. With Bowers operating the gear chain of the overhead crane and Poudrette steadying the load, the rolls were moved towards the loading bay.

The work was proceeding normally when Poudrette heard a noise and on turning saw Bowers pinned by the head between the wall of the warehouse and the top of the scoop-feeder that had tipped over. Knowing he could not lift the feeder, Poudrette ran outside and got a ride with a passing mine truck to the machine shop where he summoned several mechanics. They all rushed to the warehouse and tried to lift the scoop-feeder off Bowers but were unsuccessful. Mechanics Vic McLeod and Maynard Evans grabbed the overhead travelling chain, let the load of rolls down, put the chain block on the feeder and hoisted it to free Bowers. Gordon Cameron, safety supervisor, arrived at the scene. He had been informed of the accident and had already contacted the hospital, asking for an ambulance and a doctor. Soon both arrived and Dr. W. J. Rumball examined Bowers and pronounced him dead.

An autopsy was performed by Dr. J. Morphet. His findings were that Bowers had died instantly from a brain hemorrhage caused by a depressed fracture of the skull.

An inquest was held before Coroner J. M. Leblond. The jury returned the following verdict: "That Herbert J. Bowers came to his death by accidental means. This jury submits the following recommendations: 1) Better housekeeping of storage areas, especially the area around a loading dock; 2) More frequent inspections by the safety supervisor."

Herb Bowers was survived by his wife, a son James, and two daughters, Eleanor and Margaret.

Interment was conducted in the Kirkland Lake cemetery.

Much of the information for this account was given by James Bowers. As of 1999, he and his wife Velma live a quiet retired life in Dobie.

## 1969

- There are no fatal accidents in local mines this year
- It is calculated that the Kirkland Lake Gold Camp in the years up to 1969 produced $735,697,520 in gold, one third of that from Lake Shore Mine.
- Teck-Hughes tramway, located across Highway 66 on the west end of Kirkland Lake, is dismantled
- Lake Shore smokestack is demolished in an impressive blasting display
- The new Northern College building on the Tough-Oakes property opens and becomes a valuable community asset

*Mill building, Upper Canada Mine, Dobie. Photographer: Lawrence Chrismas*

## Chapter VII
# THE SEVENTIES

### 1970
- Welcome news again: there are no accidental fatalities in local mining activities
- Upper Canada and Upper Beaver close
- Price of gold on the free market falls below $35 an ounce

### 1971
- Rainer Stransky and Stanley Gorzalczynski are killed instantly in separate accidents at Kerr Addison
- Dofasco buys the Adams Mine from Jones and Laughlin
- A seven-ton monument made from Adams Mine ore, commemorating the players of the NHL from Kirkland Lake, is placed in a park on former Wright-Hargreaves land

### March 24, 1971
## Kerr Addison Mine
### Rainer Stransky (1932 - 1971)

Rainer Stransky lived with his wife, the former Else Egloff, and their six children at 13 28th Street in Virginiatown. The children – Gunther, Richard, Erich, Dennis, Ronald and Marieanne – were all young at the time of their father's death.

Thirty-nine-year-old Stransky was born at Kotwitz, Czechoslovakia on March 30, 1932. He had been employed at Kerr Addison for 14 years, beginning work there on May 22, 1957. He left this mine in 1964 and worked for two months at Algoma Ore Properties in Wawa and at Geco Mines in Manitouwadge. He returned to Kerr Addison in March 1964 and was a senior miner at the time of the accident.

Stransky, well known in Virginiatown, was an ardent curler and on the executive of the McGarry Curling Club. He also enjoyed fishing.

On the night-shift of March 24, he and partner Gilles Neveu were mucking and tramming broken ore from a stope to an ore pass dump on the 4,000 level using a locomotive and three ore cars. Two cars of the train were dumped by Stransky who was operating the locomotive but there was difficulty dumping the third car because a large flat slab of rock was lying across the top of it. Neveu left the operators platform and went around the cars to open the ore pass side of the train. After putting on a safety belt, he tried to bar the car door open and move the rock off the car but was unsuccessful. Stransky came around to investigate and decided to get a sledge hammer to break the rock. Unable to find a hammer in the immediate vicinity, he went around the operators platform. Neveu placed his scaling bar against the wall and was facing away from the train so he did not notice that the train began to move. He turned in time to see two cars

*Rainer Stransky. Photographer: Unknown*

and the locomotive carrying Stransky fall into the ore pass dump. The third car struck and pinned Neveau to the open dump door before it came to rest across the end of the dump opening. The crew in the immediate area rushed to his aid, released him and pulled him back from the open hole.

Supervisors were notified and a search for Stransky began. The two cars and the locomotive were found jammed 200 feet below in the ore pass. Stransky's hat and wrenches were found in the debris but he was not there. Successive ore pass openings were inspected on each level. Finally, 1,775 feet down the ore pass, his body was found.

Meanwhile, Neveu was taken to surface and treated by Dr. G. E. Hagerman who sent him to the Kirkland and District Hospital for a complete check up.

Stransky's body was removed to surface where Dr. Hagerman pronounced him dead; death was attributed to multiple fractures and internal injuries.

He was survived by his immediate family in Virginiatown and by a father, two sisters and four brothers, all living in Germany.

The body rested at Symington Funeral Home and funeral services were held in the chapel. Interment was conducted in the Kirkland Lake cemetery. The epitaph on the headstone is: BELOVED HUSBAND AND LOVING FATHER OF SIX CHILDREN – 'EVER REMEMBERED, EVER LOVED'.

Else and daughter Marieanne presently (1999) live in the Kirkland Lake area. The sons live in Edmonton, Sudbury, Thunder Bay and Timmins. Marieanne and Else supplied information and the photograph for this memorial account.

### December 17, 1971

## Kerr Addison Mine

## Stanislaw (Stanley) Gorzalczynski (1923 - 1971)

Stanley Gorzalczynski was born to John and Stanislava Gorzalczynski at Biazov in rural Poland on January 1, 1923. He had an older sister and a younger brother. As a young man, he had learned the masonry trade but at the age of 16 his native Poland was overrun and occupied by Nazi Germany. By the end of the war, six years later, he had become a displaced person. After the war, he met a girl from Czechoslovakia – Adelheid Rudisch – and in 1948, they were married in Germany.

In November of 1951, he immigrated alone to Canada to find employment in the mines at Val d'Or. His wife and their first child Richard, born in 1949, later immigrated in March 1952. In time, the family increased to four children as Danuta was born in 1952, John in 1954 and Stan in 1962.

Danuta remembers her dad as a highly disciplined, quiet hard-working man. He knew that mining was difficult and dangerous but he also believed this to be his purpose in life: to mine and provide for his family. He discouraged his sons from becoming miners and his daughter from ever marrying one.

He had been employed as a miner at Siscoe Mine, Upper Canada and the Helen Mine in Wawa. He had worked intermittently at Kerr Addison from October 1959 to August 1966 and from September

*Stanley Gorzalczynski. Photographer: Unknown*

1966 to December 1968. He was rehired in May 1969, promoted to senior miner a month later and remained employed as such until the accident.

Gorzalczynski was killed on December 17. It was a tragedy to lose a husband and a father but was especially difficult as the loss happened a week before Christmas. Left to mourn were a wife and four children ranging in age from nine to 21.

On that night-shift, he and partner Roger Boucher were instructed by Shift Boss Russell Graham to complete a slushing and cleaning operation in a stope so that the slusher could be moved away and drilling and blasting resumed. By 10:15 p.m., most of the muck was slushed and the two were cleaning fly muck around the posts and walls using Swede hoes. Boucher was about 10 feet away from Gorzalczynski when he heard a 'whoosh' and on turning around he saw many pieces of rock completely covering his partner except for his left arm and hand. Boucher pulled away what rock he could then, realizing he needed more help, he went to the 1,450 and summoned a crew for assistance.

The alarm was given and rescue operations initiated under the directions of Shift Bosses Graham and Duncan Johnson and Assistant Underground Superintendent Murray MacAuley. The body was recovered and taken by stretcher to the first aid room on surface. At 11:55, Dr. G. E. Hagerman made an examination and pronounced Gorzalczynski dead.

Another examination was performed by Dr. M. Galea, pathologist at Kirkland and District Hospital, and his findings were that the cause of death was a skull fracture, cerebral contusion and brain hemorrhage. Death would have been instantaneous.

The funeral service was held at Our Lady of the Lake Roman Catholic Church in Larder Lake and interment occurred in the Kirkland Lake cemetery. The epitaph on the headstone reads: 'REST IN PEACE'.

Some of the information and the photograph for this account was supplied by Danuta Livingston who lives in Larder Lake with her husband and children. Richard presently (1999) lives in Langley, B.C., John in Kirkland Lake and Stan in Haileybury. Mrs. Adelhied Gorzalczynski lives in Larder Lake.

Danuta recalls a powerful legacy left by her father. He always said, "A good education is something you are going to carry on your back for the rest of your life" – a lesson now being passed on to the nine grandchildren.

### 1972

- No fatal accidents in local area mines this year
- The opening of an asbestos mine in Midlothian Township near Matachewan is announced
- Price of gold is $38 (U.S.) per ounce

### 1973

- Rockburst at Macassa kills Fortune Lauzon
- Elmar Saar is killed underground at Kerr Addison
- Price of gold is $42.22 (U.S.) per ounce

### September 20, 1973

## Macassa Mine

## Fortune Lauzon (1935 - 1973)

On February 21, 1935, Fortune Lauzon was born to Mr. and Mrs. Maxime Lauzon of Englehart. The family eventually comprised 10 children; Fortune had four brothers and five sisters.

In 1971, he married Rose Anna Morin. They lived in Kirkland Lake where Fortune was a tractor-trailer driver working for City Movers. Hunting, fishing and working on old cars were his recreational interests.

He was employed intermittently at Macassa from 1968 to 1970. He was rehired on March 24, 1971 and was employed as a machine runner. He had a total of three years underground experience.

Lauzon, 38, was fatally injured on September 20 when he was struck by a rock while loading holes in a rock face. A rockburst in a wall above him occurred suddenly. J. Dupuis and John Benson found him in the muck. Dupuis went to get help while Benson and Henry Ziolkowski began to remove the rock from him. Severely injured, he was taken to surface and rushed to the Kirkland and District Hospital by ambulance but was pronounced

*Headstone photograph of Fortune Lauzon. Photographer: Author*

dead on arrival. A post-mortem conducted by Dr. G. Horner determined the cause of death as due to massive fractures at the base of the skull.

Surviving him were his wife, his nine-year-old son Fortune Lauzon Jr., his mother Marie, brothers Roland, Joseph, Herman and Maxime and sisters Dora, Florence, Annette, Delia and Rolande.

The funeral mass was held at Holy Name Church and interment took place in the local cemetery.

Some of the information for this account was given by Fortune Lauzon's sister, Mrs. Rollande Smith of Kirkland Lake.

### September 29, 1973
## Kerr Addison Mine

## Elmar Saar (1910 - 1973)

Elmar Saar was born in Estonia on May 2, 1910 and was married there to the former Maie Ratsep (b. May 30, 1910). Their eldest child, Lily, was born in Estonia and their second, Elmer, was born in Germany in 1946. During the war, Elmar worked at the war office as an accountant.

The family moved to Canada in 1950. Elmar had worked in the coal mines of Belgium three years. He was employed by Kerr Addison from December 7, 1956 and had worked as a miner, timberman, slusherman and trammer for almost 17 years.

Saar, 63, was fatally injured when he fell approximately 1,300 feet down an ore pass at Kerr Addison. Gabriel (Gabby) Houle, Saar's partner, noticed that a rock had jammed in a ramp. He was moving the train away from the ramp when he noticed Saar bending over to remove the rock. The ramp suddenly came to its vertical position, struck Saar and knocked him backwards into the open ore pass. His body was recovered by rescue operations through a chute and taken to surface where Dr. Hagerman pronounced him dead from multiple fractures and internal injuries.

Interment took place in the Kirkland Lake cemetery.

As of 1998, he is survived by his wife and daughter, both of whom reside in Southern Ontario. His son Elmer lives in Kirkland Lake with his wife and two children, Aaron and Alicia. Aaron is married to the former Lianna Dufresne, who researched information concerning the early life of Elmar Saar.

### 1974

- Leo Louko, an experienced miner, is killed in a scaling accident at Macassa
- Gold is valued at $190 (U.S.) per ounce

### April 30, 1974
## Macassa Mine

## Leo Aatos Louko (1920 - 1974)

Leo Louko was born in Finland on March 30, 1920. In January 1952, he left his homeland with his wife Niina, their daughter Marja, and his youngest sister. Louko's aunt was already living in Kirkland Lake when they arrived in search of a better life.

It did not take long for Leo and Niina to find work. They arrived on Tuesday and, by Wednesday morning, he was hired at Lake Shore; Niina had several cleaning jobs by the end of the week.

Louko worked for 13 years at Lake Shore until the mine closed; he then worked for six years at Upper Canada until its closing. In 1971, he became

*Leo Louko with his daughter Marja on her wedding day.*
*Photographer: John Scott Watson*

employed at Macassa. By 1974, 54-year-old Louko, a miner with over 22 years of experience, was living with his wife at 62 Fourth Street. Their two daughters were now on their own: Marja, her husband Tom Wakeford and their child were living in Renfrew and Margaret was living in Orillia.

On the nightshift of April 30, Louko and helper Trevor Davis were instructed by Shift Boss Fred Peddie to wash down, scale, slush and rockbolt a section of the Tegren stope. Louko was standing on a muck pile, scaling some small loose, when suddenly a 1,500-pound piece fell on him. Davis climbed the pile and tried to move the rock but was unable to do so. A crew was organized and immediately rescue operations were started. Dr. H. F. Richardson arrived, examined Louko and pronounced him dead. Later, the body was examined at the Kirkland and District Hospital; he had died of a massive skull fracture.

Louko is remembered as a kind, quiet man who loved his family. He had a special bond with new grand-daughter, Lisa: he was her 'Pappa'. She was not three when he died.

Louko enjoyed going to the cottage that he and his wife built with help from family and friends. He was an avid fisherman and during the winter enjoyed cross-country skiing and ice fishing. He was an active member in the Finnish Lutheran Church. He was a person who was always willing to help anyone.

Louko was survived by his immediate family, two sisters in Toronto and a sister and four brothers in Finland.

The funeral arrangements were handled by the Symington Funeral Home and services were held at the Finnish Lutheran Church. Interment took place in the local cemetery. The epitaph on the headstone reads: 'REST IN PEACE'.

Presently (1999), Niina, Marja and Tom live in Kirkland Lake and Margaret lives in Saskatoon. Marja provided the personal information and the photograph for this memorial to her father. Marja and Margaret credit their parents for instilling in them a strong work ethic and an appreciation of the value of education. Both are presently teachers.

## 1975

- No fatal accidents occur in Kirkland Lake area mines this year
- New Kirkland and District Hospital opens

## 1976

- For two years in a row, no fatal accidents occur in local area mines
- Midlothian, an asbestos mine west of Matachewan, begins production
- Sir Harry Oakes Chateau is vacated as a nursing home and now arouses interest as a historic building
- The Brethren in Christ Church opens on Tweedsmuir Avenue

## 1977

- For the third consecutive year, no fatal accidents occured in local area mines
- Kirkland Lake's population is about 13,000
- The new Kirkland Lake Mall is built on the former site of the Lake Shore mill

## 1978

- Stephen Nixon, 22, dies in hospital after an accident at Kerr Addison
- Stephen Rodgers, 21, is killed at Macassa
- United Asbestos Mine at Matachewan closes
- Former Kirkland Lake Gold Mine office becomes Harmony House (a halfway house)

### January 11, 1978

## Kerr Addison Mine

## Stephen Nixon (1956 - 1978)

Twenty-two-year-old Stephen Nixon died at Kirkland and District Hospital as a result of injuries sustained in a mining accident at Kerr Addison the previous day.

He worked at Kerr Addison as a mine labourer and apprentice miner from February 1975 to April 1976. He worked at Macassa as a miner from April 1976 to September 1976, then returned to Kerr Addison. Nixon was single and lived at 20 Hudson Bay Avenue.

On the day of the accident, he and Senior Miner Jocelyn Rocheleau (28) were working together underground. Nixon requested a sledgehammer from Rocheleau to break a large rock lodged in a muck slide. Wearing his safety rope, Nixon stepped into the muck slide and proceeded to break the rock with the hammer. The rock and muck began to move down the slide with him coming down along with it. He travelled a distance of 12 feet before the safety rope lost its slack, tightened, then failed. He fell a further 80 feet down the millhole to the chute area below. Rocheleau immediately went for help.

Nixon was found conscious and able to talk. He was removed by stretcher, examined underground by Dr. J. Lukezich, transported to surface and taken by ambulance to hospital where he received surgery. He died at 2:05 a.m. on January 12.

He was survived by parents Victor and Jean Nixon of Kirkland Lake, two sisters, Valerie of Ottawa and Jeanette at home, fiancée Lynn Perrier of Kirkland Lake and grandmother Mrs. Florence Hogg of Toronto.

Funeral services were held and interment occurred in the Kirkland Lake cemetery.

"Stephen was much loved, and he remains a cherished memory in the minds and hearts of his family and friends." – Jean Nixon, Stephen's mother, 1998.

### October 24, 1978

## Macassa Mine

## Stephen Wayne Rodgers (1957 - 1978)

"We loved him as one. As one with us in spirit. As part of our life." – The inscription on Stephen Rodgers' headstone at the Kirkland Lake cemetery.

Stephen Rodgers was born to Mr. and Mrs. Lorne Rodgers of Kirkland Lake on January 27, 1957. He was the fourth of five children: Dave, Larry, Suzy, Stephen and Kathy.

He was a 21-year-old miner employed at Macassa as a miner's helper from October 5, 1977. His lack of experience could have been a factor in his death.

*Stephen Rodgers. Photographer: Unknown*

Rodgers enjoyed the outdoors and fishing. According to his older brother, Dave, Stephen was somewhat apprehensive about working underground.

On October 24, while working with motorman Philip Lent, an ore car became derailed near an open ore pass. Lent, concentrating on moving the derailed car, saw Rodgers following the car then noticed that his light had disappeared. Lent immediately went to see what happened. Since he could not find Rodgers, he correctly assumed that he had fallen into the open ore pass. He was found lying on the muck almost at the next level. Dr. Ted Gardiner arrived at the 6,150 level, made an examination and pronounced Stephen dead.

An autopsy revealed that death was caused by multiple injuries including massive hemorrhage, lacerations to both lungs and a ruptured thoracic aorta.

The inquest jury made four recommendations: 1) eye bolts should be installed at all ore pass openings; 2) all gates should be closed when not in use; 3) dumping areas should have lighting; 4) more orientation and more safety instruction should be given to new employees.

Stephen Rodgers was survived by his parents, brothers and sisters.

Details of his personal history were submitted by his brother Dave who lives in King Kirkland. Lorne Rodgers, also of King Kirkland, supplied the photograph of his son for this memorial.

## 1979

- No miners are killed accidentally while at work in area mines this year
- Gold prices go from $220 to $524 an ounce and a rush of claim staking starts in the area
- Kirkland Lake train station is torn down and not replaced
- Sir Harry Oakes' widow, Eunice Lady Oakes, makes her last visit to Kirkland Lake and makes a $350,000 donation to the Museum of Northern History
- Ministry of Labour decides that the inhalation of aluminum dust by miners must not continue

*Tailings pond and mine shaft, Macassa Mine, Kirkland Lake. Photographer: Lawrence Chrismas*

## Chapter VIII
# THE EIGHTIES

### 1980

- Blasting accident claims the life of Donald Gibson at Macassa
- Gold climbs to $843 per ounce
- Old post office is refurbished to become the new town hall in Kirkland Lake
- Ecole Secondaire Jean Vanier is created
- Lake Shore office is torn down and replaced with a Canadian Tire
- Sir Harry Oakes Chateau is designated an historic site

### February 14, 1980
## Macassa Mine

## Donald Gibson (1952 - 1980)

Twenty-eight and single, Donald Gibson was a miner leader employed by Alex MacIntyre and Associates Limited, Mining Contractor, Kirkland Lake; he was hired by Willroy Mines Limited – Macassa Division. He was fatally injured when caught in his own blast. His address at the time was 270 Cedar North in Timmins, Ontario.

The accident occurred on the 5,300 level at approximately 3:30 a.m., February 14. After lighting his drift round, Gibson delayed his exit in order to relight some unlit fuses and was struck by the explosion of broken rock.

On the nightshift of February 13, the drift crew of Gibson and Lionel Landry were instructed by Shift Boss Brian Pascoe to muck out, drill off and blast an eight-foot round. The two mucked and drilled and were lighting the fuses. The easer fuse did not light and Gibson had no more matches. He went and got some igniter cord, lit it by hitting it on the track with a rock and lit the fuses. Meantime, the cut holes detonated and the broken rock hit Gibson full force, while the fringe material hit Landry.

Bleeding, Landry crawled around a corner in the drift and waited there until the remainder of the shots detonated. He then drove the motor to the station where he found help. He was taken to Kirkland and District Hospital for treatment.

The rescue team arrived and went to the scene of the accident where they found Gibson lying face down on a pile of muck, partially covered with about six inches of broken rock. There were no signs of life when he was taken to the station at the 5,300 level. Dr. R. Hughes pronounced him dead. Death was caused by severe head injuries and other multiple injuries.

### 1981

- No fatal accidents occur at Kirkland Lake and area mines this year
- Harry Oakes Chateau is restored structurally to its original condition

### 1982

- Denis de Champlain is killed at Macassa while drilling underground
- High price of gold allows LAC Minerals to put up a new headframe over the old Lake Shore workings
- Jewish synagogue on Station Road closes

### October 14, 1982
## Macassa Mine

## Denis Ivan Joseph de Champlain (1951 - 1982)

"People say that as time passes, acceptance of the death of a family member becomes easier. I have difficulty accepting the fact that my children had their father taken away from them at such an early age. The loss has had an effect on them at different times in their lives. They have coped well and have done well. In my opinion, you don't really accept the death of a loved one but as life goes on, you are able to handle it and relate to it differently." – Mrs. Denis de Champlain, 1998.

Denis de Champlain was born to Herve and Elise de Champlain in Hearst on May 12, 1951. His grandparents were Joseph and Anna de Champlain and Leon and Claudia Corriveau, all of Rimouski, Quebec.

He was married on June 20, 1972 to Dorothy Anne Dekker at the home of Mr. and Mrs. Jan Dekker in Chamberlain Township. The officiating clergy was Mrs. Shirley Jeffery.

The marriage produced two children: Carmen Marie, born on August 19, 1975 and Michael Richard, born on June 8, 1978. At the time of the accident, Carmen was seven and Michael, four.

Denis de Champlain was a devoted family man. In the fall of 1982, he was renovating the home on the Dekker farm in Englehart that he and his wife had purchased from Mrs. Dekker. He was a very hard worker and a perfectionist. For relaxation, de Champlain enjoyed motorcycling and socializing with his family and friends.

He was fatally injured on October 14 at 11:50 p.m. in the Macassa Mine at the 5,700 horizon. He died instantly when a minor rockburst dislodged a large rock that fell on him. At the time of the accident, he and Clinton Morgan were drilling on a drift face and standing about two feet apart when Morgan heard a 'bang'. He turned to his partner and saw him pinned between a large rock and the leg of the jackleg drill. Checking his partner, Morgan detected no signs of life. He then went to the station for help. De Champlain was taken to surface and to hospital where he was pronounced dead. The cause of death was a skull fracture.

The funeral service with Reverend Brad Clark and Sister Catherine McCann officiating took place at Emmanuel United Church in Englehart. The interment occurred at the Englehart cemetery.

Much of the detail of this memorial account was supplied by Dorothy de Champlain, widow of Denis, who lives near Englehart. The children are doing well: Carmen has graduated from the University of Ottawa and Michael is attending Canadore College in North Bay. Herve and Elise de Champlain presently (1998) live in Virginiatown.

### 1983

- No fatal mining accidents occur in Kirkland Lake area mines this year

### 1984

- No fatal accidents occur in Kirkland Lake area mines this year
- McBean Mine open pit operation starts up in Gauthier Township; the nearby Upper Canada Mill is used to process the ore
- Eastmaque Gold Mines builds a mill to refine once more the slimes that had filled Kirkland Lake

### 1985

- Dale Widdifield is killed in an accident at Macassa
- Three old companies become restructured: the Lake Shore, Wright-Hargreaves and Little Long Lac mines are reborn as LAC Minerals

### February 18, 1985

## Macassa Mine

## Dale Oakley Widdifield

### (1954 – 1985)

Thirty-one-year-old Dale Widdifield, employed as a shaft miner by Dynatec Mining Limited, was fatally injured on the 1,275 level of the No. 3 shaft of Macassa on February 18.

Co-worker Gerard Messier and Widdifield arrived with others at the level by means of the No. 1 compartment bucket. After disembarking, Messier accompanied Widdifield onto timbers located between the two compartments to show him the procedures for loading muck from the bin into the bucket. At this time, the crosshead and the bucket in the No. 2 compartment was ascending at normal speed when the top of the crosshead caught Widdifield, lifted him 20 feet, squeezed him against the shaft timbers and dropped him to the level station. He was pronounced dead at the scene by Dr. Ted Gardiner. The cause of death was massive hemorrhaging.

*Dale Widdifield. Photographer Unknown*

Widdifield had resided in Kirkland Lake for a year and previously in Englehart. He enjoyed sports, especially hunting and fishing. He also got a great deal of pleasure from travelling. His job had taken him to Pickle Lake, Bancroft and Yellowknife and his vacations to Newfoundland, Hawaii, Texas and Acapulco. He was a family man and particularly enjoyed his family's reunions.

He was predeceased by father Bob and three brothers. He was survived by wife Colleen, mother Mrs. Florence Widdifield, four brothers: Brian, Terry, Mike and Lee, and eight sisters: Betty, Shirley, Margie, Kelly, Cathy, Darlene, Sandra and Debbie. He was also survived by his grandmother, Mrs. Clara McKenzie, who is presently (1999) 110 years of age. (Note: she died in Sept. 1999 during the writing of this book.)

The body of Dale Widdifield rested at the McDonald Funeral Home in Englehart until the funeral mass at Holy Trinity Catholic Church, with Reverend Father Dan Sedlar officiating. Spring interment occurred in the Englehart cemetery.

Personal details and the photograph were submitted by Dale's brother and sister-in-law, Mike and Jeanette Widdifield of Englehart.

### 1986

- No miners are accidentally killed in area mines this year
- Some two million board feet of B.C. fir is used to line the new 2,200-metre-deep No. 3 shaft at Macassa, now the deepest single-lift shaft in North America. That's enough timber to build 200 average-size homes
- Kirkland Lake's movie theatre, the Strand, burns down
- New sewage treatment plant opens near the former site of the Swastika Mine

### 1987

- Mining contractor employee Siegfried Pietrzak is killed in a Lake Shore shaft operated by LAC Minerals
- Trucking company employee Robert Souriol is electrocuted when the dump box comes too close to a powerline on Eastmaque Gold property
- Employee at Adams Mine, Leonard Kazur, is killed while operating a loader
- Lake Shore tailings continue to be remilled to remove gold that was missed by earlier methods

### January 12, 1987

## Lake Shore Mine - LAC Minerals

## Siegfried (Siggy) Pietrzak

### (1952 - 1987)

Thirty-four-year-old Siggy Pietrzak of Virginiatown died underground at Lake Shore on Monday, January 12. He was survived by his wife Brenda (née Harney)* and his children, Kimberly Anne and Siggy Emmett Steven, at home. Also surviving were parents Stephan and Minna Pietrzak of Kearns.

At the time of the accident, Pietrzak was employed by mining contractor Alex McIntyre and Associates and had been employed with them for two years. He had a total of eight years of mining experience.

*Headstone photograph of Siggy Pietrzak.*
*Photographer: Author*

He fell 200 feet down a shaft manway; at the time, a three-man crew was repairing the manway. The fall occurred when the platform supporting the ladder on which he was standing collapsed. At the moment of collapse, he had disconnected himself from a safety line and was about to attach himself to another. His body was recovered by mine and ambulance crews at the 1,400 level. He was pronounced dead by Dr. J. Mutrie at the Kirkland and District Hospital.

Pietrzak will be remembered for his love of sports and children. He was a minor hockey coach for the McGarry Minor Hockey Association.

Interment was in the Kirkland Lake cemetery. The inscription on the headstone reads: PIETRZAK 'SIGGY' SIEGFRIED 1952-1987 – A LOVING AND UNDERSTANDING HUSBAND OF BRENDA – FATHER OF SIGGY AND KIMBERLEY – 'WE LOVE YOU BABE, TAKE CARE'.

** Brenda's father, Emmett Harney, was killed in a mining accident at Kerr Addison in 1968.*

## January 27, 1987
# Eastmaque Gold Mine

## Robert Frank Souriol
## (1961 - 1987)

The Souriol name is well known in Kirkland Lake. Robert's grandfather came to this area during the Depression and worked as a timber contractor in Boston Creek and South Mindoka. The family included numerous children, many of whom remained in the area to marry and raise families of their own.

Souriol was born to Arthur and Pauline Souriol in Kirkland Lake in 1961. Here he attended school and was active in scouting and spongepuck hockey. His life ambition was to become a truck driver and, when he did become one, he enjoyed his work. His marriage to the former Lorrie Nichol produced two daughters, Tammy and Bobbie-Anne. He was very proud of his daughters. He had said that when Tammy turned 16, he would buy her a red convertible so "she could allow her long hair to blow in the wind while driving." Tammy was six when he died and Bobbie-Anne only six months old.

Eastmaque Gold Mines Limited had proposed to reprocess old mine tailings from Kirkland Lake. The tailings would be taken from the lake by barge-mounted pumps and slurred to their processing facility. Before starting these mining operations, the main access road had to be constructed. A construction contractor was hired to do this work and Souriol Paving entered into an agreement with them to haul fill to various locations as required.

On January 27, Souriol Paving had five trucks hauling to the construction site; Bobby Souriol was one of the drivers. Douglas Foley, a bulldozer operator working at the site, was levelling the loads after they were dumped. Souriol was backing up his truck to the edge of the fill without the aid of a flagman. He stopped when the truck was near the bulldozer and began raising the box by using a lever inside of the cab. The box came close to an overhead powerline carrying 115 kilovolts, possibly making contact with the wires. The electricity arced from the line to the truck and the dual rear wheels ignited. Souriol fell from the cab of the truck to the ground. Foley, on the bulldozer within four feet of the truck, decided to move the dozer out of harm's way. He stood on the dozer track to plan his path of travel before starting up. At that point he 'blacked out' and also fell to the ground.

Truck driver Bill English arrived, finding both men on the ground. Unaware of the cause of the accidents, he turned both men over to see what was wrong.

Two police officers, Doug Bradley and Carl Hanson, arrived at the scene about 19 minutes after the accident and saw the two men lying on the ground between the bulldozer and the dump truck. Both the truck and the dozer were idling and the rear wheels of the truck were on fire. Souriol was not moving but Foley staggered to his feet and walked away between the vehicles. Both victims were taken by ambulance to hospital. Foley, with burns on his hands and feet, survived the mishap and was listed in fair condition. Souriol was pronounced dead on arrival.

An inquest was held with Coroner Bruce Martin presiding. Souriol had suffered cuts and bruises to his scalp and left shoulder as well as burns to his feet, thigh and abdomen. His shirt and boots were also burned. His vital organs (heart, kidneys and lungs) had suffered trauma. Cause of death was described as electric shock due to a current travelling through his body.

Gerry Donig of Ontario Hydro stated at the inquest that since the voltage was so high, no piece of equipment should be within three metres of it. He also stated that Ontario Hydro was not notified

*Bobby Souriol and wife Lorrie with daughter Tammy, nephew Adam Trudel and niece Tina Trudel (Christmas 1986). Photographer: Unknown*

of the construction and that the construction company should have called Hydro before beginning the job.

Wasyl Janischewskyj, an expert witness who teaches university-level electrical engineering, told the jury that from the evidence it was obvious Souriol was electrocuted before he hit the ground and that he had died instantly. "Souriol may not have died had he remained in the truck," he said.

The following recommendations were made by the jury: 1) signalmen should be available on every construction site to guide machine operators who are backing up; 2) signs should be erected at construction sites warning of overhead powerlines; 3) rules and regulations on construction sites should be more strictly enforced, and employees and employers should be made more aware of the hazards that exist when working near overhead powerlines.

Bobby Souriol was survived by his wife, two daughters, his parents, two brothers, Edward of North Bay and Gordon of Timmins, three sisters, Yolande (Mrs. Ivan Trudel) of South Porcupine, Annette (Mrs. Alex Cooper) of Sault Ste. Marie and Paulette (Mrs. Peter Crawford) of Kirkland Lake and grandfather Mr. Frank Lafreniere of Sudbury. He was also survived by many aunts, uncles, cousins, nieces and nephews, many of whom lived in the Kirkland Lake area.

The funeral mass was held at Holy Name Church and interment occurred in the local cemetery. A unique headstone marks the grave.

*A unique headstone marks the grave of Bobby Souriol. Photographer: Author*

Following his death, a Bobby Souriol Memorial Spongepuck League was formed.

"I believe that this was an accident that should never have happened. I feel that if proper procedures had been followed, such as the contractor informing the hydro company that they would be working in the area of such high voltage wires, my husband would still be alive." – Lorrie Souriol (1999)

"Although a man so young, Bobby had inherited the quality of making trucking an enjoyable and rewarding occupation day by day. I could always remember how his truck's engine used to roar with excellency, as each gear he shifted met with a touch of achievement. For a life so short-lived but driven day by day by his achievements and caring, it should not be forgotten or left unnoticed." – Taken from a memorial plaque given to Mom and Dad (Souriol) from their son Ed and daughter-in-law Donna Lee.

This account was compiled with the assistance of Lorrie Souriol, Pauline Souriol and Mrs. Armand Souriol, all of Kirkland Lake. Bobby's mother supplied the photograph of her son and his family.

## December 20, 1987
## Adams Mine

### Leonard Kazur (1933 - 1987)

Leonard Kazur was born in Selkirk, Manitoba on October 22, 1933 and had been a long-time resident of Kirkland Lake. He was the son of Mike and Rose Kazur of Kirkland Lake and had a sister, Janet (Mrs. Robert Barnes) of Kirkland Lake.

Kazur and his wife, the former Mary Catherine MacDonald, lived at 65 Government Road East with their children, Kimberley Anne and Christopher Leonard.

Kazur was well known and well liked in town as a family man who actively participated in family functions. A strong supporter of the Legion 87's Hockey Club, he travelled with them to out-of-town games. He chaperoned school trips in which his children participated. He was a member of the Ukrainian Club, the Numismatics Association, the Royal Canadian Legion and was a special constable

*Leonard Kazur. Photographer: Roger Morin*

for the District of Timiskaming.

In 1971, he started working at Adams in the security department and was transferred to the pellet plant in September 1986 when his job in security became redundant.

On Sunday, December 20, 54-year-old Kazur was operating a bobcat loader to clean up concentrate in the pellet plant. While dumping the concentrate over a barrier to the ground 30 feet below, the barrier gave way, and the loader and operator fell to the ground. Kazur was rushed to the Kirkland and District Hospital where Dr. Ted Gardiner pronounced him dead on arrival.

The body rested at Symington and Monette Funeral Home until the funeral mass on December 3 at Holy Name Church. Interment was in the Kirkland Lake cemetery. The epitaph on the gravestone reads: BELOVED HUSBAND – LOVING FATHER 'EVER REMEMBERED EVER LOVED'.

"A three-day inquest into the death of Leonard Kazur brought forth 11 recommendations and resulted in numerous charges being laid by the Ministry of Labour against the mine," stated Mrs. Kazur.

The personal information and the photograph were provided by Mary Kazur.

## 1988

- No miners are killed in area mines this year
- A new mill is completed at Macassa with a potential capacity of 600 tons of ore and 1,000 tons of tailings per day
- Holt-McDermott Mine, north of Kirkland Lake, begins production
- Timmins promoter refurbishes the Toburn headframe with bright red siding
- New police station is built on former Wright-Hargreaves property
- Population of Kirkland Lake is 12,126
- More old buildings on Government Road burn down
- This year, workers in mining earned the highest average weekly wage or salary of any industrial sector in Canada. The figure was $763.47 per week for metal mines, well above either forestry, manufacturing, construction, transportation or finance jobs.

## 1989

- No area miners are killed while at work this year
- Golden Shield Resources, owners of Kerr Addison Mine, declares bankruptcy

*Closed mine shaft, Lake Shore Mine - LAC Minerals, Kirkland Lake. Photographer: Lawrence Chrismas*

## Chapter IX
# THE NINETIES

### 1990

- No area miners are killed accidentally while at work this year
- Severe economic blow hits Kirkland Lake and area as Adams Mine closes and 400 jobs are lost

### 1991

- Two miners, Alex Trajkowicz and Steven Guindon, are killed in separate accidents at Macassa
- $90-million Northland Power cogeneration plant at the south end of Main Street begins to produce electricity
- New funeral home owned by Marc Monette is built using the Teck Hughes dry as its basic structure
- Another economic blow hits the area as Eastmaque Gold Mine closes in November
- A New Democratic Party MP puts forward a private member's bill to officially recognize April 28 as a remembrance day for workers killed or injured on the job. The aim of the National Day of Mourning is to remember this commitment to mourn the dead and fight for the living. The bill was passed on February 1, 1991

### November 25, 1991
## Macassa Mine

## Alexis 'Alex' Joseph Trajkowicz (1953 - 1991)

Four men were injured, one fatally, in an explosion at Macassa on November 25. Thirty-eight-year-old Alex Trajkowicz of Matheson was killed in the accident.

Robert Gagnon was transported to St. Michael's Hospital in Toronto with upper chest injuries and respiratory trauma. Dennis Perrault, with an injury to his right leg, was taken to Kirkland and District Hospital. Emile Romain was treated for injuries at emergency and then released.

The explosion happened when workers were blasting a ramp down into the mine. They had set several delayed charges, then waited for the blasts to occur before returning to the area. The final explosion was a delayed blast that killed Trajkowicz and critically injured Gagnon and Perrault.

Trajkowicz was born to Alexander and Antoinette Trajkowicz in Haileybury on August 2, 1953. He had four brothers and a sister, and was raised in Kirkland Lake where he attended school.

He was predeceased by his father and brother Teddy. Surviving Trajkowicz were his wife Anne and their child Grant, two other children, Michael and Amanda Brown and their mother Winnie, his mother, his brothers Paul of St. Catharines, Richard of North Bay, John of Gowganda, and his sister Irene Sabo of St. Catharines.

*Alex Trajkowicz. Photographer: Unknown*

The funeral service with Rector Leonard officiating was held at St. John's Anglican Church in Matheson. The funeral arrangements were entrusted to Monette and Associates of Kirkland Lake. Spring interment took place in Greenhill Cemetery at Matheson.

Some of the information and the photograph were submitted by Antoinette Trajkowicz-Laurin, Alex's mother, who presently lives in Elk Lake, Ontario (1999).

### December 20, 1991

## Macassa Mine

## Two Men are Trapped By a Rockburst

Steve Guindon and Glen Harwood, employees at Macassa, wanted to be working partners at the mine. They were good friends and had many things in common: they were married, in their thirties, patient, easy-going, and both enjoyed fishing.

Eventually, they succeeded in becoming partners and were assigned to 5635 stope. Working well together by discussing and planning their work, they formed an efficient and competent partnership.

This stope was an older one with only a small block of ore left to be mined. It had to be repaired according to the standards set by Guindon and Harwood. "One thing about Macassa," Harwood said, "if you wanted to do something to make your workplace safer, there were no questions, just do it."

During a number of consecutive shifts, they put in a cover by the track, extended a ladder above the elevation of the track, rockbolted and cabled the walls, scaled the area thoroughly and installed a bottom staging.

Glen Harwood gives the details of what happened on December 20:

*"After discussing our progress in the stope with Roger Allison, we decided to drill more holes in the face. We levelled off an area on the muck pile, set up the drilling machine, connected the air and water hoses, placed bits on the steel and then broke for lunch.*

*"At lunch, we talked about the upcoming Christmas, our families and our future. We talked about car races, a subject that Steve enjoyed. Other discussions involved bonus pay, safety and life in general. We both felt that things were going quite well.*

*"After lunch, we went back to the stope and began work by doing some light scaling before drilling. There was no abnormal loose coming down nor was there any noise coming from the rock. The work was routine and everything was progressing well. It was 2:20 p.m.*

*"I was to drill and Steve, working beside me, was to change the steels and clean the holes. I placed the steel against the face to start drilling and suddenly I realized that I was lying down – it was dark – there were rocks on my face – from a distance I heard a bump and rocks falling. Initially, I didn't feel the pressure of the rock on me.*

*"I realized that I was lying in a fetal position. My right arm was bent under my back and my left arm was extended in front of me but completely immobile. I recall a fluttering movement under my left arm but this soon stopped.*

*"'Steve, Steve,' I yelled. No answer. The effort to yell was too much as I almost became unconscious. I then realized that a rockburst must have occurred. Shockingly, I also realized that those small movements under my left arm were from Steve and had now stopped. Steve was dead.*

*"My body began to hurt. Breathing was difficult. 'Help me,' I yelled out. No answer. Of course not, I thought, who would be looking for us so soon?*

*"'Please, I don't want to die,' I yelled. I tried with all my strength to rise up, to move, to wiggle, anything but movement was impossible. I don't want to die like this, I thought. I don't like pain; I want to die quickly. 'Help me.'*

*"I heard an air hose hissing nearby. That's good, I thought but that will cause me to live longer and to suffer more.*

*"The pain was becoming unbearable; I was close to unconsciousness. I pictured my wife crying and I thought of our life together. It's over now. It will be okay; I have insurance. The house is paid off. The pain was intense. 'I don't want to die,' I yelled, 'I want to live.'*

*"I must think. Must live. Must try to move – that was impossible. I began to think of my injuries: my legs were in pain, therefore they must be attached, my arms had feeling, so they must also be attached. The pain was getting worse. Did I hear someone calling? 'Help. Help,' I yelled.*

*I began to lose energy. It was difficult to breathe and the pain was increasing. 'I want to die,' I yelled. My thoughts went to my young daughter, Veronica. She was crying for her dad at Christmas. Poor thing. Christmas is so close. Why me? Did I hurt anyone?*

*"'Get me outa here,' I shouted.*

*"Again, I almost lost consciousness. The pain caused me to feel as if my legs and hands were slowly being squeezed off my body. What good is a daddy without arms and legs, I thought.*

*What was that? – A voice? – Yes. It sounded far away. Did they hear me? 'Help. Help,' I hollered.*

*"'Hello?' called one of the rescuers. 'Hello,' I replied. I could hear rocks being moved but it sounded far away. 'Help – get the weight off,' I cried.*

*"The sound – so far away. The pain – so horrible. They won't be able to reach me. God, I love my Shirley and Veronica so much. 'Help,' I hollered.*

*Rescue workers quickly dig a tunnel towards my head. They find me. The tunnel is getting larger. The weight of rock is still on me. 'Help – get the weight off,' I yell.*

*"Ed Obradovich directs some air into the hole towards me. More rock is moved. It takes them so long to get me out. Eddy is talking to me. He is splendid. I owe him a kiss.*

*"Ed digs in and finds my hand. He holds onto it and talks to me, preventing me from slipping into unconsciousness.*

*"The rescue workers are jacking rocks away from my body. The pain never lets go. Well, not all of it anyway. As each rock is removed from my body, I feel a little better.*

*"Finally, my body is lifted free. The pain is horrible. These men are such good guys. I just don't know how to thank them. They risked their lives to rescue me. It could have bumped again. They got me out. I hope I'm worth it."*

Harwood was placed in a basket stretcher, taken through a manway up the shaft to surface where his wife was waiting for him. The ordeal to this point had lasted about six hours.

He was taken to Kirkland and District Hospital where he was listed with multiple trauma. He was flown to Sudbury on December 23.

At the Sudbury General Hospital, 75 per cent of the muscle tissue in his lower right leg was so badly damaged that the tissue had to be removed. He was later admitted into the Sudbury Laurentian Hospital where he convalesced for 30 days.

Glen recalls his ordeal. "Somehow those men, those heroes, dug down to me. After thinking a thousand times that I was going to die, I managed to force my hand out of the hole. Ed held my hand and talked to me. Without him, I may not have made it. My left arm was burning. I wiggled my head 'til I could see. I saw Steve's hat and shoulder. Poor Steve – at least he never suffered."

Guindon had died instantly in the burst. Regrettably, Harwood was unable to attend the funeral. "I wanted to go to the funeral to say goodbye to my friend," states Harwood, "I will remember him for the rest of my life." Harwood, Shirley and Veronica visit the gravesite frequently.

Today, Harwood wears a brace and lives with pain in his leg, back, chest and left arm. He has nightmares about his calamity at times. He has learned to live with his ordeal and pain, and doesn't like to talk about it. "There are people out there who are a lot worse off than I am," he says.

In talking about the support shown by the community, Shirley Harwood said, "The people were very sincere and caring. It's a good town. Kirkland Lake has always been known to pull together when there's a need. That's just the way the town is. It's like one big family."

"We will always remember Steve," adds Harwood. "The Guindons are often in our thoughts."

## Steven Allan Douglas Guindon (1960 - 1991)

Steven was born to Raymond and Geraldine Guindon on October 19, 1960, at South Porcupine; his father was employed in mining in that area at the time. In 1961, the family moved to Larder Lake as Raymond was now employed at Kerr Addison in Virginiatown.

Steven was the fourth child in a family of six children. He had three older sisters: Cathy (Mrs. Rick Nychuk) of Kenogami, Susan (Mrs. Robert Sheldon) of Larder Lake, and Ramona (Mrs. Doug Sullivan) of North Bay. His two younger brothers were Scott and Trevor.

Throughout his 30 years at Larder Lake, he was an avid camper and fisherman. Known to his friends as Duke, he loved where he lived and had no desire to leave his community. He attended elementary school in Larder Lake and graduated from Kirkland Lake CVI in 1979. He gained employment at Kerr Addison. Later, prospects looked better at Macassa and he started work there on February 16, 1987.

In August of that year, he married his long-time sweetheart Denise Lacroix. They had no children during their four-year marriage but Guindon spent many enjoyable hours with Denise and their beloved dog, Casey.

*Steve Guindon spent many enjoyable hours with Casey.*
*Photographer: Unknown*

Involved in hockey as a youngster, he renewed his passion for the game by playing in the local Old Tymers Hockey League. Unfortunately, he only played for one season after his 30th birthday.

Guindon had a renewed devotion to the Catholic Church during his life. His funeral mass was held at Our Lady of the Lake in Larder Lake with interment at the Kirkland Lake cemetery.

The tragic news of the bump at Macassa on Friday, December 20, 1991 was devastating to his family. They are left with memories of a handsome, easy-going young man who loved the outdoors and who was taken from them far too early.

This memorial account and the photograph were contributed by the family of Steven Guindon.

### 1992

- No local miners are killed at work this year
- Holloway Joint Venture is formed that includes the construction of the Holloway Mine
- Pinegar Centre, a young offenders' facility, is built near the former Tough-Oakes property
- Only two viable ethnic club halls in Kirkland Lake remain: the Polish Hall and the Italian Belvedere Hall

### 1993

- Robert Sheldon and Leonce Verrier are killed in a rockburst in Macassa
- Sudbury Contact discovers the Victoria Creek gold deposits in Gauthier Township

### November 26, 1993

## Macassa Mine

## Two Men Trapped by Rockbursts at Macassa

Two successive rockbursts occurring at 8:20:44 and 8:20:51 a.m. on November 26 displaced an estimated 15,000 tonnes of rock in stope 6723 located on the 6,600 level of Macassa.

Seismology equipment at the mine indicated that a rockburst of significant magnitude had occurred underground. It took about an hour to evacuate the mine; workers realized two men were missing. Mine rescue workers immediately began searching for them.

There were three miners preparing this stope for backfilling at the time of the burst. One miner managed to escape by climbing out on a screen. The other two, 34-year-old Robert Sheldon and 42-year-old Leonce Verrier, were trapped. Two other miners were injured during the bursts; Nelson Bourgeois received a cut on his head and Dave Kitty injured his right hand. Both were treated and released from Kirkland and District Hospital.

Two manways leading to the stope had collapsed. Searchers had to hand-muck tons of rock, piece by piece. Mine Manager Rudi Rucker said searchers had to proceed with caution as unstable ground put them at risk.

Searchers worked around the clock for three days, then had to change their approach plan due to unstable ground. The morale of the searchers was described as high but as each day progressed hope of finding the men alive was diminished. It took 34 days for the recovery team to reach the stope and 77 days to locate the bodies. The two miners had been killed instantly.

*Robert Sheldon enjoyed hunting. Photographer: Unknown*

## Robert John Elwood Sheldon

### (1957 - 1993)

Robert Sheldon was the fourth child of six; he was born to Marguerite and Fredrick Sheldon on March 7, 1957.

In the early 1900s, Sheldon's grandfather, James Elwood Sheldon, a prospector, bought property on Larder Lake to build a home. This house today remains as a home for his descendants. It was in this dwelling that Robert lived and first learned the importance of family, community and outdoor recreation.

Sheldon attended Larder Lake Public School and Kirkland Lake CVI. At the age of 16, he secured a summer job, painting mine buildings at the Kerr mine. The next summer, he worked in the mill and returned to high school in September but the steady pay cheque made him decide to return to the mine to become an underground miner.

Before long, he began dating Susan Guindon, also of Larder Lake. At the age of 21, Sheldon married Susan; their union produced two sons: Robert Aaron and Cory Adam.

On December 6, 1979, Sheldon was hired at Macassa. He decided to leave the Kerr mine because his friend Steve Nixon had died at the mine in 1978 and he thought Macassa would be a safer place to work.

Being raised in Larder Lake allowed Sheldon to become an avid hunter and fisherman. He enjoyed the outdoors and everything close to nature. He seemed to have the unique ability to enjoy every single day and every aspect of his life. Friends and family have many happy memories of Robert: organizing an annual fish derby on Estrangement Lake; spending time on the pontoon boat fishing happily with the sound of the Beach Boys blaring in the background; cruising around the lake with his family and friends.

He was closely involved with the community and was an active Legion member. Many Friday nights were reserved for darts at the Legion with his friends and he was very proud of the night he darted a perfect score. He helped out at the Larder Lake Ski Club where his two sons were active downhill skiers.

On November 26, 1993, a rockburst at Macassa trapped Robert Sheldon and Léonce Verrier on the 6,600 level. They had died instantly but for 77 agonizing days the rescue workers toiled heroically to recover the bodies, trapped under tons of fallen rock. On February 10, 1994, both bodies were found.

The family was left with only memories of a gentle family-oriented and generous man with a good sense of humour.

Cremation occurred and on February 17, 1994, a memorial service was held.

*Robert Sheldon installing flats on cable bolts with partner John Malciw (5,600-foot level of Macassa Mine). Photographer: Unknown*

*Memorial plaque placed on rock high above the shore of Larder Lake. (Two brothers and a brother-in-law are killed in mining accidents within a period of 3-1/2 years). Photographers: Denis and Jennifer Vachon*

Surviving him were his parents, his wife and two sons, sister Susan Cannon and brothers Bruce, Michael, Fred and Steven.

Sheldon was predeceased by his brother-in-law, Steven Guindon, who also died in a mining accident at Macassa on December 20, 1991. His brother Fred died later in a mining accident on June 1, 1995, at Kerr Addison.

A memorial plaque in memory of Robert Sheldon, Fred Sheldon and Steven Guindon is bolted to a rock high above the shores of Larder Lake. This memorial and its location is dedicated to these three men who spent some of the happiest times of their shortened lives on this lake.

The details for this account and the photographs of Robert were contributed by the Sheldon family.

## Léonce Verrier (1951 – 1993)

On November 26, 1993, a huge and very powerful rockburst occurred deep in the Macassa mine. The miners were evacuated and when a count was done, it was found that two, Léonce Verrier and Robert Sheldon, were still in the mine. They had died instantly but due to the extent of the damage, it would take 77 days for the rescue operation to reach their bodies.

Verrier was 42 and lived at 391 Georgina Street in Matachewan. He was born to Rolland and Helen Verrier in 1951 in Elk Lake where he was also raised. He married the former Darlene Trudel and they had three daughters: Celine, Lise and Nancy.

His surviving brothers and sisters were Huguette Rodrigue, Pauline Roy, Suzanne Roy, Juliette Norton, Lorraine D'Aoùt, Pierre and Rheal, all of Elk Lake; Lucie Parent, Rachel Lajeunesse and Hector of Cochrane; and Germain of Keswick.

He was a member of the Royal Canadian Legion and a former member of the Knights of Columbus. He was an avid fisherman. He also enjoyed doing renovations to his home and being around his grandchildren.

Verrier worked as a labourer at many places from the age of 18. His work took him to Elk Lake, Earlton, Matachewan, Notre Dame du Nord, Timmins, Swastika, Dundalk, Kirkland Lake and Cobalt. In 1986, he became employed at Macassa where he worked until his death.

Monette Funeral Home handled the funeral arrangements. On Monday, February 14, 1993, the funeral mass was held at the Church of Assumption. Interment occurred in the Kirkland Lake cemetery.

Léonce Verrier was predeceased by his father, Rolland. His mother, Helen, is presently (1999) living in Haileybury. Daughters, Celine, lives in Matachewan, Nancy (Mrs. Duncan Jodko-Narkiewicz) lives in Elmira, Ontario, and Lise Bernatchez in Dane. Lise supplied the photograph and details for this memorial to her father.

*Léonce Verrier. Photographer: Unknown*

### 1994

- No fatal mining accidents occur locally this year
- On July 23, the 'Tribute to Miners Sculpture' is unveiled as part of Kirkland Lake's 75th birthday celebration

### 1995

- Fred Sheldon is killed while working underground at Kerr Addison
- Increased emphasis is placed on exploration at Macassa
- Anticipation of shaft sinking by Sudbury Contact at its Victoria Creek Project
- Strategic exploration alliance locally develops between Queenston and Franco-Nevada
- Kirkland Lake area again becomes a hot gold camp
- Holt-McDermott Mine produces 117,621 ounces of gold this year
- Armistice Mine is reactivated; surface facilities are improved; shaft is dewatered; a large bulk sample is taken and a 20,000-foot drilling program is initiated

### June 2, 1995

## Kerr Addison Mine (A. J. Perron Gold Corporation)

## Frederic Arthur Sheldon Jr.

### (1953 - 1995)

Fred Sheldon Jr. was a life-long resident of Larder Lake with short stays in Sarnia and Fort McMurray. He was born in 1953 at Kirkland and District Hospital, the third child of six born to Fred and Marguerite Sheldon. His grandfather, James Elwood Sheldon, was one of the first prospectors to settle in Larder Lake in the early 1900s and his home remains today as the Sheldon residence. Sheldon, who was single, enjoyed golf, fishing, hunting and reading. He was an active member of the Larder Lake Legion and of the Larder Lake Lions Club.

He was a miner employed by A. J. Perron Gold Corporation in Virginiatown. On June 2, while

*Fred Sheldon. Photographer: Unknown*

mucking from a draw point on the 2,650 level, he was struck by a large rock and killed. He was 42 at the time of his death.

He was predeceased by brother Robert who was killed in a mining accident at Macassa in 1993. He was survived by his parents, a sister, Susan Cannon, and three brothers, Bruce, Michael and Steven.

This memorial account of Fred Sheldon Jr. was written by his family.

## 1996

- In July, Battle Mountain Gold begins commercial production. Also on this month, Robbie Owens is killed at this mine
- There are six producing gold mines in the Kirkland Lake area: Macassa (underground operation), Lake Shore (tailings project), Cheminis Mine, Kerr Mine, Holt-McDermott Mine and Battle Mountain Gold
- Macassa Mill is modified to accommodate 1,500 tons of ore and 1,000 tons of tailings per day
- Macassa produces 66,504 ounces of gold
- Lake Shore tailings operation produces 14,702 ounces of gold
- Cheminis, by custom milling at the Kerr Mill, produces 365.8 ounces of gold
- Throughout the year, the future of the Kerr Addison (Kerr Mine) is in doubt. In August, the mine closes after 58 years in operation. In December, the McGarry Township takes over the surface rights for back taxes. During its active time, the Kerr Addison produced 10,457,441 ounces of gold. During the same period, 45 miners died accidentally while mining this gold
- Holt-McDermott Mine experiences a slower than anticipated start-up due to a mechanical failure at a custom milling site. However, 11,000 ounces of gold are produced at the mine this year
- $8.6 million is spent on site preparation and construction of surface buildings at the Victoria Creek Project
- Queenston-Franco Nevada Joint Venture makes new discoveries below the McBean Mine and below the Anoki deposit
- Royal Oak Mines Ltd. spend $4 million on exploration drilling and underground rehabilitation at Matachewan Consolidated. Royal Oak plans to turn this mine and the Young-Davidson, past gold producers, into open pit and underground operations
- Ontario Mining Fact: Average weekly earnings in the Ontario mining industry are $1,174, one of the highest earning levels in the provincial economy

## July 30, 1996
## Battle Mountain Gold Mine

### Robert 'Robbie' Edward Owens (1962 - 1996)

It's kind of a tradition when the Owens family members get together during summer occasions and spend some of their leisure time looking for four-leafed clovers. Their Irish background tells them that finding one will bring them luck. This would have been the case in 1991 when Robbie Owens and brother Tommy were employed by a mining contractor to help sink a shaft in Malartic, Quebec. While working in the mine, Robbie slipped and fell down the shaft. Tommy, working along with him, climbed 45 feet to the bottom, expecting the worse – but Robbie was alive. He did have some lower vertebrae broken and he had to go through a year-and-a-half recuperation period but he felt lucky to be alive.

*Robbie Owens holding a lucky four-leafed clover found while at the Speed Boat Regatta at Ville Marie P.Q.*
*Photographer: Unknown*

Robbie Owens was born to Thomas and Vera (née Brookes) Owens of Virginiatown in Kirkland and District Hospital on June 20, 1962. He would be the last child in a family of eight children. Growing up in 'V-town' with his older brothers and sisters, he remained very close to them, especially to brother Kevin (Yogi). Unfortunately, at the age of 18, Yogi suddenly passed away.

Owens was always a very social person with a wide circle of friends. He had a free-spirited personality and was very active in sports, especially hockey, baseball, curling and ice fishing. He also coached a midget-aged hockey team in Virginiatown.

His father was a Kerr Addison miner and other family members were involved in mining so it was only natural that, as an 18-year-old looking for a job, he too found work at the mine. Owens worked in Kerr Addison for a period of years and later for Redpath Mining Contractors which placed him in many locations, mostly in Quebec. He eventually returned closer to home, finding work at Cheminis Mine where he remained for about two years. In 1994, he joined the work force at Battle Mountain Gold and remained there until the time of his fatal accident.

In 1996, Owens took a couple weeks off in July for personal reasons and returned to work towards the end of the month. On July 30, he was placed back on his former job, scoop tramming on the 490-metre level in a drift he thought he knew. Due to blasting the previous day, the condition of the drift had changed. While he was driving the scoop tram loaded with backfill along the drift, it fell into an unexpected hole. Owens, 34 and single, was killed in the mishap.

He was predeceased by his brother and survived by his parents, his brothers Michael of Williamstown, Ontario, Phil of Rouyn, Quebec, and Tommy of Larder Lake, his sisters Jennifer Smith of London, Ontario, Diane Bouchard of Levac, Ontario, and Kate Masching of Fort McMurray, Alberta. He was also survived by aunt and uncle

John and Joyce Owens of England and his faithful dog 'Bailey.'

The body rested at St. Paul's Anglican Church in Virginiatown and then to Monette Funeral Home for service at the chapel. A huge number of mourners assembled at the funeral of this very popular young man. Reverend Victoria Shaw officiated the service.

Interment took place in the Owens family plot at the Kirkland Lake cemetery. In memoriam contributions were made to the Kirkland Lake Miners' Memorial Foundation.

Personal information concerning this memorial to his brother was gained by an interview with Tommy Owens in 1999. He stated that Robbie is sadly missed by the family. Bailey, a continuous reminder, now lives with Tommy and his family.

## 1997

- No accidental deaths occur at local mines this year
- Following a scheduled production blast on April 12, a series of rockbursts cause damage to the Macassa No. 3 shaft between the 5,700 and 6,300 levels. The most damage occurs at the 6,025 shaft station
- Sudbury Contact complete a shaft-sinking at its Victoria Creek Project
- Barrick Gold Corporation's Holt-McDermott Mine produces 116,500 ounces of gold this year
- McGarry Township, now owner of the surface assets of the Kerr Addison mine, auctions off the surface buildings and equipment. A group of investors buy the property
- Ontario Mining Fact: Canada's mining industry is committed to continuously improving its record of safe production. Statistics show that it is twice as safe to be working in an Ontario underground mine as being a government employee

## 1998

- No miners are killed in local mines this year
- Holt-McDermott Mine employs 185 persons and is at a depth of 775 metres. The mine announces an $18-million shaft deepening project that should extend the life of the mine to the year 2003
- Low gold prices cause the Victoria Creek Project to cease operations
- Three inscribed monuments are placed near the Miners' Memorial in memory of the men who lost their lives accidentally in local mines since 1917
- A PCB recycling plant (Trans Cycle Industries) is built at the industrial park in Kirkland Lake
- Notre Development Corporation plans to develop the Adams Mine pit into a megadump
- Local Architectural Conservation Advisory Committee works on a proposal to designate the Toburn head frame as a heritage site. "The head frame is the last of the original of the once mighty Kirkland Lake gold camp." – Tom Drew-Brook. Only one other aging headframe stands as a memorial to the mining history of this area, this being the original headframe over the Upper Canada No. 2 shaft

## 1999

- Fatal accidents in local mines are eliminated again this year
- Trails 2000, a plan to create a walking trail along historic and geological sites in Kirkland Lake, is established
- Adams Mine pit could be accepting Toronto's garbage as soon as 2002
- In May, gold values hit a 20-year low
- This year marks the 70th anniversary of the creation of mine rescue in Ontario
- Kinross Gold, operators of Macassa, announce an immediate suspension of operations at the mine that operated since 1933. Low gold prices are the major reason for the closure and the loss of 160 jobs
- In September, Kinross Gold comes back with good news as it signs a deal with Exall Resources of Matheson for custom milling, saving at least 28 jobs at Macassa
- Macassa Mill continues to process the tailings from Kirkland Lake
- Battle Mountain Gold wins the Angus Campbell Safety Award for the third straight year
- The world anticipates the beginning of the year 2000

# EPILOGUE

Some miners went to work one day and never came back alive. Others were injured and died hours, days, months or even years later.

They were husbands, fathers, sons, brothers, grandfathers and friends, all of them at work and making a living. Some were very young, in their teens and twenties, just launching their independent lives. Others were experienced miners in their thirties, forties, fifties and sixties; they had family members dependent on them for love, support and a pay cheque.

Some were war veterans; others were immigrants, displaced persons and victims of war. Many came to the gold camp during the Depression and after the World Wars, looking for work. A number were recent high school graduates; some of these were raised in the Kirkland Lake area.

They came to the gold camp from England, Finland, U.S.A., Romania, Yugoslavia and many other countries. Others were Canadian born and came from Cobalt, Timmins, Sudbury, Quebec, Southern Ontario, Nova Scotia, Manitoba and Newfoundland and other localities.

Not all of these men began their working lives as miners. Some left their jobs at farming, forestry, fishing and building the railroad – all of them looked for better-paying jobs in the mines. But for others, mining was their first real job.

Not only did they work here, but they became involved in building their community and contributing to it. They were Legion members, Masons, volunteer firemen, musicians, sportsmen and members of the Loyal Order of the Moose. Many were associated with social community groups such as the Sons Of Scotland, the Croation Fraternal Society, The Polish Fraternal Society and The Italo-Canadian Club. Trinity United, Holy Name, Sacred Heart, Virginiatown United, Assumption, Our Lady of the Lake, Finnish Lutheran, St. Peter's etc., all had members in their congregations who perished in the mines.

They died here. After their death and (sometimes) a funeral service, their remains were transported back to their home towns. Many victims had no family in this country and were buried locally; their graves remain unmarked. Others were laid to rest in marked graves in the Kirkland Lake cemetery.

Could any of these accidents have been prevented? Sure, they could have. Utilizing used timber to support men on a blasting set, using a rotting rope as a safety line, walking back into a blasting area too early and jumping over an open shaft were actions that caused needless deaths. But before becoming too judgmental, have we ever exceeded the speed limit while driving, failed to buckle-up our seat belts, used an electrical device that was flawed in some way or climbed an unsafe ladder? Human error? Yes, but many of these men didn't get a second chance. And then, who can predict a rockburst, a fall of loose, a slip or a fall? Bad luck? Again, no second chances were offered to these men.

But things have changed over the years. Traditionally, mining was an occupation that was the most dangerous in the country. Years ago, mine owners were concerned with little else than production, production and more production. As one miner left in a stretcher, another was hired at the front gate to take his place. That's not the case now.

What brought on the improvements in safety? There are probably many reasons: enormous improvements in training methods for workers in the mining industry; studies done on mine safety including the Ham Commission and the Burkett Commission; implementation of new procedures; a positive attitude towards the field of safety among workers, management and unions; and an updated Health and Safety Act.

Safety has now become the joint responsibility of both employee and employer. A good example of this occurred locally on March 8, 1999. On this date, Battle Mountain Gold and its employees and families gathered to celebrate two years without a lost-time accident at the mine. The mine Health and Safety Committee organized a party that included skating, a champagne brunch, socializing, awarding of mementos and speeches. Each employee received a wrist watch and a Swiss Army pocket knife.

In his speech, Mine Manager Claude Begin said, "What is great is that 140 people decided for two years that they would accept nothing less than zero serious accidents. The goal is another year without a lost-time accident."

Mine Safety-Loss Control Officer Al Sampson in his speech said, "The mine is on a team approach to safety. The Health and Safety Committee is very active. They take walking tours of the mine and are involved in operational procedures." Mr. Sampson probably said it best when he continued, "You wonder why you have to work safely? The reasons are right in this room. You work safely for your families!"

No matter what the activity, there will always be some risk. Mining accidents are decreasing but will never be eliminated completely. We can learn from the past by not forgetting those who died accidentally. Their family and friends were devastated by the loss. Perhaps, in their memory, we should work safely everywhere so that our family members and friends will not be subjected to such tragic loss and grief.

*A parent may have lost a son…*
*A child may have lost a father…*
*A wife may have lost a husband…*
*A sibling may have lost a brother…*
*Goodbye dear miner,*
*It's sad that we must part this way.*
*You are gone, but not forgotten,*
*Your lamp is forever lit.*

*Steel sculpture of Kirkland Lake area miner by artists Sally Lawrence and Rob Moir. Photographer: Lawrence Chrismas*

# GLOSSARY

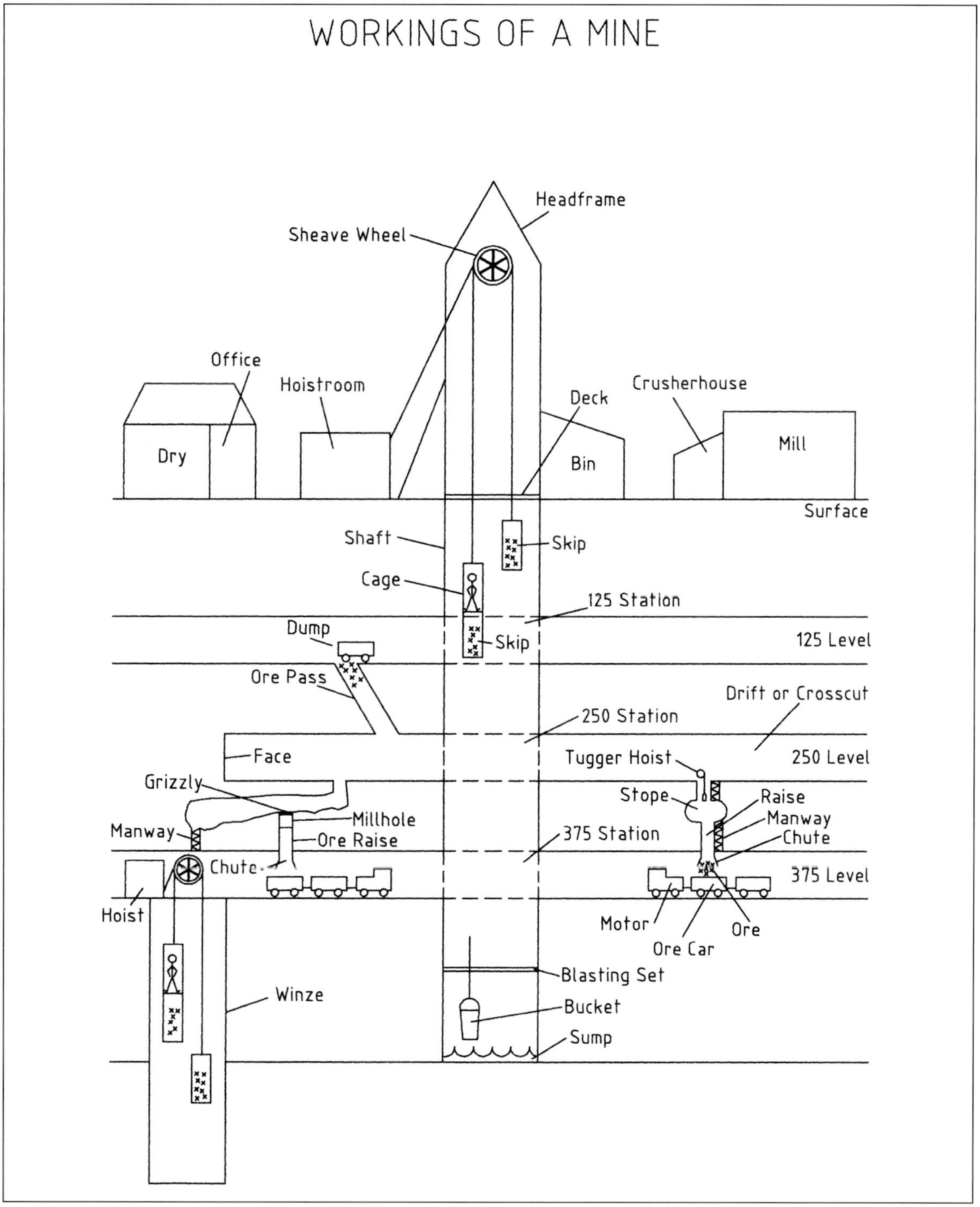


*Workings of a mine. Artist: Jan Hopkins*

assay. A chemical test performed on a sample of ore to determine the amount of valuable metals contained.

back. The roof of an underground chamber such as a stope or drift.

backfill. Waste material used to fill the void created by mining (e.g. Waste to fill a mined out stope. Waste material may be rock, sand, tailings etc.)

blasthole. A hole drilled for the purpose of blasting.

blasting pole. A pole on which a charge is placed so that it can be extended into the area to be blasted.

blasting round. A sequenced set of explosions at a rock face. Holes are loaded in such a way that each will be fired in a designated sequence in order to obtain a desired break.

blasting set. A platform supported by chain blocks and/or cable, used during shaft sinking for the purpose of storing equipment and materials.

box hole. A collection chamber for ore, made from timbers, usually located above a drift and supported by backfill.

breast. The working face of a stope.

bulkhead. A retaining structure usually made of timber.

cage. A conveyance used to transport men and equipment up and down a shaft; a mining elevator.

cant hook. A lever-like pole with a hook-like apparatus attached at one end, used to direct motion of logs.

chain block. A chain hoist.

chain chairs. The last set of timber at the bottom of the shaft.

chain hoist. A tackle employing an endless chain instead of a rope, used for hoisting heavy weights.

chute. An opening, usually constructed of timber and equipped with a gate, through which ore is drawn from a stope into ore cars.

chute blaster. A person in the mine who is responsible for placing charges in hung-up chutes.

claim. A portion of land held by a prospector or a mining company under federal or provincial law.

classifier. A machine for separating the constituents of ore to relative sizes and densities, thus facilitating concentration and treatment.

collar. The term applied to the timbering or concrete around the top of the shaft.

compressor. A machine for compressing air to a pressure sufficient to activate mining machinery.

corduroy road. A road built of logs laid side by side usually in low or swampy places.

crosscut. A horizontal underground tunnel driven to intersect an ore body.

crosshead. A heavy rectangular guide frame attached to the hoisting cable just above the bucket.

crown pillar. A body of rock of variable geometry located above the underground workings; it supports the surface above all the stopes.

crusher. A machine for crushing rock.

crusher house. A building which contains the crusher.

cut and fill. A method of stoping in which ore is removed in slices, following which the excavation is filled with backfill before the next slice is mined. The backfill supports the walls of the stope, preventing cave-ins.

cyclone (cyclone separator). Any of certain centrifugal devices for separating solid materials from liquids.

deck. The area around the shaft collar where men and materials enter or leave the cage.

deckman. An occupation in a mine whereby an employee oversees the activities of the deck area. Activities include movement of equipment, men and ore on and off the cage or bucket.

drift. A horizontal underground tunnel that follows along the length of an ore body as opposed to a crosscut which crosses the length of the ore body.

dry. A building where a miner changes his clothing before and after work.

face. The end of a drift, crosscut or stope in which work is progressing.

fill fence. A barrier intended to prevent the escape of fill.

foot wall. The wall or rock on the underside of an ore body.

gad. A short chisel-like tool.

gangway. A passageway into, through, or out of any place.

grizzly. A grating usually constructed of steel rails, placed over the top of a chute or ore pass for the purpose of stopping large pieces of rock that could get hung up in the chute or pass.

hanging wall. The wall or rock on the upper side of an ore body.

headframe. The tall slender building over the collar of the shaft. It consists of the deck and the pulley systems of the hoist and may contain storage bins.

hoist. The machine used for raising and lowering a cage, skip, crosshead, bucket or other conveyance in a shaft.

hoistman. An employee in the mine who is responsible for the proper operation of the hoist.

hoist room. A building adjacent to the headframe which houses the powerful motors that engage the hoist to move the conveyances up and down the shaft.

hung-up. A situation where rock or sand is detained in a chute, raise, ore pass etc.

jackleg drilling machine. A rock drilling machine mounted on a supporting leg for easier operation by the driller.

lagging. Planks, small timbers, poles, placed in stopes, drifts, crosscuts, etc. to prevent rocks from falling. It does not support the weight of overlying rocks.

level. The horizontal workings in a mine. Levels are usually established at regular intervals.

lifter holes. The drilled holes at the bottom of a face.

locomotive. A self-propelled vehicle operating under a single control (a motorman) running on rails, utilizing electrical energy supplied by batteries on board and used for moving ore cars.

loose. Slang for loose rock.

machine bar. A steel bar which supports a drilling machine.

machine runner. An occupation in a mine where the employee runs and operates a drilling machine.

manway. A structure allowing miners access to and from their work areas. This may consist of stairs, ladders, decks and tunnels. Also included are services such as compressed air lines, water lines and electrical wiring.

mill. A large building on the mine property where ore is treated for recovery of valuable minerals.

millhole. A passageway in which ore passes by gravity feed from a stope to a level below. The amount of ore movement is controlled by a chute at the lower end.

missed hole. A drilled hole on a rock face that contains a charge that did not detonate.

motorman. An occupation in a mine where the employee is operating a locomotive and/or an ore car train.

muck. Ore or rock that has been broken by blasting.

mucker. An employee in a mine who moves muck by machine or shovel to another position in the mine.

open pit. A surface mine, open to the daylight, such as a quarry.

ore. A mixture of rock, ore and unwanted materials from which minerals can be extracted at a profit.

ore pass. A fairly large raise, into which broken ore is dumped and then slides or falls to the desired level in the mine.

pillar. A block of solid ore or rock left in place to provide structural support to the underground workings.

pinion. A gear with a small number of teeth designed to mesh with a much larger gear.

pond. An aquatic settling area through which mill effluent flows.

primer. A highly concentrated explosive used to ignite an explosive charge and itself ignited in the process.

raise. A vertical or inclined underground working that has been driven from the bottom upward. Used for conveying ore, miners, equipment, supplies, timber or ventilation.

ram planks. The main angular located planks supporting a scaffold.

rilled. Structured in steps or stages (e.g. a rilled stope).

rock bolting. The act of supporting rock with five-foot and eight-foot-long steel bolts anchored in holes drilled specially for this purpose.

rockburst. A violent and sudden release of energy resulting in the failure of walls, backs and pillars in a mine, caused by weight or pressure of the surrounding rock.

sampler. An occupation in a mine whereby an employee selects a fractional but representative part of a mineral deposit for assay and analysis.

sand blast. A method by which a blast is set up without any previous drilling; the charge is placed on the rock and then ignited.

scaling. The act of removing loose slabs of rock from the back and walls of underground workings such as stopes and drifts, etc. This is usually done with a long hand-held scaling bar.

shaft. A vertical opening into a mine for the purpose of providing access into the mine workings. It consists of a hoist at the top and a conveyance (elevator) for handling workers and materials.

shaftman. An occupation in a mine whereby an employee is involved with the physical deepening (sinking) of a shaft.

shaft sinking. The process of making a shaft by blasting the rock and removing the broken rock by means of a bucket and cable pulley system (hoist).

sheave (wheel). A grooved wheel or pulley mounted at the top of the shaft, raise or headframe.

shrinkage stoping. A stoping method which uses part of the broken ore as a working platform and as a support for the walls of the stope.

skip. A self-dumping bucket used in a shaft for hoisting ore or rock.

slash. The process of blasting rock from the side of an underground working to widen the opening (e.g. to slash a drift wall).

slurry. A mixture of water and finely constituted waste material used as backfill in unused mine workings such as stopes. In time, this slurry will dehydrate and become a cement-like backfill material but without the strength of cement.

slush. Recovering ore or rock by using a scraper along the floor of the underground working (e.g. slushing stopes and drifts).

snatch block. A pulley that can be opened on one side to receive the middle part of a rope.

spawling. Crumbling rock due to pressure.

spit. The process of igniting a fuse which is attached to an explosive.

split lagging. Timber consisting of logs which have been sawed into two halves longitudinally.

sprag. A piece of timber serving as a prop or support.

staking. The process of marking the limits of a claim with stakes or claim posts.

stage. A platform constructed of timber onto which the drilling machine is placed.

station. An enlargement of a shaft laterally constructed for the storage and handling of equipment, supplies, ore, miners, etc. Drifts are driven from here to create levels.

steel(s). Long sharpened steel rods of various lengths used to drill holes in rock. Modern day steels have tapered ends to receive exchangeable drill bits.

stope. An excavation in a mine from which ore is being or has been extracted.

stull. A thick timber used to support the back of a mine working or wedged between two walls in a stope. It protects miners from falling rock or for supporting miners, ore or waste.

sublevel. A level or working horizon located between the main working levels.

sump. An underground excavation where water accumulates before being pumped to the surface, usually located at the bottom of the shaft.

tailings. Material rejected from a mill after most of the recoverable valuable minerals have been extracted.

thickener. A large round tank used in milling operations to separate solids from liquids. The clear fluid overflows from the tank and the heavy rock material sinks to the bottom.

timberman. An occupation in a mine whereby the employee uses lumber and timber to support and line the mine workings.

tracks. Rails for the locomotive and ore cars.

tram. To haul ore cars in a mine, manually or by a locomotive.

trammer. An occupation in a mine whereby the employee pushes an ore car manually or by means of a locomotive.

trommel. A screen usually of cylindrical or conical shape that is mounted on a revolving and slightly inclined longitudinal shaft and is used for screening or sizing ore.

truck. A steel-wheeled vehicle on rails used for moving heavy objects.

vein. A fissure, fault or crack in rock filled by minerals that have travelled upwards from some deep source.

wash down. The process of washing down a working face with water pressure after a blast to observe clearly the effect of the blasting operation.

waste. Rock that is not ore.

winze. An internal mine shaft. A shaft in a mine that has its top at an underground level, not at the surface.

# BIBLIOGRAPHY


Angus, C. and Griffin, B. *We Lived a Life and Then Some. The Life, Death and Life of a Mining Town.* Toronto: Between the Lines, 1996.

Barnes, M. *Fortunes in the Ground. Cobalt Porcupine and Kirkland Lake.* Erin: The Boston Mills Press, 1986.

Barnes, M. *Kirkland Lake On the Mile of Gold.* Erin: The Boston Mills Press, 1994.

Dufresne, V. and Thompson, C. *The Goldfields of Larder Lake. Canada's Forgotten Gold Rush.* Haileybury: Temiskaming Printing, 1999.

Hurst, T., Harvey, L., Armstrong L., Lariviere D., Jacques L., Blackburn S. *Tegousie Wabie: The Story of Larder Lake.* 1987.

*Globe and Mail.* Toronto

MacDowell, Laurel Sefton. *Remember Kirkland Lake: The History and Effects of the Kirkland Lake Gold Miners' Strike, 1941 – 1942.* Toronto: The University of Toronto Press, 1983.

*Northern Daily News.* Kirkland Lake.

*Northern News.* Cobalt.

*Ontario Department of Mines Annual Reports.* Toronto.

Ontario Mining Association. *The Miner at Work – A Story of Northern Ontario.* Toronto. 1949.

Pain, S.A. *The Way North.* Toronto: The Ryerson Press, 1964.

Pain, S.A. *Three Miles of Gold.* Toronto: The Ryerson Press, 1966.

Savage, W. S. *Mineral Resources and Mining Properties in the Kirkland Lake – Larder Lake Area.* Toronto: Frank Fogg, Queen's Printer, 1964.

*Toronto Star.* Toronto.

*Toronto Telegram.* Toronto.

Tucker, A. *Steam into Wilderness.* Don Mills: Fitzhenry and Whiteside Ltd., 1978.

Wetjen, A. and Irvine, L. H. T. *The Kirkland Lake Story: A Pictorial History.* Kirkland Lake: Town of Kirkland Lake, 1979.

# ACKNOWLEDGEMENTS/CREDITS

Many people assisted in the compilation of material for this book. Without their help, this book would not have been published.

My wife Tanya Jaworsky assisted with the typing and advised me on many occasions on the direction the book should take. I thank her for her help and patience during the four-year production period.

Cathy Blackburn has been doing genealogical studies for years. Her published works, *Kirkland Lake Cemetery* and *District of Temiskaming Monument Inscriptions*, were a valuable source of information.

In creating the magnificent monument tribute, the Miners' Memorial Committee ignited the inspiration to produce this further memorial to the miners.

Proofreaders Tom Drew-Brook and Brian Hamilton spent weeks making the compilation of information more presentable and readable, and also became important promoters of this book.

Chuck Wallsten spent many hours with the technical production, photographs and layout of the manuscript. Thank you, Chuck.

The Ministry of Northern Development and Mines directed a huge amount of information used in the research for details presented in this book. Appreciation is due to Linda Davis and Janice Brisson at the Mines Library in Sudbury, and to Dave Guindon and Gary Grabowski in Kirkland Lake.

Lydia Alexander, Curator at the Museum of Northern History, has been an advisor. She and the staff (namely, Linda Sinclair and Mary Chodoriwsky) have also contributed by assisting with the location of some photographs for this book.

Shirley Harwood, having splendid artistic abilities, designed the meaningful covers for the manuscript.

The Teck Centennial Library staff was very helpful and took a particular interest in this project.

Local Kirkland Lake historians have previously written very informative and interesting books concerning the rich history of this area. Information from the books of S. Pain, André Wetjen and L. H. T. Irvine, V. Dufresne and C. Thompson, C. Angus and B. Griffin, L.S. MacDowell, W. S. Savage, and Michael Barnes allowed me to develop a local historical perspective.

I wish to thank the many sponsors for their financial support.

There are others who made contributions to this publication. They include: Gerry O'Connor, Danny Spiegelman, Murray McGill, June Korhonen, Charlie Angus, Joe and Millie Mavrinac, Siegfried Kiessling, Fred Bukowski, Mike Chodoriwsky, Marty Koskela, Steve Yee, Duff Dufresne, Merdy Armstrong, Gib Gilchrist, Gordon Webber, Joe Laurila, Terry Croisier, Walter Franczyk, Ian MacPherson, Robin Ormerod, Jan Hopkins, Neil Culhane, Doug McPherson, Lawrence Chrismas, Steven Gerl, Doris Daigle, Eddie Duke, Rick Owen.

Special thanks go to the professional team in Calgary: Editor Kerry McArthur, Designer Sherry Mumford, Proofreader Tom Wilson,Web Master Robert Mishler and Publisher Lawrence Chrismas.

Finally, gratitude is offered to the many family members who offered their time, information and photographs. This book will be a valued memorial to their loved ones.

## CREDITS FOR USE OF PHOTOGRAPHS

(Unless otherwise indicated)

Albert Lalonde 64
Allan McNaughton 171
André Duchesne 203
Angela Hopper 116
Anthony Paver 67, 86
Antoinette Trajkowicz Laurin 235
Bellon Family (frontispiece & 145)
Berlingeri Family 161
Betty Ann Monteith 154
Bill Cameron 74
Cécile Roddick 215
Danuta Livingston 220
David & Henry Melong 43
David Yaworski 204
Denis and Jennifer Vachon 116
Donald Barkell 69
Donald Whalley 170
Edgar Harrop 132
Edward & Barbara O'Reilly 187
Esme Hurd 76, 77
Fran Haskins 199
Garry Montague 209
Gladys Wilcox 89
Guidon Family 238
Heather McDougall 185
Hilja Bently 191
Ina Pulkinen 122
Irene Clark McKay 212
Irja Porre 34
Jason Kilbride 100
Jim McCarthy 210
Jim Bowers 138
Joe Mavrinec 57
Kay Cuthbertson 68
Kazur Family 233
Laurier Soucie 133
Lawrence Chrismas (front cover)
Leo Desjardin 149
Lise Bernatchez 241
Lorne Rodgers 224
Lucille Merrick 136
Margaret Johnson 174
Marlene Pitcaithley 156, 157
Michael Barnes 31
Mike and Jeanette Widdifield 229
Morissette Family 184
Museum of Northern History 19, 20, 26, 28, 29, 38, 56, 84, 101, 106, 108, 110, 113, 128, 129, 150, 206
Nancy McFarlane 211
Northern Daily News 160, 161, 164, 169, 172, 179, 205
Olavi Huhtala 125
Ontario Mining Association 72, 78, 95, 139, 142, 143
Ontario Ministry of Mines 91
Owens Family 243
Pauline Souriol 231
Provincial Archives 25
Rebecca & Michael Starr 152
Rennie Family 154
Sandy Maguire and Brian Hamilton 155
Sheldon Family 239, 240, 242
Stan Collins 185
Stransky Family 219
Town of Cobalt 35
Violet Linney 201

# INDEX